Applied Mathematical Sciences
Volume 152

Editors
S.S. Antman J.E. Marsden L. Sirovich

Advisors
J.K. Hale P. Holmes J. Keener
J. Keller B.J. Matkowsky A. Mielke
C.S. Peskin K.R.S. Sreenivasan

Applied Mathematical Sciences

1. *John:* Partial Differential Equations, 4th ed.
2. *Sirovich:* Techniques of Asymptotic Analysis.
3. *Hale:* Theory of Functional Differential Equations, 2nd ed.
4. *Percus:* Combinatorial Methods.
5. *von Mises/Friedrichs:* Fluid Dynamics.
6. *Freiberger/Grenander:* A Short Course in Computational Probability and Statistics.
7. *Pipkin:* Lectures on Viscoelasticity Theory.
8. *Giacaglia:* Perturbation Methods in Non-linear Systems.
9. *Friedrichs:* Spectral Theory of Operators in Hilbert Space.
10. *Stroud:* Numerical Quadrature and Solution of Ordinary Differential Equations.
11. *Wolovich:* Linear Multivariable Systems.
12. *Berkovitz:* Optimal Control Theory.
13. *Bluman/Cole:* Similarity Methods for Differential Equations.
14. *Yoshizawa:* Stability Theory and the Existence of Periodic Solution and Almost Periodic Solutions.
15. *Braun:* Differential Equations and Their Applications, 3rd ed.
16. *Lefschetz:* Applications of Algebraic Topology.
17. *Collatz/Wetterling:* Optimization Problems.
18. *Grenander:* Pattern Synthesis: Lectures in Pattern Theory, Vol. I.
19. *Marsden/McCracken:* Hopf Bifurcation and Its Applications.
20. *Driver:* Ordinary and Delay Differential Equations.
21. *Courant/Friedrichs:* Supersonic Flow and Shock Waves.
22. *Rouche/Habets/Laloy:* Stability Theory by Liapunov's Direct Method.
23. *Lamperti:* Stochastic Processes: A Survey of the Mathematical Theory.
24. *Grenander:* Pattern Analysis: Lectures in Pattern Theory, Vol. II.
25. *Davies:* Integral Transforms and Their Applications, 2nd ed.
26. *Kushner/Clark:* Stochastic Approximation Methods for Constrained and Unconstrained Systems.
27. *de Boor:* A Practical Guide to Splines: Revised Edition.
28. *Keilson:* Markov Chain Models—Rarity and Exponentiality.
29. *de Veubeke:* A Course in Elasticity.
30. *Sniatycki:* Geometric Quantization and Quantum Mechanics.
31. *Reid:* Sturmian Theory for Ordinary Differential Equations.
32. *Meis/Markowitz:* Numerical Solution of Partial Differential Equations.
33. *Grenander:* Regular Structures: Lectures in Pattern Theory, Vol. III.
34. *Kevorkian/Cole:* Perturbation Methods in Applied Mathematics.
35. *Carr:* Applications of Centre Manifold Theory.
36. *Bengtsson/Ghil/Källén:* Dynamic Meteorology: Data Assimilation Methods.
37. *Saperstone:* Semidynamical Systems in Infinite Dimensional Spaces.
38. *Lichtenberg/Lieberman:* Regular and Chaotic Dynamics, 2nd ed.
39. *Piccini/Stampacchia/Vidossich:* Ordinary Differential Equations in $\mathbf{R}^n$.
40. *Naylor/Sell:* Linear Operator Theory in Engineering and Science.
41. *Sparrow:* The Lorenz Equations: Bifurcations, Chaos, and Strange Attractors.
42. *Guckenheimer/Holmes:* Nonlinear Oscillations, Dynamical Systems, and Bifurcations of Vector Fields.
43. *Ockendon/Taylor:* Inviscid Fluid Flows.
44. *Pazy:* Semigroups of Linear Operators and Applications to Partial Differential Equations.
45. *Glashoff/Gustafson:* Linear Operations and Approximation: An Introduction to the Theoretical Analysis and Numerical Treatment of Semi-Infinite Programs.
46. *Wilcox:* Scattering Theory for Diffraction Gratings.
47. *Hale/Magalhães/Oliva:* Dynamics in Infinite Dimensions, 2nd ed.
48. *Murray:* Asymptotic Analysis.
49. *Ladyzhenskaya:* The Boundary-Value Problems of Mathematical Physics.
50. *Wilcox:* Sound Propagation in Stratified Fluids.
51. *Golubitsky/Schaeffer:* Bifurcation and Groups in Bifurcation Theory, Vol. I.
52. *Chipot:* Variational Inequalities and Flow in Porous Media.
53. *Majda:* Compressible Fluid Flow and System of Conservation Laws in Several Space Variables.
54. *Wasow:* Linear Turning Point Theory.
55. *Yosida:* Operational Calculus: A Theory of Hyperfunctions.
56. *Chang/Howes:* Nonlinear Singular Perturbation Phenomena: Theory and Applications.
57. *Reinhardt:* Analysis of Approximation Methods for Differential and Integral Equations.
58. *Dwoyer/Hussaini/Voigt (eds):* Theoretical Approaches to Turbulence.
59. *Sanders/Verhulst:* Averaging Methods in Nonlinear Dynamical Systems.

(continued following index)

Helge Holden   Nils Henrik Risebro

Front Tracking for Hyperbolic Conservation Laws

With 39 Illustrations

Springer

Helge Holden
Department of Mathematical Sciences
Norwegian University of Science and
  Technology
NO-7491 Trondheim
Norway
holden@math.ntnu.no

Nils Henrik Risebro
Department of Mathematics
University of Oslo
P.O. Box 1053, Blindern
NO-0136 Oslo
Norway
nilshr@math.uio.no

Series Editors:
S. Antman
Department of Mathematics
and
Institute for Physical Science and Technology
University of Maryland
College Park, MD 20742-4015

J.E. Marsden
Control and Dynamic Systems
California Institute of Technology
Pasadena, CA 91125
USA

L. Sirovich
Division of Applied Mathematics
Brown University
Providence, RI 02912
USA

Mathematics Subject Classification (2000): 35Lxx, 35L65, 58J45

Library of Congress Cataloging-in-Publication Data
Holden, H. (Helge), 1956–
  Front tracking for hyperbolic conservation laws / Helge Holden, Nils Henrik Risebro.
    p. cm.— (Applied mathematical sciences ; 152)
  Includes bibliographical references and index.
  ISBN 3-540-43289-2 (alk. paper)
    1. Conservation laws (Mathematics) 2. Differential equations, Hyperbolic. I. Risebro,
  Nils Henrik. II. Title. III. Applied mathematical sciences (Springer-Verlag New York,
  Inc.) ; v. 152.
  QA1 .A647 vol. 152   [QA377]   510s—dc21   [515'.353]   2001057674

ISBN 3-540-43289-2   Printed on acid-free paper.

© 2002 Springer-Verlag New York, Inc.
All rights reserved. This work may not be translated or copied in whole or in part without the written permission of the publisher (Springer-Verlag New York, Inc., 175 Fifth Avenue, New York, NY 10010, USA), except for brief excerpts in connection with reviews or scholarly analysis. Use in connection with any form of information storage and retrieval, electronic adaptation, computer software, or by similar or dissimilar methodology now know or hereafter developed is forbidden.
The use in this publication of trade names, trademarks, service marks, and similar terms, even if the are not identified as such, is not to be taken as an expression of opinion as to whether or not they are subject to proprietary rights.

Printed in the United States of America.

9 8 7 6 5 4 3 2 1   SPIN 10869008

Typesetting: Pages created by authors using a Springer T_EX macro package.

www.springer-ny.com

Springer-Verlag   New York Berlin Heidelberg
A member of BertelsmannSpringer Science+Business Media GmbH

*In memory of Raphael,
who started it all*

Contents

Preface		**ix**
1	**Introduction**	**1**
	1.1 Notes	18
2	**Scalar Conservation Laws**	**23**
	2.1 Entropy Conditions	24
	2.2 The Riemann Problem	30
	2.3 Front Tracking	36
	2.4 Existence and Uniqueness	44
	2.5 Notes	56
3	**A Short Course in Difference Methods**	**63**
	3.1 Conservative Methods	63
	3.2 Error Estimates	81
	3.3 A Priori Error Estimates	92
	3.4 Measure-Valued Solutions	99
	3.5 Notes	112
4	**Multidimensional Scalar Conservation Laws**	**117**
	4.1 Dimensional Splitting Methods	117
	4.2 Dimensional Splitting and Front Tracking	127
	4.3 Convergence Rates	134
	4.4 Operator Splitting: Diffusion	147

viii Contents

	4.5	Operator Splitting: Source	154
	4.6	Notes	158

5 The Riemann Problem for Systems — 165
5.1	Hyperbolicity and Some Examples	166
5.2	Rarefaction Waves	170
5.3	The Hugoniot Locus: The Shock Curves	176
5.4	The Entropy Condition	182
5.5	The Solution of the Riemann Problem	190
5.6	Notes	202

6 Existence of Solutions of the Cauchy Problem — 207
6.1	Front Tracking for Systems	208
6.2	Convergence	220
6.3	Notes	231

7 Well-Posedness of the Cauchy Problem — 235
7.1	Stability	240
7.2	Uniqueness	267
7.3	Notes	287

A Total Variation, Compactness, etc. — 289
A.1	Notes	300

B The Method of Vanishing Viscosity — 301
B.1	Notes	314

C Answers and Hints — 317

References — 351

Index — 361

Preface

> Все счастливые семьи похожи друг на
> друга, каждая несчастливая семья
> несчастлива по-своему.[1]
>
> Лев Толстой, Анна Каренина (1875)

While it is not strictly speaking true that all linear partial differential equations are the same, the theory that encompasses these equations can be considered well developed (and these are the happy families). Large classes of linear partial differential equations can be studied using linear functional analysis, which was developed in part as a tool to investigate important linear differential equations.

In contrast to the well-understood (and well-studied) classes of linear partial differential equations, each nonlinear equation presents its own particular difficulties. Nevertheless, over the last forty years some rather general classes of nonlinear partial differential equations have been studied and at least partly understood. These include the theory of viscosity solutions for Hamilton–Jacobi equations, the theory of Korteweg–de Vries equations, as well as the theory of hyperbolic conservation laws.

The purpose of this book is to present the modern theory of hyperbolic conservation laws in a largely self-contained manner. In contrast to the modern theory of linear partial differential equations, the mathematician

---

[1] All happy families resemble one another, but each unhappy family is unhappy in its own way (Leo Tolstoy, *Anna Karenina*).

interested in nonlinear hyperbolic conservation laws does not have to cover a large body of general theory to understand the results. Therefore, to follow the presentation in this book (with some minor exceptions), the reader does not have to be familiar with many complicated function spaces, nor does he or she have to know much theory of linear partial differential equations.

The methods used in this book are almost exclusively constructive, and largely based on the front-tracking construction. We feel that this gives the reader an intuitive feeling for the nonlinear phenomena that are described by conservation laws. In addition, front tracking is a viable numerical tool, and our book is also suitable for practical scientists interested in computations.

We focus on scalar conservation laws in several space dimensions and systems of hyperbolic conservation laws in one space dimension. In the scalar case we first discuss the one-dimensional case before we consider its multidimensional generalization. Multidimensional systems will not be treated. For multidimensional equations we combine front tracking with the method of dimensional splitting. We have included a chapter on standard difference methods that provides a brief introduction to the fundamentals of difference methods for conservation laws.

This book has grown out of courses we have given over some years: full-semester courses at the Norwegian University of Science and Technology and the University of Oslo, as well as shorter courses at Universität Kaiserslautern and S.I.S.S.A., Trieste.

We have taught this material for graduate and advanced undergraduate students. A solid background in real analysis and integration theory is an advantage, but key results concerning compactness and functions of bounded variation are proved in Appendix A.

Our main audience consists of students and researchers interested in analytical properties as well as numerical techniques for hyperbolic conservation laws.

We have benefited from the kind advice and careful proofreading of various versions of this manuscript by several friends and colleagues, among them Petter I. Gustafson, Runar Holdahl, Helge Kristian Jenssen, Kenneth H. Karlsen, Odd Kolbjørnsen, Kjetil Magnus Larsen, Knut-Andreas Lie, Achim Schroll. Special thanks are due to Harald Hanche-Olsen, who has helped us on several occasions with both mathematical and TEX-nical issues. We are also grateful to Trond Iden, from Ordkommisjonen, for helping with technical issues and software for making the figures.

Our research has been supported in part by the BeMatA program of the Research Council of Norway.

A list of corrections can be found at

   www.math.ntnu.no/~holden/FrontBook/

Whenever you find an error, please send us an email about it.

Preface xi

The logical interdependence of the material in this book is depicted in the diagram below. The main line, Chapters 1, 2, 5–7, has most of the emphasis on the theory for systems of conservation laws in one space dimension. Another possible track is Chapters 1–4, with emphasis on numerical methods and theory for scalar equations in one and several space dimensions.

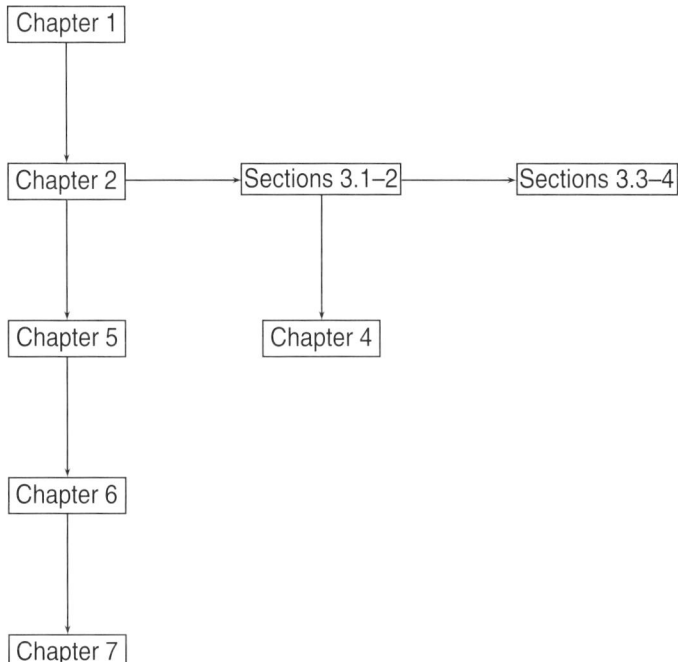

1
Introduction

> I have no objection to the use of the term "Burgers' equation" for the nonlinear heat equation (provided it is not written "Burger's equation").
>
> Letter from Burgers to Batchelor (1968)

Hyperbolic conservation laws are partial differential equations of the form

$$\frac{\partial u}{\partial t} + \nabla \cdot f(u) = 0.$$

If we write $f = (f_1, \ldots, f_m)$, $x = (x_1, x_2, \ldots, x_m) \in \mathbb{R}^m$, and introduce initial data u_0 at $t = 0$, the Cauchy problem for hyperbolic conservation laws reads

$$\frac{\partial u(x,t)}{\partial t} + \sum_{j=1}^{m} \frac{\partial}{\partial x_j} f_j\left(u(x,t)\right) = 0, \quad u|_{t=0} = u_0. \tag{1.1}$$

In applications t normally denotes the time variable, while x describes the spatial variation in m space dimensions. The unknown function u (as well as each f_j) can be a vector, in which case we say that we have a system of equations, or u and each f_j can be a scalar. This book covers the theory of scalar conservation laws in several space dimensions as well as the theory of systems of hyperbolic conservation laws in one space dimension. In the present chapter we study the one-dimensional scalar case to highlight some of the fundamental issues in the theory of conservation laws.

We use subscripts to denote partial derivatives, i.e., $u_t(x,t) = \partial u(x,t)/\partial t$. Hence we may write (1.1) when $m = 1$ as

$$u_t + f(u)_x = 0, \quad u|_{t=0} = u_0. \tag{1.2}$$

If we formally integrate equation (1.2) between two points x_1 and x_2, we obtain

$$\int_{x_1}^{x_2} u_t \, dx = -\int_{x_1}^{x_2} f(u)_x \, dx = f(u(x_1,t)) - f(u(x_2,t)).$$

Assuming that u is sufficiently regular to allow us to take the derivative outside the integral, we get

$$\frac{d}{dt}\int_{x_1}^{x_2} u(x,t) \, dx = f(u(x_1,t)) - f(u(x_2,t)). \tag{1.3}$$

This equation expresses conservation of the quantity measured by u in the sense that the rate of change in the amount of u between x_1 and x_2 is given by the difference in $f(u)$ evaluated at these points.[1] Therefore, it is natural to interpret $f(u)$ as the *flux density* of u. Often, $f(u)$ is referred to as the *flux function*.

As a simple example of a conservation law, consider a one-dimensional medium consisting of noninteracting particles, or material points, identified by their coordinates y along a line. Let $\phi(y,t)$ denote the position of material point y at a time t. The velocity and the acceleration of y at time t are given by $\phi_t(y,t)$ and $\phi_{tt}(y,t)$, respectively. Assume that for each t, $\phi(\,\cdot\,,t)$ is strictly increasing, so that two distinct material points cannot occupy the same location at the same time. Then the function $\phi(\,\cdot\,,t)$ has an inverse $\psi(\,\cdot\,,t)$, so that $y = \psi(\phi(y,t),t)$ for all t. Hence $x = \phi(y,t)$ is equivalent to $y = \psi(x,t)$. Now let u denote the velocity of the material point occupying position x at time t, i.e., $u(x,t) = \phi_t(\psi(x,t),t)$, or equivalently, $u(\phi(y,t),t) = \phi_t(y,t)$. Then the acceleration of material point y at time t is

$$\phi_{tt}(y,t) = u_t(\phi(y,t),t) + u_x(\phi(y,t),t)\phi_t(y,t)$$
$$= u_t(x,t) + u_x(x,t)u(x,t).$$

If the material particles are noninteracting, so that they exert no force on each other, and there is no external force acting on them, then Newton's second law requires the acceleration to be zero, giving

$$u_t + \left(\frac{1}{2}u^2\right)_x = 0. \tag{1.4}$$

---

[1] In physics one normally describes conservation of a quantity in integral form, that is, one starts with (1.3). The differential equation (1.2) then follows under additional regularity conditions on u.

The last equation, (1.4), is a conservation law; it expresses that u is conserved with a flux density given by $u^2/2$. This equation is often referred to as the *Burgers equation without viscosity*,[2] and is in some sense the simplest nonlinear conservation law.

Burgers' equation, and indeed any conservation law, is an example of a *quasilinear* equation, meaning that the highest derivatives occur linearly. A general inhomogeneous quasilinear equation for functions of two variables x and t can be written

$$a(x,t,u)u_t + b(x,t,u)u_x = c(x,t,u). \tag{1.5}$$

We may consider the solution as the surface $\{(t,x,u(x,t)) \,|\, (t,x) \in \mathbb{R}^2\}$ in $\mathbb{R}^3$. Let Γ be a given curve in $\mathbb{R}^3$ (which one may think of as the initial data if t is constant) parameterized by $(t(\eta), x(\eta), z(\eta))$. We want to construct a surface $S \subset \mathbb{R}^3$ parameterized by $(t,x,u(x,t))$ such that $u = u(x,t)$ satisfies (1.5) and $\Gamma \subset S$. To this end we solve the system of ordinary differential equations

$$\frac{\partial t}{\partial \xi} = a, \quad \frac{\partial x}{\partial \xi} = b, \quad \frac{\partial z}{\partial \xi} = c, \tag{1.6}$$

with

$$t(\xi_0, \eta) = t(\eta), \quad x(\xi_0, \eta) = x(\eta), \quad z(\xi_0, \eta) = z(\eta). \tag{1.7}$$

Assume that we can invert the relations $x = x(\xi, \eta)$, $t = t(\xi, \eta)$ and write $\xi = \xi(x,t)$, $\eta = \eta(x,t)$. Then

$$u(x,t) = z(\xi(x,t), \eta(x,t)) \tag{1.8}$$

satisfies both (1.5) and the condition $\Gamma \subset S$. However, there are many ifs in the above construction: The solution may only be local, and we may not be able to invert the solution of the differential equation to express (ξ, η) as functions of (x,t). These problems are intrinsic to equations of this type and will be discussed at length.

Equation (1.6) is called the *characteristic equation*, and its solutions are called *characteristics*. This can sometimes be used to find explicit solutions of conservation laws. In the homogeneous case, that is, when $c = 0$, the solution u is constant along characteristics, namely,

$$\frac{d}{d\xi} u(x(\xi, \eta), t(\xi, \eta)) = u_x x_\xi + u_t t_\xi = u_x b + u_t a = 0. \tag{1.9}$$

◇ **Example 1.1.**

Define the (quasi)linear equation

$$u_t - xu_x = -2u, \quad u(x,0) = x,$$

---

[2] Henceforth we will adhere to common practice and call it the inviscid Burgers' equation.

with associated characteristic equations
$$\frac{\partial t}{\partial \xi} = 1, \quad \frac{\partial x}{\partial \xi} = -x, \quad \frac{\partial z}{\partial \xi} = -2z.$$
The general solution of the characteristic equations reads
$$t = t_0 + \xi, \quad x = x_0 e^{-\xi}, \quad z = z_0 e^{-2\xi}.$$
Parameterizing the initial data for $\xi = 0$ by $t = 0$, $x = \eta$, and $z = \eta$, we obtain
$$t = \xi, \quad x = \eta e^{-\xi}, \quad z = \eta e^{-2\xi},$$
which can be inverted to yield
$$u = u(x,t) = z(\xi, \eta) = xe^{-t}.$$

◇

◇ **Example 1.2.**

Consider the (quasi)linear equation
$$xu_t - t^2 u_x = 0. \tag{1.10}$$
Its associated characteristic equation is
$$\frac{\partial t}{\partial \xi} = x, \quad \frac{\partial x}{\partial \xi} = -t^2.$$
This has solutions given implicitly by $x^2/2 + t^3/3 = \text{const}$, since after all, $\partial(x^2/2 + t^3/3)/\partial \xi = 0$, so the solution of (1.10) is any function φ of $x^2/2 + t^3/3$, i.e., $u(x,t) = \varphi(x^2/2 + t^3/3)$. For example, if we wish to solve the initial value problem $u(x,0) = \sin|x|$, then $u(x,0) = \varphi(x^2/2) = \sin|x|$. Consequently, $\varphi(x) = \sin\sqrt{2x}$, and the solution u is given by
$$u(x,t) = \sin\sqrt{x^2 + 2t^3/3}, \quad t \geq 0.$$

◇

◇ **Example 1.3 (Burgers' equation).**

If we apply this technique to Burgers' equation (1.4) with initial data $u(x,0) = u_0(x)$, we get that
$$\frac{\partial t}{\partial \xi} = 1, \quad \frac{\partial x}{\partial \xi} = z, \quad \text{and} \quad \frac{\partial z}{\partial \xi} = 0$$
with initial conditions $t(0,\eta) = 0$, $x(0,\eta) = \eta$, and $z(0,\eta) = u_0(\eta)$. We cannot solve these equations without knowing more about u_0, but since u (or z) is constant along characteristics, cf. (1.9), we see that

the characteristics are straight lines. In other words, the value of z is transported along characteristics, so that
$$t(\xi,\eta) = \xi, \quad x(\xi,\eta) = \eta + \xi z = \eta + \xi u_0(\eta), \quad z(\xi,\eta) = u_0(\eta).$$
We may write this as
$$x = \eta + u_0(\eta)t.$$
If we solve this equation in terms of $\eta = \eta(x,t)$, we can use η to obtain $u(x,t) = z(\xi,\eta) = u_0(\eta(x,t))$. We find that
$$\frac{\partial x}{\partial \eta} = 1 + tu_0'(\eta).$$
Thus a solution certainly exists for all $t > 0$ if $u_0' > 0$. On the other hand, if $u_0'(\tilde{x}) < 0$ for some $\tilde{x}$, then a solution cannot be found for $t > t^* = -1/u_0'(\tilde{x})$. For example, if $u_0(x) = -\arctan(x)$, there is no smooth solution for $t > 1$.

What actually happens when a smooth solution cannot be defined? From (1.3) we see that for $t > t^*$, there are several η that satisfy (1.3) for each x. In some sense, we can say that the solution u is multivalued at such points. To illustrate this, consider the surface in (t,x,u) space parameterized by ξ and η,
$$(\xi, \eta + \xi u_0(\eta), u_0(\eta)).$$
Let us assume that the initial data are given by $u_0(x) = -\arctan(x)$ and $t_0 = 0$. For each fixed t, the curve traced out by the surface is

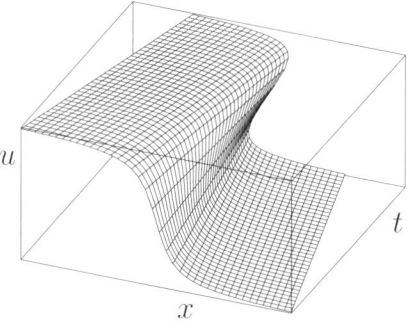

Figure 1.1. A multivalued solution.

the graph of a (multivalued) function of x. In Figure 1.1 we see how the multivaluedness starts at $t = 1$ when the surface "folds over," and that for $t > 1$ there are some x that have three associated u values. To continue the solution beyond $t = 1$ we have to choose among these three

u values. In any case, it is impossible to continue the solution and at the same time keep it continuous. ◇

Now we have seen that no matter how smooth the initial function is, we cannot expect to be able to define classical solutions of nonlinear conservation laws for all time. In this case we have to extend the concept of solution in order to allow discontinuities.

The standard way of extending the admissible set of solutions to differential equations is to look for *generalized functions*, or *distributions*, instead of smooth functions that satisfy the differential equation in the normal way. In the context of conservation laws, a distribution is a continuous linear functional on C_0^∞, the space of infinitely differentiable functions with compact support. Functions in C_0^∞ are often referred to as *test functions*.

In this book we use the following notation: $C^i(U)$ is the set of i times continuously differentiable functions on a set $U \subseteq \mathbb{R}^n$, and $C_0^i(U)$ denotes the set of such functions that have compact support in U. Then $C^\infty(U) = \bigcap_{i=0}^\infty C^i(U)$, and similarly for C_0^∞. Where there is no ambiguity, we sometimes omit the set U and write only C^0, etc.

Distributions are linear functionals $h\colon C_0^\infty \to \mathbb{R}$ with the following property: For any family $\phi_n \in C_0^\infty$ with $\operatorname{supp}\phi_n \subseteq K$ where K is compact and such that all derivatives satisfy $\phi_n^{(m)} \to \phi^{(m)}$ pointwise ($\phi^{(m)} = d^m\phi/dx^m$) for some function $\phi \in C_0^\infty$, we have that $h(\phi_n) \to h(\phi)$. Frequently we write $\langle h, \phi \rangle = h(\phi)$. Distributions are natural generalizations of functions, since if we have an integrable function h, we define the linear functional by

$$\langle h, \phi \rangle := \int_U h(x)\phi(x)\,dx, \tag{1.11}$$

for any ϕ in C_0^∞. It is not hard to see that (1.11) defines a continuous linear functional in the topology of uniform convergence of all derivatives on C_0^∞: If ϕ_k is some sequence such that $\phi_k^{(m)} \to \phi^{(m)}$ for all m, then $\langle h, \phi_k \rangle \to \langle h, \phi \rangle$. Hence h is a distribution.

Furthermore, if h is a distribution on $\mathbb{R}$, it possesses *distributional derivatives* of any order, for we define

$$\left\langle h^{(m)}, \phi \right\rangle := (-1)^m \left\langle h, \phi^{(m)} \right\rangle.$$

If h is an m times differentiable function, this definition coincides, using integration by parts, with the usual definition of $h^{(m)}(x)$. We can also define partial derivatives of distributions, e.g.,

$$\left\langle \frac{\partial h}{\partial x_i}, \phi \right\rangle = -\left\langle h, \frac{\partial \phi}{\partial x_i} \right\rangle.$$

Differentiation of distributions is linear with respect to multiplication by constants, and distributive over addition. Furthermore, if k denotes a constant distribution, $\langle k, \phi \rangle = k\int\phi$, then $k' = 0$, the zero distribution. Distributions can be multiplied by differentiable functions, and the result is a new distribution. If f is a smooth function and h a distribution, then fh is a distribution with $fh(\phi) = h(f\phi) = \langle h, f\phi \rangle$, and $(fh)' = f'h + fh'$. Two distributions, however, cannot in general be multiplied.

So, the solutions we are looking for are distributions on $C_0^\infty(\mathbb{R} \times [0, \infty))$. Thus we interpret (1.2) in the distributional sense, this means that a solution is a distribution u such that the partial derivatives u_t and $f(u)_x$ satisfy (1.2). Note that the last term $f(u)_x$ is *not* in general equal to $f'(u)u_x$; this is the case only if u is differentiable.

The above remarks also imply that $f(u)$ is not well-defined for all distributions u. We have seen that two distributions cannot always be multiplied, and in general we cannot use functions on distributions unless the distributions take pointwise values as functions. Such distributions always act on test functions as (1.11).

When considering conservation laws, we are usually interested in the initial value problem $u(x, 0) = u_0(x)$ for some given function u_0. This means that the distribution $u(x, 0)$ has to coincide with the distribution defined by the function $u_0(x)$,

$$\langle u(\cdot, 0), \phi \rangle = \langle u_0, \phi \rangle = \int_\mathbb{R} u_0(x)\phi(x)\, dx.$$

Since this is to hold for all $\phi \in C_0^\infty$, $u_0(x) = u(x, 0)$ almost everywhere with respect to the Lebesgue measure.

In this book we employ the usual notation that for any integer $p \in \mathbb{N}$, $L^p(U)$ denotes the set of all measurable functions F such that the integral

$$\int_U |F|^p\, dx$$

is finite. $L^p(U)$ is equipped with the norm

$$\|F\|_p = \|F\|_{L^p(U)} = \left(\int_U |F|^p dx \right)^{1/p}.$$

If $p = \infty$, $L^\infty(U)$ denotes the set of all measurable functions F such that

$$\operatorname{ess\,sup}_U |F|$$

is finite. The space L^∞ has the norm $\|F\|_\infty = \operatorname{ess\,sup}_U |F|$. In addition, we will frequently use the spaces

$$L^p_{\text{loc}}(U) = \{ f \in L^p(U) \mid f \in L^p(K) \text{ for every compact set } K \subseteq U \}.$$

With this notation, we see that u_0 must be equal to $u(x, 0)$ in $L^1(\mathbb{R})$. Thus, a distributional solution u to (1.2) taking the initial value $u_0(x)$ is a distribution such that $u_t + f(u)_x = 0$ as a distribution, and

$$\lim_{t \to 0} \int_\mathbb{R} u(x, t)\, dx = \int_\mathbb{R} u_0(x)\, dx.$$

Furthermore, the generalized solution u must be a function defined for almost all x and t.

Often it is convenient to use a formulation such that the initial condition is directly incorporated in the definition of a weak solution. If we consider test functions on the positive half-line $[0, \infty)$, and distributions v such that $v(0)$ makes sense, the definition of the distributional derivative is modified to

$$\langle v', \phi \rangle := -v(0)\phi(0) - \langle v, \phi' \rangle. \tag{1.12}$$

This definition is reasonable, since it must hold also if v is a function. In this case (1.12) is just the formula for partial integration. Similarly, if we consider test functions in $C_0^\infty(\mathbb{R} \times [0, \infty))$, the partial derivative of a generalized solution u with respect to t is defined to be

$$\langle u_t, \phi \rangle := -\int_\mathbb{R} u(x,0)\phi(x,0)\,dx - \left\langle u, \frac{\partial \phi}{\partial t} \right\rangle.$$

Remembering that a generalized solution is a function defined for almost all x and t, we see that

$$0 = \langle 0, \phi \rangle = \langle u_t + f(u)_x, \phi \rangle$$
$$= -\int_\mathbb{R} u(x,0)\phi(x,0)\,dx - \left\langle u, \frac{\partial \phi}{\partial t} \right\rangle - \left\langle f(u), \frac{\partial \phi}{\partial x} \right\rangle$$
$$= -\int_\mathbb{R} u(x,0)\phi(x,0)\,dx - \int_{[0,\infty)} \int_\mathbb{R} u\phi_t\,dx\,dt$$
$$\quad - \int_\mathbb{R} \int_{[0,\infty)} f(u)\phi_x\,dt\,dx$$
$$= -\int_\mathbb{R} \int_{[0,\infty)} (u\phi_t + f(u)\phi_x)\,dt\,dx - \int_\mathbb{R} u(x,0)\phi(x,0)\,dx.$$

Consequently, a solution must satisfy

$$\iint_{\mathbb{R} \times [0,\infty)} (u\phi_t + f(u)\phi_x)\,dt\,dx + \int_\mathbb{R} u_0(x)\phi(x,0)\,dx = 0 \tag{1.13}$$

for all test functions $\phi \in C_0^\infty(\mathbb{R} \times [0, \infty))$. A function satisfying (1.13) is called a *weak solution* of the initial value problem (1.2). Observe in particular that a (regular) smooth solution is a weak solution as well.

So what kind of discontinuities are compatible with (1.13)? If we assume that u is constant outside some finite interval, the remarks below (1.2) imply that

$$\frac{d}{dt} \int_{-\infty}^\infty u(x,t)\,dx = 0.$$

Hence, the total amount of u is independent of time, or the area below the graph of $u(\,\cdot\,, t)$ is constant.

◇ Example 1.4 (Burgers' equation (cont'd.)).

We now wish to determine a discontinuous function such that the graph of the function lies on the surface given earlier with $u(x,0) = -\arctan x$. Furthermore, the area under the graph of the function should be equal to the area between the x-axis and the surface. In Figure 1.2 we see a section of the surface making up the solution for $t = 3$. The curve is parameterized by x_0, and explicitly given by $u = -\arctan(x_0)$, $x = x_0 - 3\arctan(x_0)$. The function u is shown by a thick line, and the

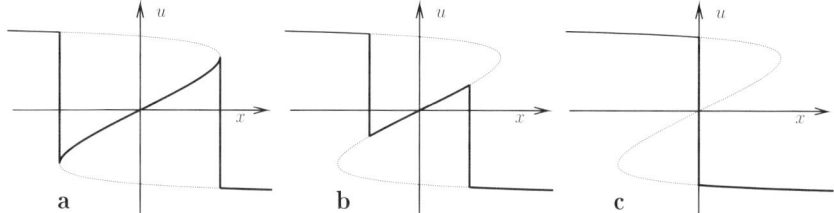

Figure 1.2. Different solutions with u conserved.

surface is shown by a dotted line. A function $u(x)$ that has the correct integral, $\int u\,dx = \int u_0\,dx$, is easily found by making any cut from the upper fold to the middle fold at some negative x_c with $x_c \geq -\sqrt{2}$, and then making a cut from the middle part to the lower part at $-x_c$. We see that in all cases, the area below the thick line is the same as the area bounded by the curve $(x(x_0), u(x_0))$. Consequently, conservation of u is not sufficient to determine a unique weak solution. ◇

Let us examine what kind of discontinuities are compatible with (1.13) in the general case. Assume that we have an isolated discontinuity that moves along a smooth curve $\Gamma \colon x = x(t)$. That the discontinuity is isolated means that the function $u(x,t)$ is differentiable in a sufficiently small neighborhood of $x(t)$ and satisfies equation (1.2) classically on each side of $x(t)$. We also assume that u is uniformly bounded in a neighborhood of the discontinuity.

Now we choose a neighborhood D around the point $(x(t),t)$ and a test function $\phi(x,t)$ whose support lies entirely inside the neighborhood. The situation is as depicted in Figure 1.3. The neighborhood consists of two parts D_1 and D_2, and is chosen so small that u is differentiable everywhere inside D except on $x(t)$. Let D_i^ε denote the set of points

$$D_i^\varepsilon = \{(x,t) \in D_i \mid \mathrm{dist}((x,t),(x(t),t)) > \varepsilon\}.$$

The function u is bounded, and hence

$$0 = \int_D (u\phi_t + f(u)\phi_x)\,dx\,dt = \lim_{\varepsilon \to 0} \int_{D_1^\varepsilon \cup D_2^\varepsilon} (u\phi_t + f(u)\phi_x)\,dx\,dt. \quad (1.14)$$

Since u is a classical solution inside each D_i^ε, we can use Green's theorem

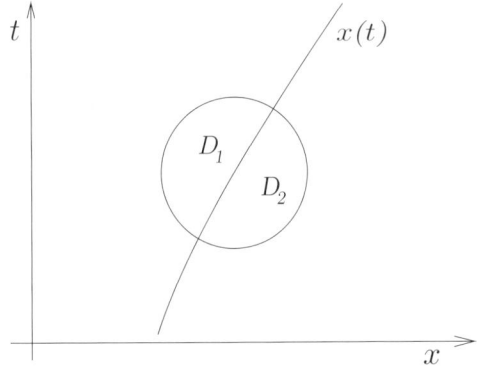

Figure 1.3. An isolated discontinuity.

and obtain

$$\int_{D_i^\varepsilon} (u\phi_t + f(u)\phi_x) \, dx \, dt = \int_{D_i^\varepsilon} (u\phi_t + f(u)\phi_x + (u_t + f(u)_x)\phi) \, dx \, dt$$

$$= \int_{D_i^\varepsilon} ((u\phi)_t + (f(u)\phi)_x) \, dx \, dt$$

$$= \int_{\partial D_i^\varepsilon} \phi \left(-u \, dx + f(u) \, dt \right). \quad (1.15)$$

But ϕ is zero everywhere on ∂D_i^ε except in the vicinity of $x(t)$. Let Γ_i^ε denote this part of ∂D_i^ε. Then

$$\lim_{\varepsilon \to 0} \int_{\Gamma_i^\varepsilon} \phi \left(-u \, dx + f(u) \, dt \right) = \pm \int_{\Gamma \cap D} \phi \left(-u_{l,r} \, dx + f_{l,r} \, dt \right).$$

Here u_l denotes the limit of $u(x,t)$ as $x \to x(t)-$, and u_r the limit as x approaches $x(t)$ from the right, i.e., $u_r = \lim_{x \to x(t)+} u(x,t)$. Similarly, $f_{l,r}$ denotes $f(u_{l,r})$. The reason for the difference in sign is that according to Green's theorem, we must integrate along the boundary counterclockwise. Therefore, the positive sign holds for $i = 1$, and the negative for $i = 2$. Using (1.14) we obtain

$$\int_{\Gamma \cap D} \phi \left[-\left(u_l(t) - u_r(t) \right) dx + \left(f_l(t) - f_r(t) \right) dt \right] = 0.$$

Since this is to hold for all test functions ϕ, we must have

$$s \left(u_l - u_r \right) = f_l - f_r, \quad (1.16)$$

where $s = x'(t)$. This equality is called the *Rankine–Hugoniot condition*, and it expresses conservation of u across jump discontinuities. It is common in the theory of conservation laws to introduce a notation for the jump in

a quantity. Write
$$[\![a]\!] = a_r - a_l \qquad (1.17)$$
for the jump in any quantity a. With this notation the Rankine–Hugoniot relation takes the form
$$s[\![u]\!] = [\![f]\!].$$

◇ **Example 1.5 (Burgers' equation (cont'd.)).**

For Burgers' equation we see that the shock speed must satisfy
$$s = \frac{[\![u^2/2]\!]}{[\![u]\!]} = \frac{(u_r^2 - u_l^2)}{2(u_r - u_l)} = \frac{1}{2}(u_l + u_r).$$
Consequently, the left shock in parts **a** and **b** in Figure 1.2 above will have greater speed than the right shock. Therefore, solutions of the type **a** or **b** cannot be isolated discontinuities moving along two trajectories starting at $t = 1$. Type **c** yields a stationary shock. ◇

◇ **Example 1.6 (Traffic flow).**

> I am ill at these numbers.
>
> W. Shakespeare, Hamlet (1603)

Rather than continue to develop the theory, we shall now consider an example of a conservation law in some detail. We will try to motivate how a conservation law can model the flow of cars on a crowded highway.

Consider a road consisting of a single lane, with traffic in one direction only. The road is parameterized by a single coordinate x, and we assume that the traffic moves in the direction of increasing x.

Suppose we position ourselves at a point x on the road and observe the number of cars $N = N(x, t, h)$ in the interval $[x, x+h]$. If some car is located at the boundary of this interval, we account for that by allowing N to take any real value. If the traffic is dense, and if h is large compared with the average length of a car, but at the same time small compared with the length of our road, we can introduce the density given by
$$\rho(x, t) = \lim_{h \to 0} \frac{N(x, t, h)}{h}.$$
Then $N(x, t, h) = \int_x^{x+h} \rho(y, t)\, dy$.

Let now the position of some vehicle be given by $r(t)$, and its velocity by $v(r(t), t)$. Considering the interval $[a, b]$, we wish to determine how the number of cars changes in this interval. Since we have assumed that there are no entries or exits on our road, this number can change only as cars are entering the interval from the left endpoint, or leaving the

interval at the right endpoint. The rate of cars passing a point x at some time t is given by
$$v(x,t)\rho(x,t).$$
Consequently,
$$-(v(b,t)\rho(b,t) - v(a,t)\rho(a,t)) = \frac{d}{dt}\int_a^b \rho(y,t)\,dy.$$
Comparing this with (1.3) and (1.2), we see that the density satisfies the conservation law
$$\rho_t + (\rho v)_x = 0. \qquad (1.18)$$

In the simplest case we assume that the velocity u is given as a function of the density ρ only. This may be a good approximation if the road is uniform and does not contain any sharp bends or similar obstacles that force the cars to slow down. It is also reasonable to assume that there is some maximal speed $v_{\max}$ that any car can attain. When traffic is light, a car will drive at this maximum speed, and as the road gets more crowded, the cars will have to slow down, until they come to a complete standstill as the traffic stands bumper to bumper. Hence, we assume that the velocity v is a monotone decreasing function of ρ such that $v(0) = v_{\max}$ and $v(\rho_{\max}) = 0$. The simplest such function is a linear function, resulting in a flux function given by
$$f(\rho) = v\rho = \rho v_{\max}\left(1 - \frac{\rho}{\rho_{\max}}\right). \qquad (1.19)$$
For convenience we normalize by introducing $u = \rho/\rho_{\max}$ and $\tilde{x} = v_{\max}x$. The resulting normalized conservation law reads
$$u_t + (u(1-u))_x = 0. \qquad (1.20)$$
Setting $\tilde{u} = \frac{1}{2} - u$, we recover Burgers' equation, but this time with a new interpretation of the solution.

Let us solve an initial value problem explicitly by the method of characteristics described earlier. We wish to solve (1.20), with initial function $u_0(x)$ given by
$$u_0(x) = u(x,0) = \begin{cases} \frac{3}{4} & \text{for } x \leq -a, \\ \frac{1}{2} - x/(4a) & \text{for } -a < x < a, \\ \frac{1}{4} & \text{for } a \leq x. \end{cases}$$

The characteristics satisfy $t'(\xi) = 1$ and $x'(\xi) = 1 - 2u(x(\xi), t(\xi))$. The solution of these equations is given by $x = x(t)$, where

$$x(t) = \begin{cases} x_0 - t/2 & \text{for } x_0 < -a, \\ x_0 + x_0 t/(2a) & \text{for } -a \le x_0 \le a, \\ x_0 + t/2 & \text{for } a < x_0. \end{cases}$$

Inserting this into the solution $u(x,t) = u_0(x_0(x,t))$, we find that

$$u(x,t) = \begin{cases} \frac{3}{4} & \text{for } x \le -a - t/2, \\ \frac{1}{2} - x/(4a + 2t) & \text{for } -a - t/2 < x < a + t/2, \\ \frac{1}{4} & \text{for } a + t/2 \le x. \end{cases}$$

This solution models a situation where the traffic density initially is small for positive x, and high for negative x. If we let a tend to zero, the solution reads

$$u(x,t) = \begin{cases} \frac{3}{4} & \text{for } x \le -t/2, \\ \frac{1}{2} - x/(2t) & \text{for } -t/2 < x < t/2, \\ \frac{1}{4} & \text{for } t/2 \le x. \end{cases}$$

As the reader may check directly, this is also a classical solution everywhere except at $x = \pm t/2$. It takes discontinuous initial values:

$$u(x,0) = \begin{cases} \frac{3}{4} & \text{for } x < 0, \\ \frac{1}{4} & \text{otherwise.} \end{cases} \tag{1.21}$$

This initial function may model the situation when a traffic light turns green at $t = 0$. Facing the traffic light the density is high, while on the other side of the light there is a small constant density.

Initial value problems of the kind (1.21), where the initial function consists of two constant values, are called *Riemann problems*. We will discuss Riemann problems at great length in this book.

If we simply insert $u_l = \frac{3}{4}$ and $u_r = \frac{1}{4}$ in the Rankine–Hugoniot condition (1.16), we find another weak solution to this initial value problem. These left and right values give $s = 0$, so the solution found here is simply $u_2(x,t) = u_0(x)$. A priori, this solution is no better or worse than the solution computed earlier. But when we examine the situation the equation is supposed to model, the second solution u_2 is unsatisfactory, since it describes a situation where the traffic light is green, but the density of cars facing the traffic light does not decrease!

In the first solution the density decreased. Examining the model a little more closely we find, perhaps from experience of traffic jams, that the allowable discontinuities are those in which the density is increasing. This corresponds to the situation where there is a traffic jam ahead, and we suddenly have to slow down when we approach it.

When we emerge from a traffic jam, we experience a gradual decrease in the density of cars around us, not a sudden jump from a bumper to bumper situation to a relatively empty road.

We have now formulated a condition, in addition to the Rankine–Hugoniot condition, that allows us to reduce the number of weak solutions to our conservation law. This condition says, *Any weak solution u has to increase across discontinuities.* Such conditions are often called *entropy conditions*. This terminology comes from gas dynamics, where similar conditions state that the physical entropy has to increase across any discontinuity.

Let us consider the opposite initial value problem, namely,

$$u_0(x) = \begin{cases} \frac{1}{4} & \text{for } x < 0, \\ \frac{3}{4} & \text{for } x \geq 0. \end{cases}$$

Now the characteristics starting at negative x_0 are given by $x(t) = x_0 + t/2$, and the characteristics starting on the positive half-line are given by $x(t) = x_0 - t/2$. We see that these characteristics immediately will run into each other, and therefore the solution is multivalued for any positive time t. Thus there is no hope of finding a continuous solution to this initial value problem for any time interval $\langle 0, \delta \rangle$, no matter how small δ is. When inserting the initial $u_l = \frac{1}{4}$ and $u_r = \frac{3}{4}$ into the Rankine–Hugoniot condition, we see that the initial function is already a weak solution. This time, the solution increases across the discontinuity, and therefore satisfies our entropy condition. Thus, an admissible solution is given by $u(x,t) = u_0(x)$.

Now we shall attempt to solve a more complicated problem in some detail. Assume that we have a road with a uniform density of cars initially. At $t = 0$ a traffic light placed at $x = 0$ changes from green to red. It remains red for some time interval Δt, then turns green again and stays green thereafter. We assume that the initial uniform density is given by $u = \frac{1}{2}$, and we wish to determine the traffic density for $t > 0$.

When the traffic light initially turns red, the situation for the cars to the left of the traffic light is the same as when the cars stand bumper to bumper to the right of the traffic light. So in order to determine the situation for t in the interval $[0, \Delta t \rangle$, we must solve the Riemann problem with the initial function

$$u_0^l(x) = \begin{cases} \frac{1}{2} & \text{for } x < 0, \\ 1 & \text{for } x \geq 0. \end{cases} \tag{1.22}$$

For the cars to the right of the traffic light, the situation is similar to the situation when the traffic abruptly stopped at $t = 0$ behind the car located at $x = 0$. Therefore, to determine the density for $x > 0$ we have to solve the Riemann problem given by

$$u_0^r(x) = \begin{cases} 0 & \text{for } x < 0, \\ \frac{1}{2} & \text{for } x \geq 0. \end{cases} \tag{1.23}$$

Returning to (1.22), here u is increasing over the initial discontinuity, so we can try to insert this into the Rankine–Hugoniot condition. This gives

$$s = \frac{f_r - f_l}{u_r - u_l} = \frac{\frac{1}{4} - 0}{\frac{1}{2} - 1} = -\frac{1}{2}.$$

Therefore, an admissible solution for $x < 0$ and t in the interval $[0, \Delta t)$ is given by

$$u^l(x,t) = \begin{cases} \frac{1}{2} & \text{for } x < -t/2, \\ 1 & \text{for } x \geq -t/2. \end{cases}$$

This is indeed close to what we experience when we encounter a traffic light. We see the discontinuity approaching as the brake lights come on in front of us, and the discontinuity has passed us when we have come to a halt. Note that although each car moves only in the positive direction, the discontinuity moves to the left.

In general, there are three different speeds when we study conservation laws: the particle speed, in our case the speed of each car, the characteristic speed; and the speed of a discontinuity. These three speeds are not equal if the conservation law is nonlinear. In our case, the speed of each car is nonnegative, but both the characteristic speed and the speed of a discontinuity may take both positive and negative values. Note that the speed of an admissible discontinuity is less than the characteristic speed to the left of the discontinuity, and larger than the characteristic speed to the right. This is a general feature of admissible discontinuities.

It remains to determine the density for positive x. The initial function given by (1.23) also has a positive jump discontinuity, so we obtain an admissible solution if we insert it into the Rankine–Hugoniot condition. Then we obtain $s = \frac{1}{2}$, so the solution for positive x is

$$u^r(x,t) = \begin{cases} 0 & \text{for } x < t/2, \\ \frac{1}{2} & \text{for } x \geq t/2. \end{cases}$$

Piecing together u^l and u^r we find that the density u in the time interval $[0, \Delta t\rangle$ reads

$$u(x,t) = \begin{cases} \frac{1}{2} & \text{for } x \leq -t/2, \\ 1 & \text{for } -t/2 < x \leq 0, \\ 0 & \text{for } 0 < x \leq t/2, \\ \frac{1}{2} & \text{for } t/2 < x, \end{cases} \quad t \in [0, \Delta t\rangle.$$

What happens for $t > \Delta t$? To find out, we have to solve the Riemann problem

$$u(x, \Delta t) = \begin{cases} 1 & \text{for } x < 0, \\ 0 & \text{for } x \geq 0. \end{cases}$$

Now the initial discontinuity is not acceptable according to our entropy condition, so we have to look for some other solution. We can try to mimic the example above where we started with a nonincreasing initial function that was linear on some small interval $\langle -a, a \rangle$. Therefore, let $v(x,t)$ be the solution of the initial value problem

$$v_t + (v(1-v))_x = 0,$$

$$v(x,0) = v_0(x) = \begin{cases} 1 & \text{for } x < -a, \\ \frac{1}{2} - x/(2a) & \text{for } -a \leq x < a, \\ 0 & \text{for } a \leq x. \end{cases}$$

As in the above example, we find that the characteristics are not overlapping, and they fill out the positive half-plane exactly. The solution is given by $v(x,t) = v_0(x_0(x,t))$:

$$v(x,t) = \begin{cases} 1 & \text{for } x < -a - t, \\ \frac{1}{2} - x/(2a + 2t) & \text{for } -a - t \leq x < a + t, \\ 0 & \text{for } a + t \leq x. \end{cases}$$

Letting $a \to 0$, we obtain the solution to the Riemann problem with a left value 1 and a right value 0. For simplicity we also denote this function by $v(x,t)$.

This type of solution can be depicted as a "fan" of characteristics emanating from the origin, and it is called a *centered rarefaction wave*, or sometimes just a *rarefaction wave*. The origin of this terminology lies in gas dynamics.

We see that the rarefaction wave, which is centered at $(0, \Delta t)$, does not immediately influence the solution away from the origin. The leftmost part of the wave moves with a speed -1, and the front of the wave moves with speed 1. So for some time after Δt, the density is obtained by piecing together three solutions, $u^l(x,t)$, $v(x, t - \Delta t)$, and $u^r(x,t)$.

The rarefaction wave will of course catch up with the discontinuities in the solutions u^l and u^r. Since the speeds of the discontinuities are $\mp\frac{1}{2}$, and the speeds of the rear and the front of the rarefaction wave are ∓ 1, and the rarefaction wave starts at $(0, \Delta t)$, we conclude that this will happen at $(\mp \Delta t, 2\Delta t)$.

It remains to compute the solution for $t > 2\Delta t$. Let us start with examining what happens for positive x. Since the u values that are transported along the characteristics in the rarefaction wave are less than $\frac{1}{2}$, we can construct an admissible discontinuity by using the Rankine–Hugoniot condition (1.16). Define a function that has a discontinuity moving along a path $x(t)$. The value to the right of the discontinuity is $\frac{1}{2}$, and the value to the left is determined by $v(x, t - \Delta t)$. Inserting this into (1.16) we get

$$x'(t) = s = \frac{\frac{1}{4} - \left(\frac{1}{2} + \frac{x}{2(t-\Delta t)}\right)\left(\frac{1}{2} - \frac{x}{2(t-\Delta t)}\right)}{\frac{1}{2} - \left(\frac{1}{2} - \frac{x}{2(t-\Delta t)}\right)} = \frac{x}{2(t-\Delta t)}.$$

Since $x(2\Delta t) = \Delta t$, this differential equation has solution

$$x_+(t) = \sqrt{\Delta t(t - \Delta t)}.$$

The situation is similar for negative x. Here, we use the fact that the u values in the left part of the rarefaction fan are larger than $\frac{1}{2}$. This gives a discontinuity with a left value $\frac{1}{2}$ and right values taken from the rarefaction wave. The path of this discontinuity is found to be $x_-(t) = -x_+(t)$.

Now we have indeed found a solution that is valid for all positive time. This function has the property that it is a classical solution at all points x and t where it is differentiable, and it satisfies both the Rankine–Hugoniot condition and the entropy condition at points of discontinuity. We show this weak solution in Figure 1.4, both in the (x, t) plane, where we show characteristics and discontinuities, and u as a function of x for various times. The characteristics are shown as gray lines, and the discontinuities as thicker black lines. This concludes our example. Note that we have been able to find the solution to a complicated initial value problem by piecing together solutions from Riemann problems. This is indeed the main idea behind front tracking, and a theme to which we shall give considerable attention in this book. ◇

18   1. Introduction

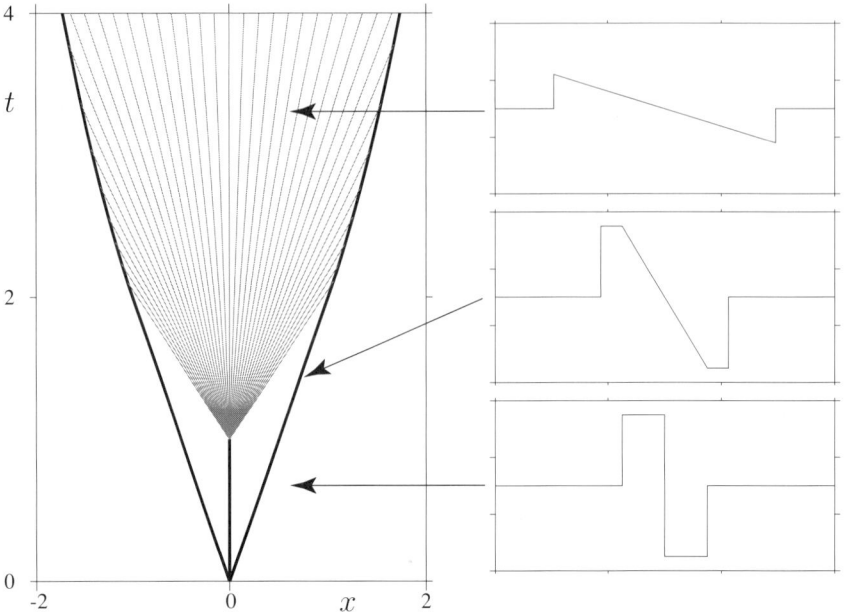

Figure 1.4. A traffic light on a single road. To the left we show the solution in (x,t), and to the right the solution $u(x,t)$ at three different times t.

1.1 Notes

> Never any knowledge was delivered in the same order it was invented.[3]
>
> Sir Francis Bacon (1561–1626)

The simplest nontrivial conservation law, the inviscid Burgers' equation, has been extensively analyzed. Burgers introduced the "nonlinear diffusion equation"

$$u_t + \frac{1}{2}(u^2)_x = u_{xx}, \qquad (1.24)$$

which is currently called (the viscous) Burgers' equation, in 1940 [25] (see also [26]) as a model of turbulence. Burgers' equation is linearized, and thereby solved, by the Cole–Hopf transformation [33], [73]. Both the equation and the Cole–Hopf transformation were, however, known already in 1906; see Forsyth [48, p. 100]. See also Bateman [10].

The most common elementary example of application of scalar conservation laws is the model of traffic flow called "traffic hydrodynamics" that was

---

[3]in *Valerius Terminus: Of the Interpretation of Nature*, c. 1603.

introduced independently by Lighthill and Whitham [101] and Richards [118]. A modern treatment can be found in Haberman [61].

The jump condition, or the Rankine–Hugoniot condition, was derived heuristically from the conservation principle independently by Rankine in 1870 [116] and Hugoniot in 1886 [76, 77, 78]. Our presentation of the Rankine–Hugoniot condition is taken from Smoller [130].

The notion of "Riemann problem" is fundamental in the theory of conservation laws. It was introduced by Riemann in 1859 [119, 120] in the context of gas dynamics. He studied the situation where one initially has two gases with different (constant) pressures and densities separated by a thin membrane in a one-dimensional cylindrical tube. See [72] for a historical discussion.

There are by now several books on various aspects of hyperbolic conservation laws, starting with the classical book by Courant and Friedrichs [37]. Nice treatments with emphasis on the mathematical theory can be found in books by Lax [96], Chorin and Marsden [29], Roždestvenskiĭ and Janenko [126], Smoller [130], Málek et al. [107], Hörmander [74], Liu [104], Serre [128, 129], Bressan [18], and Dafermos [42]. The books by Godlewski and Raviart [58, 59], LeVeque [98], Kröner [86], Toro [140], and Thomas [139] focus more on the numerical theory.

Exercises

1.1 Determine characteristics for the following quasilinear equations:
$$u_t + \sin(x)u_x = u,$$
$$\sin(t)u_t + \cos(x)u_x = 0,$$
$$u_t + \sin(u)u_x = u,$$
$$\sin(u)u_t + \cos(u)u_x = 0.$$

1.2 Use characteristics to solve the following initial value problems:

a.
$$uu_x + xu_y = 0, \quad u(0,s) = 2s \text{ for } s > 0.$$

b.
$$e^y u_x + uu_y + u^2 = 0, \quad u(x,0) = 1/x \text{ for } x > 0.$$

c.
$$xu_y - yu_x = u, \quad u(x,0) = h(x) \text{ for } x > 0.$$

d.
$$(x+1)^2 u_x + (y-1)^2 u_y = (x+y)u, \quad u(x,0) = -1 - x.$$

e.
$$u_x + 2xu_y = x + xu, \quad u(1,y) = e^y - 1.$$

f.
$$u_x + 2xu_y = x + xu, \quad u(0,y) = y^2 - 1.$$

g.
$$xuu_x + u_y = 2y, \quad u(x,0) = x.$$

1.3 Find the shock condition (i.e., the Rankine–Hugoniot condition) for one-dimensional systems, i.e., the unknown u is a vector $u = (u_1, \ldots, u_n)$ for some $n > 1$, and also $f(u) = (f_1(u), \ldots, f_n(u))$.

1.4 Consider a scalar conservation law in two space dimensions,
$$u_t + \frac{\partial f(u)}{\partial x} + \frac{\partial g(u)}{\partial y} = 0,$$
where the flux functions f and g are continuously differentiable. Now the unknown u is a function of x, y, and t. Determine the Rankine–Hugoniot condition across a jump discontinuity in u, assuming that u jumps across a regular surface in (x, y, t). Try to generalize your answer to a conservation law in n space dimensions.

1.5 We shall consider a linearization of Burgers' equation. Let
$$u_0(x) = \begin{cases} 1 & \text{for } x < -1, \\ -x & \text{for } -1 \leq x \leq 1, \\ -1 & \text{for } 1 < x. \end{cases}$$

a. First determine the maximum time that the solution of the initial value problem
$$u_t + \frac{1}{2}\left(u^2\right)_x = 0, \quad u(x,0) = u_0(x),$$
will remain continuous. Find the solution for t less than this time.

b. Then find the solution v of the linearized problem
$$v_t + u_0(x)v_x = 0, \quad v(x,0) = u_0(x).$$
Determine the solution also in the case where $v(x,0) = u_0(\alpha x)$ where α is nonnegative.

c. Next, we shall determine a procedure for finding u by solving a sequence of linearized equations. Fix $n \in \mathbb{N}$. For t in the interval $\langle m/n, (m+1)/n]$ and $m \geq 0$, let v_n solve
$$(v_n)_t + v_n(x, m/n)(v_n)_x = 0,$$

and set $v_n(x, 0) = u_0(x)$. Then show that
$$v_n\left(x, \frac{m}{n}\right) = u_0\left(\alpha_{m,n} x\right)$$
and find a recursive relation (in m) satisfied by $\alpha_{m,n}$.

d. Assume that
$$\lim_{n \to \infty} \alpha_{m,n} = \bar{a}(t),$$
for some continuously differentiable $\bar{a}(t)$, where $t = m/n < 1$. Show that $\bar{a}(t) = 1/(1 - t)$, and thus $v_n(x) \to u(x)$ for $t < 1$. What happens for $t \geq 1$?

1.6   a. Solve the initial value problem for Burgers' equation
$$u_t + \frac{1}{2}\left(u^2\right)_x = 0, \quad u(x, 0) = \begin{cases} -1 & \text{for } x < 0, \\ 1 & \text{for } x \geq 0. \end{cases} \quad (1.25)$$

b. Then find the solution where the initial data are
$$u(x, 0) = \begin{cases} 1 & \text{for } x < 0, \\ -1 & \text{for } x \geq 0. \end{cases}$$

c. If we multiply Burgers' equation by u, we formally find that u satisfies
$$\frac{1}{2}\left(u^2\right)_t + \frac{1}{3}\left(u^3\right)_x = 0. \quad (1.26)$$
Are the solutions to (1.25) you found in parts **a** and **b** weak solutions to (1.26)? If not, then find the corresponding weak solutions to (1.26). *Warning: This shows that manipulations valid for smooth solutions are not necessarily true for weak solutions.*

1.7 ([130, p. 250]) Show that
$$u(x, t) = \begin{cases} 1 & \text{for } x \leq (1 - \alpha)t/2, \\ -\alpha & \text{for } (1 - \alpha)t/2 < x \leq 0, \\ \alpha & \text{for } 0 < x \leq (\alpha - 1)t/2, \\ -1 & \text{for } x \geq (\alpha - 1)t/2 \end{cases}$$
is a weak solution of
$$u_t + \left(\frac{1}{2}u^2\right)_x = 0, \quad u(x, 0) = \begin{cases} 1 & \text{for } x \leq 0, \\ -1 & \text{for } x > 0, \end{cases}$$
for all $\alpha \geq 1$. *Warning: Thus we see that weak solutions are not necessarily unique.*

2
Scalar Conservation Laws

> It is a capital mistake to theorise before one has data. Insensibly one begins to twist facts to suit theories, instead of theories to suit facts.
>
> *Sherlock Holmes, A Scandal in Bohemia (1891)*

In this chapter we consider the Cauchy problem for a scalar conservation law. Our goal is to show that, subject to certain conditions, there exists a unique solution to the general initial value problem. Our method will be completely constructive, and we shall exhibit a procedure by which this solution can be constructed. This procedure is, of course, front tracking. The basic ingredient in the front-tracking algorithm is the solution of the Riemann problem.

Already in the example on traffic flow, we observed that conservation laws may have several weak solutions, and that some principle is needed to pick out the correct ones. The problem of lack of uniqueness for weak solutions is intrinsic in the theory of conservation laws. There are by now several different approaches to this problem, and they are commonly referred to as "entropy conditions."

Thus the solution of Riemann problems requires some mechanism to choose one of possibly several weak solutions. Therefore, before we turn to front tracking, we will discuss entropy conditions.

2.1 Entropy Conditions

We study the conservation law[1]

$$u_t + f(u)_x = 0, \tag{2.1}$$

whose solutions $u = u(x,t)$ are to be understood in the distributional sense; see (1.13). We will not state any continuity properties of f, but tacitly assume that f is sufficiently smooth for all subsequent formulas to make sense.

One of the most common entropy conditions is so-called viscous regularization, where the scalar conservation law $u_t + f(u)_x = 0$ is replaced by $u_t + f(u)_x = \epsilon u_{xx}$. The idea is that the physical problem has some diffusion, and that the conservation law represents a limit model when the diffusion is small. Based on this, one is looking for solutions of the conservation law that are limits of the regularized equation when $\epsilon \to 0$.

Therefore, we are interested in the *viscous regularization* of the conservation law (2.1),

$$u_t^\varepsilon + f(u^\varepsilon)_x = \varepsilon u_{xx}^\varepsilon, \tag{2.2}$$

as $\varepsilon \to 0$. In order for this equation to be well posed, ε must be nonnegative. Equations such as (2.2) are called viscous, since the right-hand side u_{xx}^ε models the effect of viscosity or diffusion. We then demand that the distributional solutions of (2.1) should be limits of solutions of the more fundamental equation (2.2) as the viscous term disappears.

This has some interesting consequences. Assume that (2.1) has a solution consisting of constant states on each side of a discontinuity moving with a speed s, i.e.,

$$u(x,t) = \begin{cases} u_l & \text{for } x < st, \\ u_r & \text{for } x \geq st. \end{cases} \tag{2.3}$$

We say that $u(x,t)$ satisfies a *traveling wave entropy condition* if $u(x,t)$ is the pointwise limit almost everywhere of some $u^\varepsilon(x,t) = U((x-st)/\varepsilon)$ as $\varepsilon \to 0$, where u^ε solves (2.2) in the classical sense.

Inserting $U((x-st)/\varepsilon)$ into (2.2), we obtain

$$-s\dot{U} + \frac{df(U)}{d\xi} = \ddot{U}. \tag{2.4}$$

Here $U = U(\xi)$, $\xi = (x-st)/\varepsilon$, and $\dot{U}$ denotes the derivative of U with respect to ξ. This equation can be integrated once, yielding

$$\dot{U} = -sU + f(U) + A, \tag{2.5}$$

---

[1] The analysis up to and including (2.7) could have been carried out for systems on the line as well.

2.1. Entropy Conditions 25

where A is a constant of integration. We see that as $\varepsilon \to 0$, ξ tends to plus or minus infinity, depending on whether $x - st$ is positive or negative.

If u should be the limit of u^ε, we must have that

$$\lim_{\varepsilon \to 0} u^\varepsilon = \lim_{\varepsilon \to 0} U(\xi) = \begin{cases} u_l & \text{for } x < st, \\ u_r & \text{for } x > st, \end{cases} = \begin{cases} \lim_{\xi \to -\infty} U(\xi), \\ \lim_{\xi \to +\infty} U(\xi). \end{cases}$$

It follows that

$$\lim_{\xi \to \pm\infty} \dot{U}(\xi) = 0.$$

Inserting this into (2.5), we obtain (recall that $f_r = f(u_r)$, etc.)

$$A = su_l - f_l = su_r - f_r, \tag{2.6}$$

which again gives us the Rankine–Hugoniot condition

$$s(u_l - u_r) = f_l - f_r.$$

Summing up, the traveling wave U must satisfy the following boundary value problem:

$$\dot{U} = -s(U - u_l) + (f(U) - f_l), \quad U(\pm\infty) = \begin{cases} u_r, \\ u_l. \end{cases} \tag{2.7}$$

Using the Rankine–Hugoniot condition, we see that both u_l and u_r are fixed points for this equation. What we want is an orbit of (2.7) going from u_l to u_r. If the triplet (s, u_l, u_r) has such an orbit, we say that the discontinuous solution (2.3) satisfies a traveling wave entropy condition, or that the discontinuity has a *viscous profile*. (For the analysis so far in this section we were not restricted to the scalar case, and could as well have worked with the case of systems where u is a vector in $\mathbb{R}^n$ and $f(u)$ is some function $\mathbb{R}^n \to \mathbb{R}^n$.)

From now on we say that any isolated discontinuity satisfies the traveling wave entropy condition if (2.7) holds locally across the discontinuity.

Let us examine this in more detail. First we assume that $u_l < u_r$. Observe that $\dot{U}$ can never be zero. Assuming otherwise, namely that $\dot{U}(\xi_0) = 0$ for some ξ_0, the constant $U(\xi_0)$ would be the unique solution, which contradicts that $U(-\infty) = u_l < u_r = U(\infty)$. Thus $\dot{U}(\xi) > 0$ for all ξ, and hence

$$f_l + s(u - u_l) < f(u), \tag{2.8}$$

for all $u \in \langle u_l, u_r \rangle$. Remembering that according to the Rankine–Hugoniot conditions $s = (f_l - f_r)/(u_l - u_r)$, this means that the graph of $f(u)$ has to lie *above* the straight line segment joining the points (u_l, f_l) and (u_r, f_r). On the other hand, if the graph of $f(u)$ is above the straight line, then (2.8) is satisfied, and we can find a solution of (2.7). Similarly, if $u_l > u_r$, $\dot{U}$ must be negative in the whole interval $\langle u_r, u_l \rangle$. Consequently, the graph of $f(u)$ must be *below* the straight line.

By combining the two cases we conclude that the viscous profile or traveling wave entropy condition is equivalent to

$$s |k - u_l| < \text{sign}\,(k - u_l)\,(f(k) - f(u_l)), \tag{2.9}$$

for all k strictly between u_l and u_r. Note that an identical inequality holds with u_l replaced by u_r.

The inequality (2.9) motivates another entropy condition, the *Kružkov entropy condition*. This condition is often more convenient to work with in the sense that it combines the definition of a weak solution with that of the entropy condition.

Choose a smooth convex function $\eta = \eta(u)$ and a nonnegative test function ϕ in $C_0^\infty(\mathbb{R} \times \langle 0, \infty \rangle)$. (Such a test function will be supported away from the x-axis, and thus we get no contribution from the initial data.) Then we find

$$\begin{aligned}
0 &= \iint (u_t + f(u)_x - \epsilon u_{xx})\,\eta'(u)\phi\,dx\,dt \\
&= \iint \eta(u)_t \phi\,dx\,dt + \iint q'(u) u_x \phi\,dx\,dt \\
&\quad - \varepsilon \iint \left(\eta(u)_{xx} - \eta''(u)(u_x)^2\right) \phi\,dx\,dt \\
&= -\iint \eta(u)\phi_t\,dx\,dt - \iint q(u)\phi_x\,dx\,dt \\
&\quad - \varepsilon \iint \eta(u)\phi_{xx}\,dx\,dt + \varepsilon \iint \eta''(u)(u_x)^2 \phi\,dx\,dt \\
&\geq -\iint \left(\eta(u)\phi_t + q(u)\phi_x + \varepsilon \eta \phi_{xx}\right) dx\,dt,
\end{aligned} \tag{2.10}$$

where we first introduced q such that

$$q'(u) = f'(u)\eta'(u) \tag{2.11}$$

and subsequently used the convexity of η, i.e., $\eta'' \geq 0$. Interpreted in a distributional sense we may write this as

$$\frac{\partial}{\partial t}\eta + \frac{\partial}{\partial x}q \leq \varepsilon \eta_{xx}.$$

If this is to hold as $\varepsilon \to 0$, we need

$$\frac{\partial}{\partial t}\eta + \frac{\partial}{\partial x}q \leq 0. \tag{2.12}$$

Consider now the case with

$$\eta(u) = \left((u-k)^2 + \delta^2\right)^{1/2}, \quad \delta > 0,$$

for some constant k. By taking $\delta \to 0$ we can extend the analysis to the case where

$$\eta(u) = |u - k|. \tag{2.13}$$

In this case we find that
$$q(u) = \operatorname{sign}(u-k)\left(f(u)-f(k)\right).$$

Remark 2.1. Consider a fixed bounded weak solution u and a nonnegative test function ϕ, and define the linear functional
$$\Lambda(\eta) = \iint \left(\eta(u)\phi_t + q(u)\phi_x\right) dx\, dt. \tag{2.14}$$
(The function q depends linearly on η; cf. (2.11).) Assume that $\Lambda(\eta) \geq 0$ when η is convex. Introduce
$$\eta_i(u) = \alpha_i \left|u-k_i\right|, \quad k_i \in \mathbb{R},\ \alpha_i \geq 0.$$
Clearly,
$$\Lambda\left(\sum_i \eta_i\right) \geq 0.$$
Since u is a weak solution, we have
$$\Lambda(\alpha u + \beta) = 0, \quad \alpha, \beta \in \mathbb{R},$$
and hence the convex piecewise linear function
$$\eta(u) = \alpha u + \beta + \sum_i \eta_i(u) \tag{2.15}$$
satisfies $\Lambda(\eta) \geq 0$. On the other hand, any convex piecewise linear function η can be written in the form (2.15). This can be proved by induction on the number of breakpoints for η, where by breakpoints we mean those points where η' is discontinuous. The induction step goes as follows. Consider a breakpoint for η, which we without loss of generality can assume is at the origin. Near the origin we may write η as
$$\eta(u) = \begin{cases} \sigma_1 u & \text{for } u \leq 0, \\ \sigma_2 u & \text{for } u > 0, \end{cases}$$
for $|u|$ small. Since η is convex, $\sigma_1 < \sigma_2$. Then the function
$$\tilde\eta(u) = \eta(u) - \frac{1}{2}(\sigma_2-\sigma_1)|u| - \frac{1}{2}(\sigma_1+\sigma_2)u \tag{2.16}$$
is a convex piecewise linear function with one breakpoint fewer than η for which one can use the induction hypothesis. Hence we infer that $\Lambda(\eta) \geq 0$ for all convex, piecewise linear functions η. Consider now any convex function η. By sampling points, we can approximate η with convex, piecewise linear functions η_j such that $\eta_j \to \eta$ in L^∞. Thus we find that
$$\Lambda(\eta) \geq 0.$$
We conclude that if $\Lambda(\eta) \geq 0$ for the *Kružkov function* $\eta(u) = |u-k|$ for all $k \in \mathbb{R}$, then this inequality holds for all convex functions.

We say that a function is a *Kružkov entropy solution* to (2.1) if the inequality

$$\frac{\partial}{\partial t}|u-k| + \frac{\partial}{\partial x}\operatorname{sign}(u-k)(f(u)-f(k)) \leq 0 \qquad (2.17)$$

holds in the sense of distributions for all constants k. Thus, in integral form, (2.17) appears as

$$\iint (|u-k|\phi_t + \operatorname{sign}(u-k)(f(u)-f(k))\phi_x)\,dx\,dt \geq 0, \qquad (2.18)$$

which is to hold for all constants k and all nonnegative test functions ϕ in $C_0^\infty(\mathbb{R} \times \langle 0,\infty\rangle)$.

If we consider solutions on a time interval $[0,T]$, and thus use nonnegative test functions $\phi \in C_0^\infty(\mathbb{R} \times [0,T])$, we find that

$$\int_0^T \!\!\int [\,|u-k|\phi_t + \operatorname{sign}(u-k)(f(u)-f(k))\phi_x\,]\,dx\,dt$$
$$-\int |u(x,T)-k|\phi(x,T)\,dx + \int |u_0(x)-k|\phi(x,0)\,dx \geq 0 \qquad (2.19)$$

should hold for all $k \in \mathbb{R}$ and for all nonnegative test functions ϕ in $C_0^\infty(\mathbb{R} \times [0,T])$.

If we assume that u is bounded, and set $k \leq -\|u\|_\infty$, (2.18) gives

$$0 \leq \iint ((u-k)\phi_t + (f(u)-f(k))\phi_x)\,dx\,dt = \iint (u\phi_t + f(u)\phi_x)\,dx\,dt.$$

Similarly, setting $k \geq \|u\|_\infty$ gives

$$0 \geq \iint (u\phi_t + f(u)\phi_x)\,dx\,dt.$$

These two inequalities now imply that

$$\iint (u\phi_t + f(u)\phi_x)\,dx\,dt = 0 \qquad (2.20)$$

for all nonnegative ϕ. By considering test functions ϕ in $C_0^\infty(\mathbb{R} \times \langle 0,\infty\rangle)$ of the form $\phi_+ - \phi_-$, with $\phi_\pm \in C_0^\infty(\mathbb{R} \times \langle 0,\infty\rangle)$ nonnegative, equation (2.20) implies the usual definition (1.13) of a weak solution.

For sufficiently regular weak solutions we can show that the traveling wave entropy condition and the Kružkov entropy condition are equivalent. Assume that u is a classical solution away from isolated jump discontinuities along piecewise smooth curves, and that it satisfies the Kružkov entropy condition (2.18). We may apply the argument used to derive the Rankine–Hugoniot condition to (2.18). This gives us the following inequality to be satisfied for all k:

$$s[\![\,|u-k|\,]\!] \geq [\![\,\operatorname{sign}(u-k)(f(u)-f(k))\,]\!] \qquad (2.21)$$

(recall that $[\![a]\!] = a_r - a_l$ for any quantity a). Conversely, if (2.21) holds, we can multiply by a nonnegative test function ϕ with support in a neighborhood of the discontinuity and integrate to obtain (2.18).

If we assume that $u_l < u_r$, and choose k to be between u_l and u_r, we obtain
$$s\left(-(u_l - k) - (u_r - k)\right) \leq -(f_r - f(k)) - (f_l - f(k)),$$
or
$$f(k) - sk \geq \bar{f} - s\bar{u}. \tag{2.22}$$
Here, $\bar{f}$ denotes $(f_l + f_r)/2$, and similarly, $\bar{u} = (u_l + u_r)/2$. If we choose coordinates[2] such that $\bar{u} = \bar{f} = 0$, we see that $f(k) \geq sk$ for all $k \in \langle u_l, u_r \rangle$. In other words, the graph of $f(u)$ must lie above the straight line segment between (u_l, f_l) and (u_r, f_r). Similarly, if $u_r < u_l$, we find that the graph has to lie below the line segment. Hence the Kružkov entropy condition implies (2.9).

Assume now that u is a piecewise continuous weak solution whose discontinuities satisfy (2.9), and that u is differentiable, and consequently a classical solution of (2.1) in some neighborhood of a point (x, t). In this neighborhood we have
$$0 = u_t + f(u)_x = (u - k)_t + (f(u) - f(k))_x,$$
for any constant k. If we choose a nonnegative test function ϕ with support in a small neighborhood of (x, t), and a constant k such that $u > k$ where $\phi \neq 0$, we have
$$\iint \left(|u - k|\phi_t + \operatorname{sign}(u - k)(f(u) - f(k))\phi_x\right) dx\,dt = 0. \tag{2.23}$$
If $k > u$ where $\phi \neq 0$, we obtain the same equality. Therefore, since u is continuous around (x, t) and ϕ can be chosen to have arbitrarily small support, (2.23) holds for all nonnegative ϕ having support where u is a classical solution.

If u has an isolated discontinuity, with limits u_l and u_r that are such that the Rankine–Hugoniot equality holds, then
$$s[\![|u - k|]\!] = [\![\operatorname{sign}(u - k)(f(u) - f(k))]\!] \tag{2.24}$$
for any constant k *not* between u_l and u_r. For constants k between u_l and u_r, we have seen that if $f(k) - sk \geq \bar{f} - s\bar{u}$, i.e., the viscous profile entropy condition holds, then
$$s[\![|u - k|]\!] \geq [\![\operatorname{sign}(u - k)(f(u) - f(k))]\!], \tag{2.25}$$
and thus the last two equations imply that the Kružkov entropy condition will be satisfied.

---
[2]Replace u and f by $v = u - \bar{u}$ and $g(v) = f(u) - \bar{f}$, respectively.

Hence, for sufficiently regular solutions, these two entropy conditions are equivalent. We shall later see that Kružkov's entropy condition implies that for sufficiently regular initial data, the solution does indeed possess the necessary regularity; consequently, these two entropy conditions "pick" the same solutions. We will therefore in the following use whichever entropy condition is more convenient to work with.

2.2 The Riemann Problem

> With my two algorithms one can solve all problems—without error, if God will!
>
> Al-Khwarizmi (c. 780–850)

For conservation laws, the Riemann problem is the initial value problem

$$u_t + f(u)_x = 0, \quad u(x,0) = \begin{cases} u_l & \text{for } x < 0, \\ u_r & \text{for } x \geq 0. \end{cases} \tag{2.26}$$

Assume temporarily that $f \in C^2$ with finitely many inflection points. We have seen examples of Riemann problems and their solutions in the previous chapter, in the context of traffic flow. Since both the equation and the initial data are invariant under the transformation $x \mapsto kx$ and $t \mapsto kt$, it is reasonable to look for solutions of the form $u = u(x,t) = w(x/t)$. We set $z = x/t$ and insert this into (2.26) to obtain

$$-\frac{x}{t^2} w' + \frac{1}{t} f'(w) w' = 0, \text{ or } z = f'(w). \tag{2.27}$$

If f' is strictly monotone, we can simply invert this relation to obtain the solution $w = (f')^{-1}(z)$. In the general case we have to replace f' by a monotone function on the interval between u_l and u_r. In the example of traffic flow, we saw that it was important whether $u_l < u_r$ or vice versa. Assume first that $u_l < u_r$. Now we claim that the solution of (2.27) is given by

$$u(x,t) = w(z) = \begin{cases} u_l & \text{for } x \leq f'_\smile(u_l)t, \\ (f'_\smile)^{-1}(x/t) & \text{for } f'_\smile(u_l)t \leq x \leq f'_\smile(u_r)t, \\ u_r & \text{for } x \geq f'_\smile(u_r)t, \end{cases} \tag{2.28}$$

for $u_l < u_r$. Here $f_\smile$ denotes the *lower convex envelope* of f in the interval $[u_l, u_r]$, and $\left((f_\smile)'\right)^{-1}$, or, to be less pedantic, $(f'_\smile)^{-1}$, denotes the inverse of its derivative. The lower convex envelope is defined to be the largest convex function that is smaller than or equal to f in the interval $[u_l, u_r]$, i.e.,

$$f_\smile(u) = \sup\left\{g(u) \mid g \leq f \text{ and } g \text{ convex on } [u_l, u_r]\right\}. \tag{2.29}$$

To picture the envelope of f, we can imagine the graph of f cut out from a board, so that the lower boundary of the board has the shape of the graph. An elastic rubber band stretched from $(u_l, f(u_l))$ to $(u_r, f(u_r))$ will then follow the graph of $f_\smile$. Note that $f_\smile$ depends on the interval $[u_l, u_r]$, and thus is a highly nonlocal function of f.

Since $f''_\smile \geq 0$, we have that $f'_\smile$ is nondecreasing, and hence we can form its inverse, denoted by $(f'_\smile)^{-1}$, permitting jump discontinuities where $f'_\smile$ is constant. Hence formula (2.28) at least makes sense. In Figure 2.1 we see a flux function and the envelope between two points u_l and u_r. If

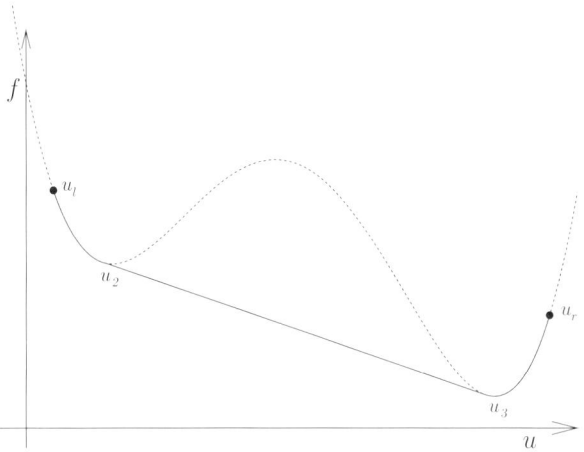

Figure 2.1. In a series of figures we will illustrate the solution of an explicit Riemann problem. We start by giving the flux function f, two states u_l and u_r (with $u_l < u_r$), and the convex envelope of f relative to the interval $[u_l, u_r]$.

$f \in C^2$ with finitely many inflection points, there will be a finite number of intervals with endpoints $u_l = u_1 < u_2 < \cdots < u_n = u_r$ such that $f_\smile = f$ on every other interval. That is, if $f_\smile(u) = f(u)$ for $u \in [u_i, u_{i+1}]$, then $f_\smile(u) < f(u)$ for $u \in \langle u_{i+1}, u_{i+2}\rangle \cup \langle u_{i-1}, u_i\rangle$. In this case the solution $u(\,\cdot\,, t)$ consists of finitely many intervals where u is a regular solution given by $u(x,t) = (f')^{-1}(x/t)$ separated by jump discontinuities at points x such that $x = f'(u_j)t = t(f(u_{j+1}) - f(u_j))/(u_{j+1} - u_j) = f'(u_{j+1})t$ that clearly satisfy the Rankine–Hugoniot relation. In Figure 2.1 we have three intervals, where $f_\smile < f$ on the middle interval.

To show that (2.28) defines an entropy solution, we shall need some notation. For $i = 1, \ldots, n$ set $\sigma_i = f'_\smile(u_i)$ and define $\sigma_0 = -\infty$, $\sigma_{n+1} = \infty$. By discarding identical σ_i's and relabeling if necessary, we can and will assume that $\sigma_0 < \sigma_1 < \cdots < \sigma_{n+1}$. Then for $i = 2, \ldots, n$ define

$$v_i(x,t) = (f'_\smile)^{-1}\left(\frac{x}{t}\right), \quad \sigma_{i-1} \leq \frac{x}{t} \leq \sigma_i$$

and set $v_1(x,t) = u_l$ for $x \leq \sigma_1 t$ and $v_{n+1}(x,t) = u_r$ for $x \geq \sigma_n t$. Let Ω_i denote the set

$$\Omega_i = \{(x,t) \mid 0 \leq t \leq T, \quad \sigma_{i-1} t < x < \sigma_i t\}$$

for $i = 1, \ldots, n+1$. Using this notation, u defined by (2.28) can be written

$$u(x,t) = \sum_{i=1}^{n+1} \chi_{\Omega_i}(x,t) v_i(x,t), \tag{2.30}$$

where χ_{Ω_i} denotes the characteristic function of the set Ω_i. For $i = 1, \ldots, n$ we then define

$$\underline{u}_i = \lim_{x \to \sigma_i t-} u(x,t) \quad \text{and} \quad \bar{u}_i = \lim_{x \to \sigma_i t+} u(x,t).$$

The values $\underline{u}_i$ and $\bar{u}_i$ are the left and right limits of the discontinuities of u.

With this notation at hand, we show that u defined by (2.30) is an entropy solution in the sense of (2.19) of the initial value problem (2.26). Note that u is also continuously differentiable in each Ω_i. First we use Green's theorem (similarly as in proving the Rankine–Hugoniot relation (1.15)) on each Ω_i to show that

$$\int_0^T \int (\eta \varphi_t + q \varphi_x) \, dx \, dt = \sum_{i=1}^{n+1} \iint_{\Omega_i} (\eta_i \varphi_t + q_i \varphi_x) \, dx \, dt$$

$$= \sum_{i=1}^{n+1} \iint_{\Omega_i} \left((\eta_i \varphi)_t + (q_i \varphi)_x \right) dx \, dt$$

$$= \sum_{i=1}^{n+1} \int_{\partial \Omega_i} \varphi \left(-\eta_i \, dx + q_i \, dt \right)$$

$$= \int \left(\eta(x,T) \varphi(x,T) - \eta(x,0) \varphi(x,0) \right) dx$$

$$+ \sum_{i=1}^{n} \int_0^T \varphi(\sigma_i t, t) \left[\sigma_i (\bar{\eta}_i - \underline{\eta}_i) - (\bar{q}_i - \underline{q}_i) \right] dt.$$

Here

$$\eta = \eta(u,k) = |u - k|,$$
$$\eta_i = \eta(u_i(x,t), k), \quad \bar{\eta}_i = \eta(\bar{u}_i, k), \quad \underline{\eta}_i = \eta(\underline{u}_i, k),$$
$$q = q(u,k) = \text{sign}(u-k)(f(u) - f(k)),$$
$$q_i = q(u_i(x,t), k), \quad \bar{q}_i = q(\bar{u}_i, k), \quad \text{and} \quad \underline{q}_i = q(\underline{u}_i, k).$$

Now we observe that by (2.24) and (2.25) we have

$$\sigma_i (\bar{\eta}_i - \underline{\eta}_i) - (\bar{q}_i - \underline{q}_i) \geq 0,$$

for all constants k. Hence

$$\int_0^T \int \left(\eta \varphi_t + q \varphi_x \right) dx\, dt + \int \left(\eta(x,0)\varphi(x,0) - \eta(x,T)\varphi(x,T) \right) dx \geq 0,$$

i.e., u satisfies (2.19). Now we have found an entropy-satisfying solution to

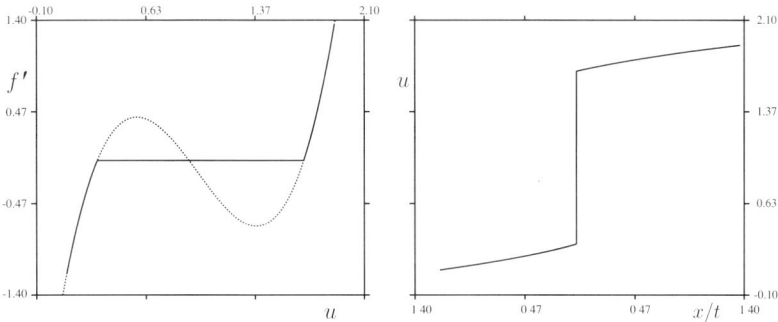

Figure 2.2. The function $f'_\smile$ (left) and its inverse (right).

the Riemann problem if $u_l < u_r$.

If $u_l > u_r$, we can transform the problem to the case discussed above by sending $x \mapsto -x$. Then we obtain the Riemann problem

$$u_t - f(u)_x = 0, \qquad u(x,0) = \begin{cases} u_r & \text{for } x < 0, \\ u_l & \text{for } x \geq 0. \end{cases}$$

In order to solve this we have to take the lower convex envelope of $-f$ from u_r to u_l. But this envelope is exactly the negative of the *upper concave envelope* from u_l to u_r. The upper concave envelope is defined to be

$$f_\frown(u) = \inf \left\{ g(u) \;\middle|\; g \geq f \text{ and } g \text{ concave on } [u_r, u_l] \right\}. \tag{2.31}$$

In this case the weak solution is given by

$$u(x,t) = w(z) = \begin{cases} u_l & \text{for } x \leq f'_\frown(u_l)t, \\ (f'_\frown)^{-1}(z) & \text{for } f'_\frown(u_l)t \leq x \leq f'_\frown(u_r)t, \\ u_r & \text{for } x \geq f'_\frown(u_r)t, \end{cases} \tag{2.32}$$

for $u_l > u_r$, where $z = x/t$.

This construction of the solution is valid as long as the envelope consists of a finite number of intervals where $f_{\smile,\frown} \neq f$, alternating with intervals where the envelope and the function coincide. We will later extend the solution to the case where f is only *Lipschitz continuous*.

We have now proved a theorem about the solution of the Riemann problem for scalar conservation laws.

Theorem 2.2. *The initial value problem*

$$u_t + f(u)_x = 0, \qquad u(x,0) = \begin{cases} u_l & \text{for } x < 0, \\ u_r & \text{for } x \geq 0, \end{cases}$$

with a flux function $f(u)$ such that $f_{\smile,\frown} \neq f$ on finitely many intervals, alternating with intervals where they coincide, has a weak solution given by equation (2.28) if $u_l < u_r$, or by (2.32) if $u_r < u_l$. This solution satisfies the Kružkov entropy condition (2.19).

The solution $u(x,t)$ given by (2.28) and (2.32) consists of a finite number of discontinuities separated by "wedges" (i.e., intervals $\langle z_i, z_{i+1}\rangle$) inside which u is a classical solution. A discontinuity that satisfies the entropy condition is called a *shock wave* or simply a *shock*, and the continuous parts of the solution of the Riemann problem are called *rarefaction waves*. This terminology, as well as the term "entropy condition," comes from gas dynamics. Thus we may say that the solution of a Riemann problem consists of a finite sequence of rarefaction waves alternating with shocks.

◇ **Example 2.3 (Traffic flow (cont'd.)).**

In the conservation law model of traffic flow, we saw that the flux function was given as

$$f(u) = u(1-u).$$

This is a concave function. Consequently, any upper envelope will be the function f itself, whereas any lower envelope will be the straight line segment between its endpoints. Any Riemann problem with $u_l > u_r$ will be solved by a rarefaction wave, and if $u_l < u_r$, the solution will consist of a single shock. This is, of course, in accordance with our earlier results, and perhaps also with our experience. ◇

The solution of a Riemann problem is frequently depicted in (x,t) space as a collection of rays emanating from the origin. The slope of these rays is the reciprocal of $f'(u)$ for rarefaction waves, and if the ray illustrates a shock, the reciprocal of $[\![f]\!]/[\![u]\!]$. In Figure 2.3 we illustrate the solution of the previous Riemann problem in this way; broken lines indicate rarefaction waves, and the solid line the shock. Note that Theorem 2.2 does *not* require the flux function f to be differentiable. Assume now that the flux function is *continuous* and *piecewise linear* on a finite number of intervals. Thus f' will then be a step function taking a finite number of values. The discontinuity points of f' will hereafter be referred to as *breakpoints*.

Making this approximation is reasonable in many applications, since the precise form of the flux function is often the result of some measurements. These measurements are taken for a discrete set of u values, and a piecewise linear flux function is the result of a linear interpolation between these values.

2.2. The Riemann Problem   35

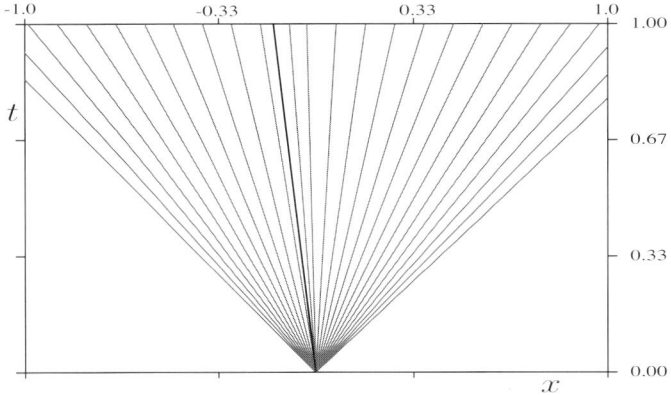

Figure 2.3. The solution of a Riemann problem, shown in (x,t) space.

Both upper concave and lower convex envelopes will also be piecewise linear functions with a finite number of breakpoints. This means that $f'_\smile$ and $f'_\frown$ will be step functions, as will their inverses. Furthermore, the inverses of the derivatives will take their values among the breakpoints of $f_\smile$ (or $f_\frown$), and therefore also of f. If the initial states in a Riemann problem are breakpoints, then the entire solution will take values in the set of breakpoints.

If we assume that $u_l < u_r$, and label the breakpoints $u_l = u_0 < u_1 < \cdots < u_n = u_r$, then $f_\smile$ will have breakpoints in some subset of this, say $u_l < u_{i_1} < \cdots < u_{i_k} < u_r$. The solution will be a step function in $z = x/t$, monotonically nondecreasing between u_l and u_r. The discontinuities will be located at z_{i_k}, given by

$$z_{i_k} = \frac{f(u_{i_{k-1}}) - f(u_{i_k})}{u_{i_{k-1}} - u_{i_k}}.$$

Thus the following corollary of Theorem 2.2 holds.

Corollary 2.4. *Assume that f is a continuous piecewise linear function $f\colon [-K, K] \to \mathbb{R}$ for some constant K. Denote the breakpoints of f by $-K = u_0 < u_1 < \cdots < u_{n-1} < u_n = K$. Then the Riemann problem*

$$u_t + f(u)_x = 0, \qquad u(x,0) = \begin{cases} u_j & \text{for } x < 0, \\ u_k & \text{for } x \geq 0, \end{cases} \qquad (2.33)$$

has a piecewise constant (in $z = x/t$) solution. If $u_j < u_k$, let $u_j = v_1 < \cdots < v_m = u_k$ denote the breakpoints of $f_\smile$, and if $u_j > u_k$, let $u_k = v_m < \cdots < v_1 = u_j$ denote the breakpoints of $f_\frown$. The weak solution of the

Riemann problem is then given by

$$u(x,t) = \begin{cases} v_1 & \text{for } x \leq s_1 t, \\ v_2 & \text{for } s_1 t < x \leq s_2 t, \\ \vdots & \\ v_i & \text{for } s_{i-1} t < x \leq s_i t, \\ \vdots & \\ v_m & \text{for } s_{m-1} t < x. \end{cases} \quad (2.34)$$

Here, the speeds s_i are computed from the derivative of the envelope, that is,

$$s_i = \frac{f(v_{i+1}) - f(v_i)}{v_{i+1} - v_i}.$$

Note that this solution is an admissible solution in the sense that it satisfies the Kružkov entropy condition. The viscous profile entropy condition is somewhat degenerate in this case. Across discontinuities over which $f(u)$ differs from the envelope, it is satisfied. But across those discontinuities over which the envelope and the flux function coincide, the right-hand side of the defining ordinary differential equation (2.7) collapses to zero. The conservation law is called *linearly degenerate* in each such interval $\langle v_i, v_{i+1} \rangle$. Nevertheless, these discontinuities are also limits of the viscous regularization, as can be seen by changing to Lagrangian coordinates $x \mapsto x - s_i t$; see Exercise 2.3.

With this we conclude our discussion of the Riemann problem, and in the next section we shall see how the solutions of Riemann problems may be used as a building block to solve more general initial value problems.

2.3 Front Tracking

> This algorithm is admittedly complicated, but no simpler mechanism seems to do nearly as much.
>
> D. E. Knuth, The TEXbook (1984)

We begin this section with an example that illustrates the ideas of front tracking for scalar conservation laws, as well as some of the properties of solutions.

◇ **Example 2.5.**

In this example we shall study a piecewise linear approximation of Burgers' equation, $u_t + (u^2/2)_x = 0$. This means that we study a conservation law with a flux function that is piecewise linear and agrees with $u^2/2$

at its breakpoints. To be specific, we choose intervals of unit length. We shall be interested in the flux function only in the interval $[-1, 2]$, where we define it to be

$$f(u) = \begin{cases} -u/2 & \text{for } u \in [-1, 0], \\ u/2 & \text{for } u \in [0, 1], \\ 3u/2 - 1 & \text{for } u \in [1, 2]. \end{cases} \quad (2.35)$$

This flux function has two breakpoints, and is convex.

We wish to solve the initial value problem

$$u_t + f(u)_x = 0, \quad u_0(x) = \begin{cases} 2 & \text{for } x \le x_1, \\ -1 & \text{for } x_1 < x \le x_2, \\ 1 & \text{for } x_2 < x, \end{cases} \quad (2.36)$$

with f given by (2.35). Initially, the solution must consist of the solutions of the two Riemann problems located at x_1 and x_2. This is so, since the waves from these solutions move with a finite speed, and will not interact until some positive time.

This feature, sometimes called *finite speed of propagation*, characterizes hyperbolic, as opposed to elliptic or parabolic, partial differential equations. It implies that if we change the initial condition locally around some point, it will not immediately influence the solution "far away." Recalling the almost universally accepted assumption that nothing moves faster than the speed of light, one can say that hyperbolic equations are more fundamental than the other types of partial differential equations.

Returning to our example, we must then solve the two initial Riemann problems. We commence with the one at x_1. Since f is convex, and $u_l = 2 > -1 = u_r$, the solution will consist of a single shock wave with speed $s_1 = \frac{1}{2}$ given from the Rankine–Hugoniot condition of this Riemann problem. For small t and x near x_1 the solution reads

$$u(x, t) = \begin{cases} 2 & \text{for } x < s_1 t + x_1, \\ -1 & \text{for } x \ge s_1 t + x_1. \end{cases} \quad (2.37)$$

The other Riemann problem has $u_l = -1$ and $u_r = 1$, so we must use the lower convex envelope, which in this case coincides with the flux function f. The flux function has two linear segments and one breakpoint $u = 0$ in the interval $\langle -1, 1 \rangle$. Hence, the solution will consist of two discontinuities moving apart. The speeds of the discontinuities are computed from $f'(u)$, or equivalently from the Rankine–Hugoniot condition, since f is linearly degenerate over each discontinuity. This gives $s_2 = -\frac{1}{2}$ and $s_3 = \frac{1}{2}$. The solution equals

$$u(x,t) = \begin{cases} -1 & \text{for } x < s_2 t + x_2, \\ 0 & \text{for } s_2 t + x_2 \leq x < s_3 t + x_2, \\ 1 & \text{for } s_3 t + x_2 \leq x, \end{cases} \quad (2.38)$$

for small t and x near x_2.

It remains to connect the two solutions (2.37) and (2.38). This is easily done for sufficiently small t:

$$u(x,t) = \begin{cases} 2 & \text{for } x < x_1 + s_1 t, \\ -1 & \text{for } x_1 + s_1 t \leq x \leq x_2 + s_2 t, \\ 0 & \text{for } x_2 + s_2 t \leq x < x_2 + s_3 t, \\ 1 & \text{for } x_2 + s_3 t \leq x. \end{cases} \quad (2.39)$$

The problem now is that the shock wave located at $x_1(t) = x_1 + t/2$ will collide with the discontinuity $x_2(t) = x_2 - t/2$. Then equation (2.39) is no longer valid, since the middle interval has collapsed. This will happen at time $t = t_1 = (x_2 - x_1)$ and position $x = x_4 = (x_1 + x_2)/2$.

To continue the solution we must solve the interaction between the shock and the discontinuity. Again, using finite speed of propagation, we have that the solution away from (x_4, t_1) will not be directly influenced by the behavior here. Consider now the solution at time t_1 and in a vicinity of x_4. Here u takes two constant values, 2 for $x < x_4$ and 0 for $x > x_4$. Therefore, the interaction of the shock wave $x_1(t)$ and the discontinuity $x_2(t)$ is determined by solving the Riemann problem with $u_l = 2$ and $u_r = 0$.

Again, this Riemann problem is solved by a single shock, since the flux function is convex and $u_l > u_r$. The speed of this shock is $s_4 = 1$. Thus, for t larger than $x_2 - x_1$, the solution consists of a shock located at $x_4(t)$ and a discontinuity located at $x_3(t)$. The locations are given by

$$x_4(t) = \frac{1}{2}(x_1 + x_2) + 1\,(t - (x_2 - x_1)) = t + \frac{1}{2}(3x_1 - x_2),$$

$$x_3(t) = x_2 + \frac{1}{2}t.$$

We can then write the solution $u(x,t)$ as

$$u(x,t) = 2 + [\![u\,(x_4(t))]\!]H\,(x - x_4(t)) + [\![u\,(x_3(t))]\!]H\,(x - x_3(t)), \quad (2.40)$$

where H is the Heaviside function.

Indeed, any function $u(x,t)$ that is piecewise constant in x with discontinuities located at $x_j(t)$ can be written in the form

$$u(x,t) = u_l + \sum_j [\![u\,(x_j(t))]\!]H\,(x - x_j(t)), \quad (2.41)$$

where u_l now denotes the value of u to the left of the leftmost discontinuity.

Since the speed of $x_4(t)$ is greater than the speed of $x_3(t)$, these two discontinuities will collide. This will happen at $t = t_2 = 3(x_2 - x_1)$ and $x = x_5 = (5x_2 - 3x_1)/2$. In order to resolve the interaction of these two discontinuities, we have to solve the Riemann problem with $u_l = 2$ and $u_r = 1$.

In the interval $[1, 2]$, $f(u)$ is linear, and hence the solution of the Riemann problem will consist of a single discontinuity moving with speed $s_5 = \frac{3}{2}$. Therefore, for $t > t_2$ the solution is defined as

$$u(x,t) = \begin{cases} 2 & \text{for } x < 3t/2 - 3x_1 - 2x_2, \\ 1 & \text{for } x \geq 3t/2 - 3x_1 - 2x_2. \end{cases} \quad (2.42)$$

Since the solution now consists of a single moving discontinuity, there will be no further interactions, and we have found the solution for all positive t. Figure 2.4 depicts this solution in (x, t) plane; the discontinuities are shown as solid lines. We call the method that we have used to obtain the solution *front tracking*. Front tracking consists in *tracking* all discontinuities in the solution, whether they represent shocks or

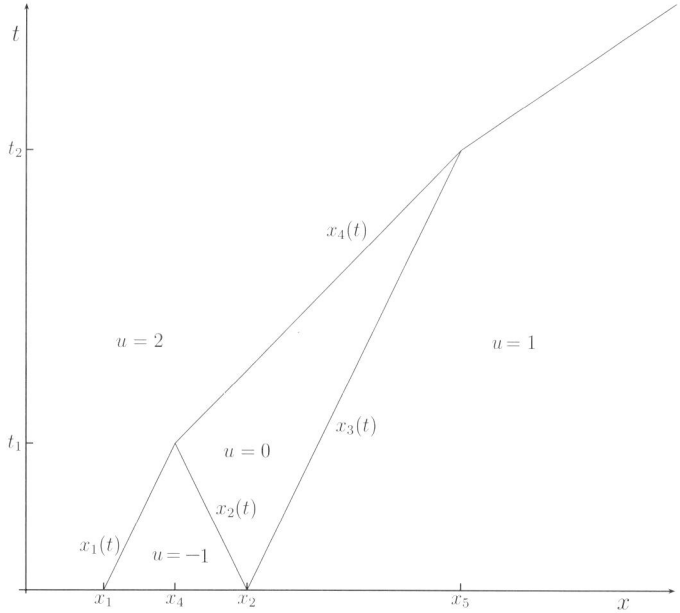

Figure 2.4. The solution of (2.36) with the piecewise linear continuous flux function (2.35).

not. Hereafter, if the flux function is continuous and piecewise linear, all discontinuities in the solution will be referred to as *fronts*.

Notice that if the flux function is continuous and piecewise linear, the Rankine–Hugoniot condition can be used to calculate the speed of any front. So from a computational point of view, all discontinuities are equivalent. ◇

With this example in mind we can define a general front-tracking algorithm for scalar conservation laws. Loosely speaking, front tracking consists in making a step function approximation to the initial data, and a piecewise linear approximation to the flux function. The approximate initial function will define a series of Riemann problems, one at each step. One can solve each Riemann problem, and since the solutions have finite speed of propagation, they will be independent of each other until waves from neighboring solutions interact. Front tracking should then resolve this interaction in order to propagate the solution to larger times.

By considering flux functions that are continuous and piecewise linear we are providing a method for resolving interactions.

(**i**) We are given a scalar one-dimensional conservation law
$$u_t + f(u)_x = 0, \quad u|_{t=0} = u_0. \tag{2.43}$$

(**ii**) Approximate f by a continuous, piecewise linear flux function f^δ.

(**iii**) Approximate initial data u_0 by a piecewise constant function u_0^η.

(**iv**) Solve initial value problem
$$u_t + f^\delta(u)_x = 0, \quad u|_{t=0} = u_0^\eta$$
exactly. Denote solution by $u_{\delta,\eta}$.

(**v**) As f^δ and u_0^η approach f and u_0, respectively, the approximate solution $u_{\delta,\eta}$ will converge to u, the solution of (2.43).

Front tracking in a box (scalar case).

We have seen that the solution of a Riemann problem always is a monotone function taking values between u_l and u_r. Another way of stating this is to say that the solution of a Riemann problem obeys a *maximum principle*. This means that if we solve a collection of Riemann problems, the solutions (all of them) will remain between the minimum and the maximum of the left and right states.

Therefore, fix a large positive number M and let $u_i = i\delta$, for $-M \le i\delta \le M$. In this section we hereafter assume, unless otherwise stated, that the flux function $f(u)$ is continuous and piecewise linear, with breakpoints u_i.

We assume that u_0 is some piecewise constant function taking values in the set $\{u_i\}$ with a finite number of discontinuities, and we wish to solve the initial value problem

$$u_t + f(u)_x = 0, \qquad u(x,0) = u_0(x). \tag{2.44}$$

As remarked above, the solution will initially consist of a number of non-interacting solutions of Riemann problems. Each solution will be a piecewise constant function with discontinuities traveling at constant speed. Hence, at some later time $t_1 > 0$, two discontinuities from neighboring Riemann problems will interact.

At $t = t_1$ we can proceed by considering the initial value problem with solution $v(x,t)$:

$$v_t + f(v)_x = 0, \qquad v(x,t_1) = u(x,t_1).$$

Since the solutions of the initial Riemann problems will take values among the breakpoints of f, i.e., $\{u_i\}$, the initial data $u(x,t_1)$ is the same type of function as $u_0(x)$. Consequently, we can proceed as we did initially, by solving the Riemann problems at the discontinuities of $u(x,t_1)$. However, except for the Riemann problem at the interaction point, these Riemann problems have all been solved initially, and their solution merely consists in continuing the discontinuities at their present speed. The Riemann problem at the interaction point has to be solved, giving a new fan of discontinuities. In this fashion the solution can be calculated up to the next interaction at t_2, say. Note that what we calculate in this way is *not* an approximation to the entropy weak solution of (2.44), but the exact solution.

It is clear that we can continue this process for any number of interactions occurring at times t_n where $0 < t_1 \le t_2 \le t_3 \le \cdots \le t_n \le \cdots$. However, we cannot a priori be sure that $\lim t_n = \infty$, or in other words, that we can calculate the solution up to any predetermined time. One might envisage that the number of discontinuities grow for each interaction, and that their number increases without bound at some finite time. The next lemma assures us that this does not happen.

Lemma 2.6. *For each fixed δ, and for each piecewise constant function u_0 taking values in the set $\{u_i\}$, there is only a finite number of interactions between discontinuities of the weak solution to (2.44) for t in the interval $[0,\infty)$.*

Remark 2.7. In particular, this means that we can calculate the solution by front tracking up to infinite time using only a *finite* number of operations. In connection with front tracking used as a numerical method, this property is called *hyperfast*. In the rest of this book we call a discontinuity in a front-tracking solution a *front*. Thus a front can represent either a shock or a discontinuity over which the flux function is linearly degenerate.

Proof of Lemma 2.6. Let $N(t)$ denote the total number of fronts in the front-tracking solution $u(x,t)$ at time t.

If a front represents a jump from u_l to u_r, we say that the front contains ℓ linear segments if the flux function has $\ell - 1$ breakpoints between u_l and

u_r. We use the notation $[\![u]\!]$ to denote the jump in u across a front. In this notation $\ell = |[\![u]\!]|/\delta$.

Let $L(t)$ be the total number of linear segments present in all fronts of $u(x,t)$ at time t. Thus, if we number the fronts from left to right, and the ith front contains ℓ_i linear segments, then

$$L(t) = \sum_i \ell_i = \frac{1}{\delta} \sum_i |[\![u]\!]_i|.$$

Let Q denote the number of linear segments in the piecewise linear flux function $f(u)$ for u in the interval $[-M, M]$. Now we claim that the functional

$$T(t) = QL(t) + N(t)$$

is strictly decreasing for each collision of fronts. Since $T(t)$ takes only integer values, this means that we can have at most $T(0)$ collisions.

It remains to prove that $T(t)$ is strictly decreasing for each collision. Assume that a front separating values u_l and u_m collides from the left with a front separating u_m and u_r. We will first show that T is decreasing if u_m is between u_l and u_r.

We assume that $u_l < u_m < u_r$. If $u_r < u_m < u_l$, the situation is analogous, and the statement can be proved with the same arguments. The situation is as depicted in Figure 2.5. Since a single front connects u_l with u_m, the graph of the flux function cannot cross the straight line segment connecting the points $(u_l, f(u_l))$ and $(u_m, f(u_m))$. The entropy condition also implies that the graph of the flux function must be above this segment. The same holds for the front on the right separating u_m and u_r. As the two fronts are colliding, the speed of the left front must be larger than the speed of the right front. This means that the slope of the segment from $(u_l, f(u_l))$ to $(u_m, f(u_m))$ is greater than the slope of the segment from $(u_m, f(u_m))$ to $(u_r, f(u_r))$. Therefore, the lower convex envelope from u_l to u_r consists of the line from $(u_l, f(u_l))$ to $(u_r, f(u_r))$. Accordingly, the solution of the Riemann problem consists of a single front separating u_l and u_r. See Figure 2.5. Consequently, L does not change at the interaction, and N decreases by one. Thus, in the case where u_m is between u_r and u_l, T decreases.

It remains to show that T also decreases if u_m is not between u_l and u_r. We will show this for the case $u_m < u_l < u_r$. The other cases are similar, and can be proved by analogous arguments.

Since the Riemann problem with a left state u_l and right state u_m is solved by a single discontinuity, the graph of the flux function cannot lie above the straight line segment connecting the points $(u_l, f(u_l))$ and $(u_m, f(u_m))$. Similarly, the graph of the flux function must lie entirely above the straight line segment connecting $(u_m, f(u_m))$ and $(u_r, f(u_r))$. Also, the slope of the latter segment must be smaller than that of the former, since the fronts are colliding. This means that the Riemann problem

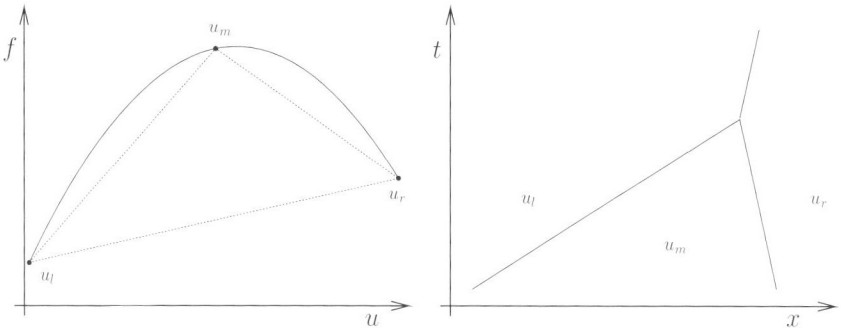

Figure 2.5. An interaction of fronts where $u_l < u_m < u_r$.

with left state u_l and right state u_r defined at the collision of the fronts will have a solution consisting of fronts with speed smaller than or equal to the speed of the right colliding front. See Figure 2.6. The maximal number

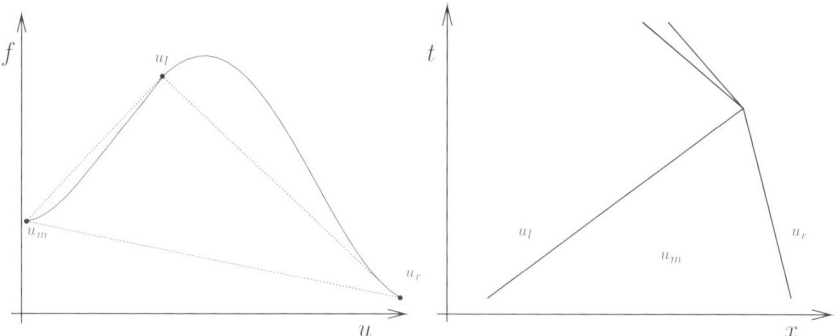

Figure 2.6. An interaction of fronts where $u_m < u_l < u_r$.

of fronts resulting from the collision is $|u_l - u_r|/\delta$. This is strictly less than Q. Hence N increases by at most $Q - 1$. At the same time, L decreases by at least one. Consequently, T must decrease by at least one. This concludes the proof of Lemma 2.6. □

As a corollary of Lemma 2.6, we know that for a piecewise constant initial function with a finite number of discontinuities, and for a continuous and piecewise linear flux function with a finite number of breakpoints, the initial value problem has a weak solution satisfying both Kružkov entropy conditions (2.17), as well as the viscous entropy condition for every discontinuity. It is not difficult to prove the following slight generalization of what we have already shown:

Corollary 2.8. *Let $f(u)$ be a continuous and piecewise linear function with a finite number of breakpoints for u in the interval $[-M, M]$, where M is some constant. Assume that u_0 is a piecewise constant function with a finite number of discontinuities, $u_0 \colon \mathbb{R} \to [-M, M]$. Then the initial value problem*

$$u_t + f(u)_x = 0, \qquad u|_{t=0} = u_0 \qquad (2.45)$$

has a weak solution $u(x, t)$. The function $u(x, t)$ is a piecewise constant function of x for each t, and $u(x, t)$ takes values in the finite set $\{u_0(x)\} \cup \{\text{the breakpoints of } f\}$. Furthermore, there are only a finite number of interactions between the fronts of u. The distribution u also satisfies the Kružkov entropy condition (2.19).

This is all well and fine, but we could wish for more. For instance, is this solution the only one? And piecewise linear flux functions and piecewise constant initial functions seem more like an approximation than what we would expect to see in "real life." So what happens when the piecewise constant initial function and the piecewise linear flux function converge to general initial data and flux functions, respectively?

It turns out that these two questions are connected and can be answered by elegant, but indirect, analysis starting from the Kružkov formulation (2.18).

2.4 Existence and Uniqueness

> Det var en ustyrtelig mængde lag!
> Kommer ikke kærnen snart for en dag?[3]
>
> *Henrik Ibsen, Peer Gynt (1875)*

By a clever choice of the test function ϕ, we shall use the Kružkov formulation to show stability with respect to the initial value function, and thereby uniqueness.

The approach used in this section is also very useful in estimating the error in numerical methods. We shall return to this in a later chapter.

Let therefore $u = u(x, t)$ and $v = v(x, t)$ be two weak solutions to

$$u_t + f(u)_x = 0,$$

with initial data

$$u|_{t=0} = u_0, \qquad v|_{t=0} = v_0,$$

---

[3] What an enormous number of swathings! Isn't the kernel soon coming to light?

2.4. Existence and Uniqueness

respectively, satisfying the Kružkov entropy condition. Equivalently,

$$\iint \left(|u-k|\,\phi_t + \operatorname{sign}(u-k)\,(f(u)-f(k))\phi_x\right)\,dx\,dt$$
$$+ \int |u_0 - k|\,\phi|_{t=0}\,dx \geq 0 \tag{2.46}$$

for every nonnegative test function ϕ with compact support (and similarly for the function v).

We assume that f is *Lipschitz continuous*, that is, that there is a constant L such that

$$\|f\|_{\operatorname{Lip}} := \sup_{u \neq v} \left|\frac{f(u)-f(v)}{u-v}\right| \leq L, \tag{2.47}$$

and we denote by $\|f\|_{\operatorname{Lip}}$ the Lipschitz constant, or seminorm[4], of f.

If ϕ is compactly supported in $t > 0$, then (2.46) reads

$$\iint \left(|u-k|\,\phi_t + \operatorname{sign}(u-k)\,(f(u)-f(k))\phi_x\right)\,dx\,dt \geq 0. \tag{2.48}$$

For simplicity we shall in this section use the notation

$$q(u,k) = \operatorname{sign}(u-k)\,(f(u)-f(k)).$$

For functions of two variables we define the Lipschitz constant by

$$\|q\|_{\operatorname{Lip}} = \sup_{(u_1,v_1) \neq (u_2,v_2)} \frac{|q(u_1,v_1) - q(u_2,v_2)|}{|u_1-u_2|+|v_1-v_2|}.$$

Since $q_u(u,k) = \operatorname{sign}(u-k)\,f'(u)$ and $q_k(u,k) = -\operatorname{sign}(u-k)\,f'(k)$, it follows that if $\|f\|_{\operatorname{Lip}} \leq L$, then also $\|q\|_{\operatorname{Lip}} \leq L$.

Now let $\phi = \phi(x,t,y,s)$ be a nonnegative test function both in (x,t) and (y,s) with compact support in $t > 0$ and $s > 0$. Using that both u and v satisfy (2.48), we can set $k = v(y,s)$ in the equation for u, and set $k = u(x,t)$ in the equation for $v = v(y,s)$. We integrate the equation for $u(x,t)$ with respect to y and s, and the equation for $v(y,s)$ with respect to x and t, and add the two resulting equations. We then obtain

$$\iint \iint \left(|u(x,t)-v(y,s)|\,(\phi_t+\phi_s) + q(u,v)(\phi_x+\phi_y)\right)\,dx\,dt\,dy\,ds \geq 0. \tag{2.49}$$

Now we temporarily leave the topic of conservation laws in order to establish some facts about "approximate δ distributions," or mollifiers. This is a sequence of smooth functions ω_ε such that the corresponding distributions tend to the δ_0 distribution, i.e., $\omega_\varepsilon \to \delta_0$ as $\varepsilon \to 0$. There are several ways of defining these distributions. We shall use the following: Let $\omega(\sigma)$ be a

---

[4] $\|f\|_{\operatorname{Lip}}$ is *not* a norm; after all, constants k have $\|k\|_{\operatorname{Lip}} = 0$.

46   2. Scalar Conservation Laws

C^∞ function such that

$$0 \leq \omega(\sigma) \leq 1, \quad \operatorname{supp}\omega \subseteq [-1,1], \quad \omega(-\sigma) = \omega(\sigma), \quad \int_{-1}^{1} \omega(\sigma)\,d\sigma = 1.$$

Now define

$$\omega_\varepsilon(\sigma) = \frac{1}{\varepsilon}\omega\left(\frac{\sigma}{\varepsilon}\right). \tag{2.50}$$

It is not hard to verify that ω_ε has the necessary properties such that $\lim_{\varepsilon \to 0} \omega_\varepsilon = \delta_0$ as a distribution.

We will need the following result:

Lemma 2.9. *Let* $h \in L^\infty(\mathbb{R}^2)$ *with compact support. Assume that for almost all* $x_0 \in \mathbb{R}$ *the function* $h(x,y)$ *is continuous at* (x_0, x_0). *Then*

$$\lim_{\varepsilon \to 0} \int\int h(x,y)\omega_\varepsilon(x-y)\,dy\,dx = \int h(x,x)\,dx. \tag{2.51}$$

Proof. Observe first that

$$\int h(x,y)\omega_\varepsilon(x-y)\,dy - h(x,x)$$
$$= \int (h(x,y) - h(x,x))\,\omega_\varepsilon(x-y)\,dy$$
$$= \int_{|z| \leq 1} (h(x, x+\varepsilon z) - h(x,x))\,\omega(z)\,dz \to 0 \text{ as } \varepsilon \to 0,$$

for almost all x, using the continuity of h. Furthermore,

$$\left|\int h(x,y)\omega_\varepsilon(x-y)\,dy\right| \leq \|h\|_\infty,$$

and hence we can use Lebesgue's bounded convergence theorem to obtain the result. □

Returning now to conservation laws and (2.49), we must make a smart choice of a test function $\phi(x,t,y,s)$. Let $\psi(x,t)$ be a test function that has support in $t > \epsilon_0$. We then define

$$\phi(x,t,y,s) = \psi\left(\frac{x+y}{2}, \frac{t+s}{2}\right)\omega_{\varepsilon_0}(t-s)\omega_\varepsilon(x-y),$$

where ε_0 and ε are (small) positive numbers. In this case[5]

$$\phi_t + \phi_s = \frac{\partial \psi}{\partial t}\left(\frac{x+y}{2}, \frac{t+s}{2}\right)\omega_{\varepsilon_0}(t-s)\omega_\varepsilon(x-y),$$

---

[5] Beware! Here $\frac{\partial \psi}{\partial t}\left(\frac{x+y}{2}, \frac{t+s}{2}\right)$ means the partial derivative of ψ with respect to the second variable, and this derivative is evaluated at $\left(\frac{x+y}{2}, \frac{t+s}{2}\right)$.

and[6]
$$\phi_x + \phi_y = \frac{\partial \psi}{\partial x}\left(\frac{x+y}{2}, \frac{t+s}{2}\right) \omega_{\varepsilon_0}(t-s)\omega_\varepsilon(x-y).$$

Now apply Lemma 2.9 to $h(x,y) = |u(x) - v(y)|\omega_{\varepsilon_0}\partial\psi/\partial t$ and $h(x,y) = q(x,y)\omega_{\varepsilon_0}\partial\psi/\partial x$. Then, letting ε_0 and ε tend to zero, (2.49) and Lemma 2.9 give

$$\iint (|u(x,t) - v(x,t)|\psi_t + q(u,v)\psi_x)\, dt\, dx \geq 0 \qquad (2.52)$$

for any two weak solutions u and v and any nonnegative test function ψ with support in $t > \epsilon$.

If we considered (2.17) in the strip $t \in [0,T]$ and test functions whose support included 0 and T, the Kružkov formulation would imply

$$\iint \iint (|u(x,t) - v(y,s)|(\phi_t + \phi_s) + q(u,v)(\phi_x + \phi_y))\, dx\, dt\, dy\, ds$$
$$- \iint |u(x,T) - v(y,T)|\phi(x,T,y,T)\, dx\, dy$$
$$+ \iint |u_0(x) - v_0(y)|\phi(x,0,y,0)\, dx\, dy \geq 0.$$

We can make the same choice of test function as before, now ending up with

$$\iint (|u(x,t) - v(x,t)|\psi_t + q(u,v)\psi_x)\, dt\, dx$$
$$- \int |u(x,T) - v(x,T)|\psi(x,T)\, dx \qquad (2.53)$$
$$+ \int |u_0(x) - v_0(x)|\psi(x,0)\, dx \geq 0.$$

In order to exploit (2.53) we define ψ as

$$\psi(x,t) = \left(\chi_{[-M+Lt+\varepsilon, M-Lt-\varepsilon]} * \omega_\varepsilon\right)(x), \qquad (2.54)$$

for $t \in [0,T]$. Here L denotes the Lipschitz constant of f, $\chi_{[a,b]}$ the characteristic function of the interval $[a,b]$, and $*$ the convolution product. We make the constant M so large that $M - Lt - \varepsilon > -M + Lt + \varepsilon$ for $t < T$. In order to make ψ an admissible test function, we modify it to go smoothly to zero for $t > T$.

We can compute for $t < T$,

$$\psi_t = \frac{d}{dt}\int_{-M+Lt+\varepsilon}^{M-Lt-\varepsilon} \omega_\varepsilon(x-y)\, dy \qquad (2.55)$$

---

[6] As in the previous equation, $\frac{\partial \psi}{\partial x}\left(\frac{x+y}{2}, \frac{t+s}{2}\right)$ means the partial derivative of ψ with respect to the first variable, and this derivative is evaluated at $\left(\frac{x+y}{2}, \frac{t+s}{2}\right)$.

$$= -L\left(\omega_\varepsilon\left(x - M + Lt + \varepsilon\right) + \omega_\varepsilon\left(x + M - Lt - \varepsilon\right)\right) \leq 0,$$

and

$$\psi_x = -\left(\omega_\varepsilon\left(x - M + Lt + \varepsilon\right) - \omega_\varepsilon\left(x + M - Lt - \varepsilon\right)\right). \tag{2.56}$$

With our choice of M, the two functions on the right-hand side of (2.56) have nonoverlapping support. Therefore,

$$0 = \psi_t + L\left|\psi_x\right| \geq \psi_t + \frac{q(u,v)}{|u-v|}\psi_x,$$

and hence

$$|u - v|\,\psi_t + q(u,v)\psi_x \leq 0.$$

Using this in (2.53), and letting ε go to zero, we find that

$$\int_{-M+Lt}^{M-Lt} |u(x,t) - v(x,t)|\,dx \leq \int_{-M}^{M} |u_0(x) - v_0(x)|\,dx. \tag{2.57}$$

If we assume that $u_0(x) = v_0(x)$ for $|x|$ sufficiently large (or $u_0 - v_0 \in L^1$), we have that also $u(x,t) = v(x,t)$ for large $|x|$. Consequently,

$$\|u(\,\cdot\,,t) - v(\,\cdot\,,t)\|_1 \leq \|u_0 - v_0\|_1 \tag{2.58}$$

in this case. Thus we have proved the following result.

Proposition 2.10. *Let f be Lipschitz continuous. Let u and v be two weak solutions of the initial value problems*

$$u_t + f(u)_x = 0, \quad u|_{t=0} = u_0,$$
$$v_t + f(v)_x = 0, \quad v|_{t=0} = v_0,$$

respectively, satisfying the Kružkov entropy condition. Assume that $u_0 - v_0$ is integrable. Then

$$\|u(\,\cdot\,,t) - v(\,\cdot\,,t)\|_1 \leq \|u_0 - v_0\|_1. \tag{2.59}$$

In particular, if $u_0 = v_0$, then $u = v$.

In other words, we have shown, starting from the Kružkov formulation of the entropy condition, that the initial value problem is stable in L^1, assuming existence of solutions.

The idea is now to obtain existence of solutions using front tracking; for Riemann initial data and continuous, piecewise linear flux functions we already have existence from Corollary 2.8. For given initial data and flux function we show that the solution can be obtained by approximating with front-tracking solutions.

First we consider general initial data $u_0 \in L^1(\mathbb{R})$ but with a continuous, piecewise linear flux function f. Equation (2.58) shows that if $u_0^i(x)$ is a sequence of step functions converging in L^1 to some $u_0(x)$, then the corresponding front-tracking solutions $u^i(x,t)$ will also converge in L^1 to some function $u(x,t)$. What is the equation satisfied by $u(x,t)$?

To answer this question, let $\phi(x,t)$ be a fixed test function, and let C be a constant such that
$$C > \max\{\|\phi\|_\infty, \|\phi_t\|_\infty, \|\phi_x\|_\infty\}.$$
Let also T be such that $\phi(x,t) = 0$ for all $t \geq T$. Since $u^i(x,t)$ all are weak solutions, we compute

$$\left| \int\int_0^\infty (u(x,t)\phi_t + f(u(x,t))\phi_x)\, dt\, dx + \int \phi(x,0) u_0(x)\, dx \right|$$

$$= \left| \int\int_0^\infty \left((u - u^i)\phi_t + (f(u) - f(u^i))\phi_x \right) dt\, dx \right.$$

$$\left. + \int \phi(x,0) \left(u_0 - u_0^i\right) dx \right|$$

$$\leq \int_0^T \left(C \|u(\,\cdot\,,t) - u^i(\,\cdot\,,t)\|_1 + LC \|u(\,\cdot\,,t) - u^i(\,\cdot\,,t)\|_1 \right) dt$$

$$+ C \|u_0 - u_0^i\|_1. \qquad (2.60)$$

Here L is a constant given by the Lipschitz condition (2.47). This implies

$$\left| \int_{-\infty}^\infty \int_0^\infty (u(x,t)\phi_t + f(u(x,t))\phi_x)\, dt\, dx + \int_{-\infty}^\infty \phi(x,0) u_0(x)\, dx \right|$$

$$\leq C\left((1+L)T + 1\right) \|u_0 - u_0^i\|_1. \qquad (2.61)$$

Since the right-hand side of (2.61) can be made arbitrarily small, $u(x,t)$ is a weak solution to the initial value problem
$$u_t + f(u)_x = 0, \qquad u(x,0) = u_0(x),$$
where f is piecewise linear and continuous.

Now, since we have eliminated the restriction of using step functions as initial functions, we would like to be able to consider conservation laws with more general flux functions. That is, if f^i is some sequence of piecewise linear continuous functions converging to f in Lipschitz norm, we would like to conclude that the corresponding front-tracking solutions converge as well.

That is, if f^i is some sequence of piecewise linear continuous functions such that $f^i \to f$ in the Lipschitz seminorm, and $u_0(x)$ is a function with finite total variation,[7] that is, T.V. (u_0) is bounded, then the corresponding solutions $u^i(x,t)$ converge in L^1 to some function $u(x,t)$.

To this end we need to consider stability with respect to the flux function. We start by studying two Riemann problems with the same initial data,

---

[7]See Appendix A for definitions and properties of the total variation T.V. (u) of a function u.

but with different flux functions. Let u and v be the weak solutions of

$$u_t + f(u)_x = 0, \quad v_t + g(v)_x = 0 \tag{2.62}$$

with initial data

$$u(x,0) = v(x,0) = \begin{cases} u_l & \text{for } x < 0, \\ u_r & \text{for } x > 0. \end{cases}$$

We assume that both f and g are continuous and piecewise linear with a finite number of breakpoints. The solutions u and v of (2.62) will be piecewise constant functions of x/t that are equal outside a finite interval in x/t. We need to estimate the difference in L^1 between the two solutions.

Lemma 2.11. *The following inequality holds:*

$$\frac{d}{dt}\|u - v\|_1 \leq \sup_u |f'(u) - g'(u)| \, |u_l - u_r|, \tag{2.63}$$

where the supremum is over all u between u_l and u_r.

Proof. Assume that $u_l \leq u_r$; the case $u_l \geq u_r$ is similar. Consider first the case where f and g both are convex. Without loss of generality we may assume that f and g have common breakpoints $u_l = w_1 < w_2 < \cdots < w_n = u_r$, and let the speeds be denoted by

$$f'|_{\langle w_j, w_{j+1} \rangle} = s_j \text{ and } g'|_{\langle w_j, w_{j+1} \rangle} = \tilde{s}_j.$$

Then

$$\int_{u_l}^{u_r} |f'(u) - g'(u)| \, du = \sum_{j=1}^{n-1} |s_j - \tilde{s}_j| (w_{j+1} - w_j).$$

Let σ_j be an ordering, that is, $\sigma_j < \sigma_{j+1}$, of all the speeds $\{s_j, \tilde{s}_j\}$. Then we may write

$$u(x,t)|_{x \in \langle \sigma_j t, \sigma_{j+1} t \rangle} = u_{j+1},$$
$$v(x,t)|_{x \in \langle \sigma_j t, \sigma_{j+1} t \rangle} = v_{j+1},$$

where both u_{j+1} and v_{j+1} are from the set of all possible breakpoints, namely $\{w_1, w_2, \ldots, w_n\}$, and $u_j \leq u_{j+1}$ and $v_j \leq v_{j+1}$. Thus

$$\|u(\,\cdot\,,t) - v(\,\cdot\,,t)\|_1 = t \sum_{j=1}^{m} |u_{j+1} - v_{j+1}| (\sigma_{j+1} - \sigma_j).$$

We easily see that

$$\frac{d}{dt}\|u(\,\cdot\,,t) - v(\,\cdot\,,t)\|_1 = \int_{u_l}^{u_r} |f'(u) - g'(u)| \, du$$
$$\leq \sup_u |f'(u) - g'(u)| \, |u_l - u_r|.$$

2.4. Existence and Uniqueness

The case where f and g are not necessarily convex is more involved. We will show that

$$\int_{u_l}^{u_r} |f'_\smile(u) - g'_\smile(u)|\, du \le \int_{u_l}^{u_r} |f'(u) - g'(u)|\, du \qquad (2.64)$$

when the convex envelopes are taken on the interval $[u_l, u_r]$. To this end we use the following general lemma:

Lemma 2.12 (Crandall–Tartar). *Let D be a subset of $L^1(\Omega)$, where Ω is some measure space. Assume that if ϕ and ψ are in D, then also $\phi \vee \psi = \max\{\phi, \psi\}$ is in D. Assume furthermore that there is a map $T \colon D \to L^1(\Omega)$ such that*

$$\int_\Omega T(\phi) = \int_\Omega \phi, \quad \phi \in D.$$

Then the following statements, valid for all $\phi, \psi \in D$, are equivalent:

(i) *If $\phi \le \psi$, then $T(\phi) \le T(\psi)$.*

(ii) $\int_\Omega (T(\phi) - T(\psi))^+ \le \int_\Omega (\phi - \psi)^+$, *where $\phi^+ = \phi \vee 0$.*

(iii) $\int_\Omega |T(\phi) - T(\psi)| \le \int_\Omega |\phi - \psi|$.

Proof of Lemma 2.12. For completeness we include a proof of the lemma. Assume (i). Then $T(\phi \vee \psi) - T(\phi) \ge 0$, which trivially implies $T(\phi) - T(\psi) \le T(\phi \vee \psi) - T(\psi)$, and thus $(T(\phi) - T(\psi))^+ \le T(\phi \vee \psi) - T(\psi)$. Furthermore,

$$\int_\Omega (T(\phi) - T(\psi))^+ \le \int_\Omega (T(\phi \vee \psi) - T(\psi)) = \int_\Omega (\phi \vee \psi - \psi) = \int_\Omega (\phi - \psi)^+,$$

proving (ii). Assume now (ii). Then

$$\int_\Omega |T(\phi) - T(\psi)| = \int_\Omega (T(\phi) - T(\psi))^+ + \int_\Omega (T(\psi) - T(\phi))^+$$

$$\le \int_\Omega (\phi - \psi)^+ + \int_\Omega (\psi - \phi)^+$$

$$= \int_\Omega |\phi - \psi|,$$

which is (iii). It remains to prove that (iii) implies (i). Let $\phi \le \psi$. For real numbers we have $x^+ = (|x| + x)/2$. This implies

$$\int_\Omega (T(\phi) - T(\psi))^+ = \frac{1}{2} \int_\Omega |T(\phi) - T(\psi)| + \frac{1}{2} \int_\Omega (T(\phi) - T(\psi))$$

$$\le \int_\Omega |\phi - \psi| + \int_\Omega (\phi - \psi) = 0.$$

$\square$

To apply this lemma in our context, we let D be the set of all piecewise constant functions on $[u_l, u_r]$. For any piecewise linear and continuous flux

function f its derivative f' is in D, and we define
$$T(f') = (f_{\smile})',$$
where the convex envelope is taken on the full interval $[u_l, u_r]$. Then
$$\int_{u_l}^{u_r} T(f')\, du = \int_{u_l}^{u_r} (f_{\smile})'(u)\, du = f_{\smile}(u_r) - f_{\smile}(u_l)$$
$$= f(u_r) - f(u_l) = \int_{u_l}^{u_r} f'(u)\, du.$$

To prove (2.64), it suffices to prove that **(i)** holds, that is,
$$f' \leq g' \text{ implies } T(f') \leq T(g')$$
for another piecewise linear and continuous flux function g. Assume otherwise, and let $[u_1, u_2]$ be the interval with left point u_1 nearest u_l such that $(f_{\smile})' > (g_{\smile})'$. Assume furthermore that u_2 is chosen maximal. By construction the point u_1 has to be a breakpoint of $f_{\smile}$. Thus $f_{\smile}(u_1) = f(u_1)$, while u_2 has to be a breakpoint of $g_{\smile}$. Thus $g_{\smile}(u_2) = g(u_2)$. Using this and the fact that the lower convex envelope never exceeds the function, we obtain
$$g(u_2) - g(u_1) \leq g_{\smile}(u_2) - g_{\smile}(u_1) < f_{\smile}(u_2) - f_{\smile}(u_1) \leq f(u_2) - f(u_1),$$
which is a contradiction. Hence we conclude that property **(i)** holds, which implies **(iii)** and subsequently (2.64). $\square$

If we then let $u_0(x)$ be an arbitrary piecewise constant function with a finite number of discontinuities, and let u and v be the solutions of the initial value problem (2.62), but $u(x,0) = v(x,0) = u_0(x)$, then the following inequality holds for all t until the first front collision:
$$\frac{d}{dt}\|u(t) - v(t)\|_1 \leq \|f - g\|_{\mathrm{Lip}} \sum_{i=1}^{n} |[\![u_0]\!]_i|. \tag{2.65}$$
where we used the Lipschitz seminorm $\|\cdot\|_{\mathrm{Lip}}$ defined in (2.47). At this point, it is also convenient to consider the *total variation of* u_0.

In this notation (2.65) gives
$$\|u(t) - v(t)\|_1 \leq t\,\|f - g\|_{\mathrm{Lip}}\,\mathrm{T.V.}(u_0). \tag{2.66}$$

This estimate holds until the first interaction of fronts for either u or v. Let t_1 be this first collision time, and let w be the weak solution constructed by front tracking of
$$w_t + f(w)_x = 0, \qquad w(x, t_1) = v(x, t_1).$$
Then for $t_1 < t < t_2$, with t_2 denoting the next time two fronts of either v or u interact,
$$\|u(t) - v(t)\|_1 \leq \|u(t) - w(t)\|_1 + \|w(t) - v(t)\|_1$$

$$\leq \|u(t_1) - w(t_1)\|_1 + (t - t_1) \|f - g\|_{\text{Lip}} \text{T.V.} (v(t_1)). \tag{2.67}$$

However,

$$\|u(t_1) - w(t_1)\|_1 = \|u(t_1) - v(t_1)\|_1 \leq t_1 \|f - g\|_{\text{Lip}} \text{T.V.} (u_0). \tag{2.68}$$

Observe furthermore that front-tracking solutions have the property that the total variation of the initial function T.V. (u_0) is larger than or equal to the total variation of the front-tracking solution T.V. $(u(\,\cdot\,, t))$. This holds because as two fronts interact, the solution of the resulting Riemann problem is always a monotone function. Thus, no new extrema are created.

When this and (2.68) are used in (2.67), we obtain

$$\|u(t) - v(t)\|_1 \leq t \|f - g\|_{\text{Lip}} \text{T.V.} (u_0). \tag{2.69}$$

This now holds for $t_1 < t < t_2$, but we can repeat the above argument inductively for every collision time t_i. Consequently, (2.69) holds for all positive t.

Consider now a Lipschitz continuous function f, and let f^i be a sequence of continuous, piecewise linear functions converging to f in the Lipschitz seminorm. Then the corresponding solutions u^i converge to some function u from (2.69).

Next, we show that the limit $u(x,t)$ satisfies the Kružkov entropy condition (2.18), thereby showing that it is indeed a weak solution that satisfies the entropy condition. By the Lebesgue bounded convergence theorem both

$$|u^i - k| \to |u - k|,$$

and

$$\text{sign}\,(u^i - k)\,(f^i\,(u^i) - f^i(k)) \to \text{sign}\,(u - k)\,(f(u) - f(k))$$

in $L^1_{\text{loc}}(\mathbb{R} \times [0, \infty))$. Hence, the limit $u(x,t)$ satisfies (2.18), and we have shown that the front-tracking method indeed converges to the unique solution of conservation law.

Let u be the limit obtained by front tracking by letting $f^i \to f$ in the Lipschitz seminorm, and similarly let v be the limit obtained by letting $g^i \to g$. Assume that $u_0 = v_0$. Using (2.69) we then obtain

$$\|u(t) - v(t)\|_1 \leq t \|f - g\|_{\text{Lip}} \text{T.V.} (u_0). \tag{2.70}$$

We can combine (2.58) and the above equation (2.70) to show a comparison result. Let u, v, and w be the solutions of

$$\begin{aligned} u_t + f(u)_x &= 0, & u(x,0) &= u_0(x), \\ v_t + g(v)_x &= 0, & v(x,0) &= v_0(x), \\ w_t + f(w)_x &= 0, & w(x,0) &= v_0(x). \end{aligned}$$

54   2. Scalar Conservation Laws

Then
$$\|u(\,\cdot\,,t) - v(\,\cdot\,,t)\|_1 \leq \|u(\,\cdot\,,t) - w(\,\cdot\,,t)\|_1 + \|w(\,\cdot\,,t) - v(\,\cdot\,,t)\|_1$$
$$\leq \|u_0 - w(\,\cdot\,,0)\|_1 + t\,\text{T.V.}\,(v_0)\,\|f - g\|_{\text{Lip}},$$
by (2.58) and (2.70). If we had defined w to be the solution of
$$w_t + g(w)_x = 0, \quad w(x,0) = u_0(x),$$
we would have obtained
$$\|u(\,\cdot\,,t) - v(\,\cdot\,,t)\|_1 \leq \|v_0 - w(\,\cdot\,,0)\|_1 + t\,\text{T.V.}\,(u_0)\,\|f - g\|_{\text{Lip}}.$$
Thus we have proved the following theorem.

Theorem 2.13. *Let u_0 be a function of bounded variation that is also in L^1, and let $f(u)$ be a Lipschitz continuous function. Then there exists a unique weak solution $u = u(x,t)$ to the initial value problem*
$$u_t + f(u)_x = 0, \quad u(x,0) = u_0(x),$$
which also satisfies the Kružkov entropy condition (2.46). Furthermore, if v_0 is another function in $BV \cap L^1$, $g(v)$ is Lipschitz continuous, and v is the unique weak Kružkov entropy solution to
$$v_t + g(v)_x = 0, \quad v(x,0) = v_0(x),$$
then
$$\|u(\,\cdot\,,t) - v(\,\cdot\,,t)\|_1 \leq \|u_0 - v_0\|_1 + t\min\{\text{T.V.}\,(u_0),\text{T.V.}\,(v_0)\}\,\|f - g\|_{\text{Lip}}. \tag{2.71}$$

In the special case where $g = f_\delta$, the piecewise linear and continuous approximation to f obtained by taking linear interpolation between points $f_\delta(j\delta) = f(j\delta)$, we have, assuming that f is piecewise C^2, that (see Exercise 2.11)
$$\|f - f_\delta\|_{\text{Lip}} \leq \delta\,\|f''\|_\infty, \tag{2.72}$$
and hence we obtain, when u_δ denotes the solution with flux f_δ,
$$\|u(\,\cdot\,,t) - u_\delta(\,\cdot\,,t)\|_1 \leq \|u_0 - u_{0,\delta}\|_1 + t\,\delta\,\text{T.V.}\,(u_0)\,\|f''\|_\infty,$$
showing that the convergence is of first order in the approximation parameter δ.

We end this section by summarizing some of the fundamental properties of solutions of scalar conservation laws in one dimension.

Theorem 2.14. *Let u_0 be an integrable function of bounded variation, and let $f(u)$ be a Lipschitz continuous function. Then the unique weak entropy solution $u = u(x,t)$ to the initial value problem*
$$u_t + f(u)_x = 0, \quad u(x,0) = u_0(x),$$
satisfies the following properties for all $t \in [0,\infty\rangle$:

(i) *Maximum principle*:
$$\|u(\,\cdot\,,t)\|_\infty \le \|u_0\|_\infty.$$

(ii) *Total variation diminishing (TVD)*:
$$\text{T.V.}\,(u(\,\cdot\,,t)) \le \text{T.V.}\,(u_0).$$

(iii) L^1-*contractive*: If v_0 is a function in $BV \cap L^1$ and $v = v(x,t)$ denotes the entropy solution with v_0 as initial data, then
$$\|u(\,\cdot\,,t) - v(\,\cdot\,,t)\|_1 \le \|u_0 - v_0\|_1.$$

(iv) *Monotonicity preservation*:
$$u_0 \text{ monotone implies } u(\,\cdot\,,t) \text{ monotone}.$$

(v) *Monotonicity*: Let v_0 be a function in $BV \cap L^1$, and let $v = v(x,t)$ denote the entropy solution with v_0 as initial data. Then
$$u_0 \le v_0 \text{ implies } u(\,\cdot\,,t) \le v(\,\cdot\,,t).$$

(vi) *Lipschitz continuity in time*:
$$\|u(\,\cdot\,,t) - u(\,\cdot\,,s)\|_1 \le \|f\|_{\text{Lip}}\,\text{T.V.}\,(u_0)\,|t-s|,$$
for all $s,t \in [0,\infty)$.

Proof. The maximum principle and the monotonicity preservation properties are all easily seen to be true for the front-tracking approximation by checking the solution of isolated Riemann problems, and the properties carry over in the limit.

Monotonicity holds by the Crandall–Tartar lemma, Lemma 2.12, applied with the solution operator $u_0 \mapsto u(x,t)$ as T, and the L^1 contraction property.

The fact that the total variation is nonincreasing follows using Theorem 2.13 (with $g = f$ and $v_0 = u_0(\,\cdot\, + h)$) and
$$\text{T.V.}\,(u(\,\cdot\,,t)) = \lim_{h\to 0} \frac{1}{h}\int |u(x+h,t) - u(x,t)|\,dx$$
$$\le \lim_{h\to 0} \frac{1}{h}\int |u_0(x+h) - u_0(x)|\,dx = \text{T.V.}\,(u_0).$$

The L^1-contractivity is a special case of (2.71). Finally, to prove the Lipschitz continuity in time of the spatial L^1-norm, we first consider the solution of a single Riemann problem for a continuous piecewise linear flux function f. From Corollary 2.4 we see that
$$\|u(\,\cdot\,,t) - u(\,\cdot\,,s)\|_1 \le (|s_1|\,|v_2 - v_1| + \cdots + |s_{m-1}|\,|v_m - v_{m-1}|)\,|t-s|$$
$$\le \|f\|_{\text{Lip}}\,\text{T.V.}\,(u_0)\,|t-s|.$$

This will carry over to the general case by taking appropriate approximations (of the initial data and the flux function). For an alternative argument, see Theorem 7.10. □

2.5 Notes

> Ofte er det jo sådan, at når man kigger det nye efter
> i sømmene, så er det bare sømmene, der er nye.
>
> Kaj Munk, En Digters Vej og andre Artikler (*1948*)

The "viscous regularization" as well as the weak formulation of the scalar conservation law were studied in detail by Hopf [73] in the case of Burgers' equation where $f(u) = u^2/2$. Hopf's paper initiated the rigorous analysis of conservation laws. Oleĭnik, [110], [111], gave a systematic analysis of the scalar case, proving existence and uniqueness of solutions using finite differences. See also the survey by Gel'fand [50].

Kružkov's approach, which combines the notion of weak solution and uniqueness into one equation, (2.23), was introduced in [88], in which he studied general scalar conservation laws in many dimensions with explicit time and space dependence in flux functions and a source term.

The solution of the Riemann problem in the case where the flux function f has one or more inflection points was given by Gel'fand [50], Cho-Chun [28], and Ballou [7].

It is quite natural to approximate any flux function by a continuous and piecewise linear function. This method is frequently referred to as "Dafermos' method" [41]. Dafermos used it to derive existence of solutions of scalar conservation laws. Prior to that a similar approach was studied numerically by Barker [9]. Further numerical work based on Dafermos' paper can be found in Hedstrom [63, 64] and Swartz and Wendroff [133]. Applications of front tracking to hyperbolic conservation laws on a half-line appeared in [136].

Unaware of this earlier development, Holden, Holden, and Høegh-Krohn rediscovered the method [67], [66] and proved L^1-stability and that the method in fact can be used as a numerical method. We here use the name "front-tracking method" as a common name for this approach and an analogous method that works for systems of hyperbolic conservation laws. We combine the front-tracking method with Kružkov's ingenious method of "doubling the variables"; Kružkov's method shows stability (and thereby uniqueness) of the solution, and we use front tracking to construct the solution.

The original argument in [66] followed a direct but more cumbersome analysis. An alternative approach to show convergence of the front-tracking approximation is first to establish boundedness of the approximation both in L^∞ and total variation, and then to use Helly's theorem to deduce convergence. Subsequently one has to show that the limit is a Kružkov entropy solution, and finally invoke Kuznetsov's theory to conclude stability in the sense of Theorem 2.13. We will use this argument in Chapters 3 and 4.

Lemma 2.12 is due to Crandall and Tartar; see [40] and [38].

The L^1-contractivity of solutions of scalar conservation laws is due to Volpert [143]; see also Keyfitz [115, 62].

The uniqueness result, Theorem 2.13, was first proved by Lucier [106], using an approach due to Kutznetsov [90]. Our presentation here is different in that we avoid Kutznetsov's theory; see Section 3.2. For an alternative proof of (2.58) we refer to Málek et al. [107, pp. 92 ff].

The term "front tracking" is also used to denote other approaches to hyperbolic conservation laws. Glimm and coworkers [55, 56, 53, 54] have used a front-tracking method as a computational tool. In their approach the discontinuities, or shocks, are introduced as independent computational objects and moved separately according to their own dynamics. Away from the shocks, traditional numerical methods can be employed. This method yields sharp fronts. The name "front tracking" is also used in connection with level set methods, in particular in connection with Hamilton–Jacobi equations; see, e.g., [114]. Here one considers the dynamics of interfaces or fronts. These methods are distinct from those treated in this book.

Exercises

> Our problems are manmade;
> therefore they may be solved by man.
>
> John F. Kennedy (1963)

2.1 Let
$$f(u) = \frac{u^2}{u^2 + (1-u)^2}.$$

Find the solution of the Riemann problem for the scalar conservation law $u_t + f(u)_x = 0$ where $u_l = 0$ and $u_r = 1$. This equation is an example of the so-called Buckley–Leverett equation and represents a simple model of two-phase fluid flow in a porous medium. In this case u is a number between 0 and 1 and denotes the saturation of one of the phases.

2.2 Consider the initial value problem for Burgers' equation,
$$u_t + \frac{1}{2}\left(u^2\right)_x = 0, \quad u(x,0) = u_0(x) = \begin{cases} -1 & \text{for } x < 0, \\ 1 & \text{for } x \geq 0. \end{cases}$$

a. Show that $u(x,t) = u(x,0)$ is a weak solution.
b. Let
$$u_0^\varepsilon(x) = \begin{cases} -1 & \text{for } x < -\varepsilon, \\ x/\varepsilon & \text{for } -\varepsilon \le x \le \varepsilon, \\ 1 & \text{for } \varepsilon < x. \end{cases}$$

Find the solution, $u^\varepsilon(x,t)$, of Burgers' equation if $u(x,0) = u_0^\varepsilon(x)$.
c. Find $\bar{u}(x,t) = \lim_{\varepsilon \downarrow 0} u^\varepsilon(x,t)$.
d. Since $\bar{u}(x,0) = u_0(x)$, why do we not have $\bar{u} = u$?

2.3 For $\varepsilon > 0$, consider the linear viscous regularization
$$u_t^\varepsilon + a u_x^\varepsilon = \varepsilon u_{xx}^\varepsilon, \qquad u^\varepsilon(x,0) = u_0(x) = \begin{cases} u_l, & \text{for } x \le 0, \\ u_r, & \text{for } x > 0, \end{cases}$$

where a is a constant. Show that
$$\lim_{\varepsilon \downarrow 0} u^\varepsilon(x,t) = \begin{cases} u_l, & \text{for } x < at, \\ u_r, & \text{for } x > at, \end{cases}$$

and thus that in $L^1(\mathbb{R} \times [0,T])$, $u^\varepsilon \to u_0(x - at)$.

2.4 This exercise outlines another way to prove monotonicity. If u and v are entropy solutions, then we have
$$\iint [(u-v)\psi_t + (f(u) - f(v))\psi_x] \, dt \, dx - \int (u-v)\psi \big|_0^T \, dx = 0.$$

Set $\Phi(\sigma) = |\sigma| + \sigma$, and use (2.53) to conclude that
$$\iint \left(\Phi(u-v)\psi_t + \Psi(u,v)\psi_x \right) dx \, dt - \int \Phi(u-v)\psi \big|_0^T \, dx \ge 0 \quad (2.73)$$

for a Lipschitz continuous Ψ. Choose a suitable test function ψ to show that (2.73) implies the monotonicity property.

2.5 Let $c(x)$ be a continuous and locally bounded function. Consider the conservation law with "coefficient" c,
$$u_t + c(x) f(u)_x = 0, \quad u(x,0) = u_0(x). \tag{2.74}$$

a. Define the characteristics for (2.74).
b. What is the Rankine–Hugoniot condition in this case?
c. Set $f(u) = u^2/2$, $c(x) = 1 + x^2$, and
$$u_0 = \begin{cases} -1 & \text{for } x < 0, \\ 1 & \text{for } x \ge 0. \end{cases}$$

Find the solution of (2.74) in this case.

d. Formulate a front-tracking algorithm for the general case of (2.74).
e. What is the entropy condition for (2.74)?

2.6 Consider the conservation law where the x dependency is "inside the derivation,"

$$u_t + (c(x)f(u))_x = 0. \qquad (2.75)$$

The coefficient c is assumed to be continuously differentiable.

a. Define the characteristics for (2.75).
b. What is the entropy condition for this problem?
c. Modify the proof of Proposition 2.10 to show that if u and v are entropy solutions of (2.75) with initial data u_0 and v_0, respectively, then

$$\|u(\,\cdot\,,t) - v(\,\cdot\,,t)\|_{L^1(\mathbb{R})} \leq \|u_0 - v_0\|_{L^1(\mathbb{R})}.$$

2.7 Let η and q be an entropy/entropy flux pair as in (2.12). Assume that u is a piecewise continuous solution (in the distributional sense) of

$$\eta(u)_t + q(u)_x \leq 0.$$

Show that across any discontinuity u satisfies

$$(\eta_l - \eta_r) - \sigma\,(q_l - q_r) \leq 0,$$

where σ is the speed of the discontinuity, and $q_{l,r}$ and $\eta_{l,r}$ are the values to the left and right of the discontinuity.

2.8 Consider the initial value problem for (the inviscid) Burgers' equation

$$u_t + \frac{1}{2}\left(u^2\right)_x = 0, \quad u(x,0) = u_0(x),$$

and assume that the entropy solution is bounded. Set $\eta = \frac{1}{2}u^2$, and find the corresponding entropy flux $q(u)$. Then choose a test function $\psi(x,t)$ to show that

$$\|u(\,\cdot\,,t)\|_{L^2(\mathbb{R})} \leq \|u_0\|_{L^2(\mathbb{R})}.$$

If v is another bounded entropy solution of Burgers' equation with initial data v_0, is $\|u - v\|_{L^2(\mathbb{R})} \leq \|u_0 - v_0\|_{L^2(\mathbb{R})}$?

2.9 Consider the scalar conservation law with a zeroth-order term

$$u_t + f(u)_x = g(u), \qquad (2.76)$$

where $g(u)$ is a locally bounded and Lipschitz continuous function.

a. Determine the Rankine–Hugoniot relation for (2.76).
b. Find the entropy condition for (2.76).

2.10 The initial value problem
$$v_t + H(v_x) = 0, \quad v(x,0) = v_0(x), \tag{2.77}$$
is called a Hamilton–Jacobi equation. One is interested in solving (2.77) for $t > 0$, and the initial function v_0 is assumed to be bounded and uniformly continuous. Since the differentiation is inside the nonlinearity, we cannot define solutions in the distributional sense as for conservation laws. A *viscosity solution* of (2.77) is a bounded and uniformly continuous function v such that for all test functions φ, the following hold:

$$\text{subsolution} \begin{cases} \text{if } v - \varphi \text{ has a local maximum at } (x,t) \text{ then} \\ \varphi(x,t)_t + H(\varphi(x,t)_x) \leq 0, \end{cases}$$

$$\text{supsolution} \begin{cases} \text{if } v - \varphi \text{ has a local minimum at } (x,t) \text{ then} \\ \varphi(x,t)_t + H(\varphi(x,t)_x) \geq 0. \end{cases}$$

If we set $p = v_x$, then formally p satisfies the conservation law
$$p_t + H(p)_x = 0, \quad p(x,0) = \partial_x v_0(x). \tag{2.78}$$

Assume that
$$v_0(x) = v_0(0) + \begin{cases} p_l x & \text{for } x \leq 0, \\ p_r x & \text{for } x > 0, \end{cases}$$
where p_l and p_r are constants. Let p be an entropy solution of (2.78) and set
$$v(x,t) = v_0(0) + xp(x,t) - tH(p(x,t)).$$
Show that v defined in this way is a viscosity solution of (2.77).

2.11 Let f be piecewise C^2. Show that if we define the continuous, piecewise linear interpolation f_δ by $f_\delta(j\delta) = f(j\delta)$, then we have
$$\|f - f_\delta\|_{\text{Lip}} \leq \delta \|f''\|_\infty,$$
cf. (2.72).

2.12 ([113]) Consider the Riemann problem
$$u_t + f(u)_x = 0, \quad u|_{t=0} = u_0,$$
where
$$u_0 = \begin{cases} u_l & \text{for } x < 0, \\ u_r & \text{for } x > 0. \end{cases}$$
Show that the solution can be written
$$u(x,t) = \begin{cases} -\frac{d}{d\xi} \min_{u \in [u_l, u_r]} (f(u) - \xi u)\big|_{\xi=x/t} & \text{for } u_l < u_r, \\ -\frac{d}{d\xi} \max_{u \in [u_r, u_l]} (f(u) - \xi u)\big|_{\xi=x/t} & \text{for } u_r < u_l. \end{cases}$$

In particular, the formula is valid also for nonconvex flux functions.

2.13 Find the unique weak entropy solution of the initial value problem (cf. Exercise 2.9)
$$u_t + \left(\frac{1}{2}u^2\right)_x = -u,$$
$$u|_{t=0} = \begin{cases} 1 & \text{for } x \leq -\frac{1}{2}, \\ -2x & \text{for } -\frac{1}{2} < x < 0, \\ 0 & \text{for } x \geq 0. \end{cases}$$

2.14 Find the weak entropy solution of the initial value problem
$$u_t + (e^u)_x = 0, \quad u(x,0) = \begin{cases} 2 & \text{for } x < 0, \\ 0 & \text{for } x \geq 0. \end{cases}$$

2.15 Find the weak entropy solution of the initial value problem
$$u_t + \left(u^3\right)_x = 0$$
with initial data

a.
$$u(x,0) = \begin{cases} 1 & \text{for } x < 2, \\ 0 & \text{for } x \geq 2, \end{cases}$$

b.
$$u(x,0) = \begin{cases} 0 & \text{for } x < 2, \\ 1 & \text{for } x \geq 2. \end{cases}$$

2.16 Find the weak entropy solution of the initial value problem
$$u_t + \frac{1}{2}\left(u^2\right)_x = 0$$
with initial data
$$u(x,0) = \begin{cases} 1 & \text{for } 0 < x < 1, \\ 0 & \text{otherwise}. \end{cases}$$

2.17 Redo Example 2.5 with the same flux function but initial data
$$u_0(x) = \begin{cases} -1 & \text{for } x \leq x_1, \\ 1 & \text{for } x_1 < x < x_2, \\ -1 & \text{for } x \geq x_2. \end{cases}$$

3
A Short Course in Difference Methods

> Computation will cure what ails you.
>
> *Clifford Truesdell, The Computer, Ruin of Science and Threat to Mankind, 1980/1982*

Although front tracking can be thought of as a numerical method, and has indeed been shown to be excellent for one-dimensional conservation laws, it is not part of the standard repertoire of numerical methods for conservation laws. Traditionally, difference methods have been central to the development of the theory of conservation laws, and the study of such methods is very important in applications.

This chapter is intended to give a brief introduction to difference methods for conservation laws. The emphasis throughout will be on methods and general results rather than on particular examples. Although difference methods and the concepts we discuss can be formulated for systems, we will exclusively concentrate on scalar equations. This is partly because we want to keep this chapter introductory, and partly due to the lack of general results for difference methods applied to systems of conservation laws.

3.1 Conservative Methods

We are interested in numerical methods for the scalar conservation law in one dimension. (We will study multidimensional problems in Chapter 4.)

Thus we consider

$$u_t + f(u)_x = 0, \qquad u|_{t=0} = u_0. \tag{3.1}$$

A difference method is created by replacing the derivatives by finite differences, e.g.,

$$\frac{\Delta u}{\Delta t} + \frac{\Delta f(u)}{\Delta x} = 0. \tag{3.2}$$

Here Δt and Δx are small positive numbers. We shall use the notation

$$U_j^n = u\left(j\Delta x, n\Delta t\right) \quad \text{and} \quad U^n = \left(U_{-K}^n, \ldots, U_j^n, \ldots, U_K^n\right),$$

where u now is our numerical approximation to the solution of (3.1). Normally, since we are interested in the initial value problem (3.1), we know the initial approximation

$$U_j^0, \quad -K \le j \le K,$$

and we want to use (3.2) to calculate U^n for $n \in \mathbb{N}$. We will not say much about boundary conditions in this book. Often one assumes that the initial data is periodic, i.e.,

$$U_{-K+j}^0 = U_{K+j}^0, \quad \text{for } 0 \le j \le 2K,$$

which gives $U_{-K+j}^n = U_{K+j}^n$. Another commonly used device is to assume that $\partial_x f(u) = 0$ at the boundary of the computational domain. For a numerical scheme this means that

$$f\left(U_{-K-j}^n\right) = f\left(U_{-K}^n\right) \quad \text{and} \quad f\left(U_{K+j}^n\right) = f\left(U_K^n\right) \quad \text{for } j > 0.$$

For nonlinear equations, explicit methods are most common. These can be written

$$U^{n+1} = G\left(U^n, \ldots, U^{n-l}\right) \tag{3.3}$$

for some function G.

◇ **Example 3.1 (A nonconservative method).**

If $f(u) = u^2/2$, then we can define an explicit method

$$U_j^{n+1} = U_j^n - \frac{\Delta t}{\Delta x} U_j^n \left(U_{j+1}^n - U_j^n\right). \tag{3.4}$$

If U^0 is given by

$$U_j^0 = \begin{cases} 0 & \text{for } j \le 0, \\ 1 & \text{for } j > 0, \end{cases}$$

then $U^n = U^0$ for all n. So the method produces a nicely converging sequence, but the limit is not a solution to the original problem. The difference method (3.4) is based on a nonconservative formulation. Henceforth, we will not discuss nonconservative schemes. ◇

We call a difference method *conservative* if it can be written in the form
$$U_j^{n+1} = G(U_{j-1-p}^n, \ldots, U_{j+q}^n)$$
$$= U_j^n - \lambda \left(F\left(U_{j-p}^n, \ldots, U_{j+q}^n\right) - F\left(U_{j-1-p}^n, \ldots, U_{j-1+q}^n\right) \right), \quad (3.5)$$
where
$$\lambda = \frac{\Delta t}{\Delta x}.$$
The function F is referred to as the *numerical flux*. For brevity, we shall often use the notation
$$G(U; j) = G\left(U_{j-1-p}, \ldots, U_{j+q}\right),$$
$$F(U; j) = F\left(U_{j-p}, \ldots, U_{j+q}\right),$$
so that (3.5) reads
$$U_j^{n+1} = G(U^n; j) = U_j^n - \lambda \left(F\left(U^n; j\right) - F\left(U^n; j-1\right) \right).$$
Conservative methods have the property that U is conserved, since
$$\sum_{j=-K}^{K} U_j^{n+1} \Delta x = \sum_{j=-K}^{K} U_j^n \Delta x - \Delta t \left(F\left(U^n; K\right) - F\left(U^n; -K-1\right) \right).$$
If we set U_j^0 equal to the average of u_0 over the jth grid cell, i.e.,
$$U_j^0 = \frac{1}{\Delta x} \int_{j\Delta x}^{(j+1)\Delta x} u_0(x)\, dx,$$
and for the moment assume that $F\left(U^n; K\right) = F\left(U^n; -K-1\right)$, then
$$\int U^n(x)\, dx = \int u_0(x)\, dx. \quad (3.6)$$
A conservative method is said to be *consistent* if
$$F(u, \ldots, u) = f(u). \quad (3.7)$$
In addition we demand that F be Lipschitz continuous in all its variables.

◇ **Example 3.2 (Some conservative methods).**

The simplest conservative method is the *upwind scheme*
$$F(U; j) = f(U_j). \quad (3.8)$$
Another common method is the *Lax–Friedrichs scheme*, usually written
$$U_j^{n+1} = \frac{1}{2}\left(U_{j+1}^n + U_{j-1}^n\right) - \frac{1}{2}\lambda\left(f\left(U_{j+1}^n\right) - f\left(U_{j-1}^n\right)\right). \quad (3.9)$$
In conservation form, this reads
$$F(U^n; j) = \frac{1}{2\lambda}\left(U_j^n - U_{j+1}^n\right) + \frac{1}{2}\left(f\left(U_j^n\right) + f\left(U_{j+1}^n\right)\right).$$

Also, two-step methods are used. One is the *Richtmyer two-step Lax–Wendroff scheme*:

$$F(U;j) = f\left(\frac{1}{2}\left(U_{j+1}^n + U_j^n\right) - \lambda\left(f\left(U_{j+1}^n\right) - f\left(U_j^n\right)\right)\right). \qquad (3.10)$$

Another two-step method is the *MacCormack scheme*:

$$F(U;j) = \frac{1}{2}\left(f\left(U_j^n - \lambda\left(f\left(U_{j+1}^n\right) - f\left(U_j^n\right)\right)\right) + f\left(U_j^n\right)\right). \qquad (3.11)$$

The *Godunov scheme* is a generalization of the upwind method. Let $\tilde{u}_j$ be the solution of the Riemann problem with initial data

$$\tilde{u}_j(x,0) = \begin{cases} U_j^n & \text{for } x \leq 0, \\ U_{j+1}^n & \text{for } x > 0. \end{cases}$$

The numerical flux is given by

$$F(U;j) = f\left(\tilde{u}(0,\Delta t)\right). \qquad (3.12)$$

To avoid that waves from neighboring grid cells start to interact before the next time step, we cannot take too long time steps Δt. Since the maximum speed is bounded by $\max|f'(u)|$, we need to enforce the requirement that

$$\lambda|f'(u)| < 1. \qquad (3.13)$$

The condition (3.13) is called the *Courant–Friedrichs–Lewy (CFL) condition*. If all characteristic speeds are nonnegative (nonpositive), Godunov's method reduces to the upwind (downwind) method.

The Lax–Friedrichs and Godunov schemes are both of first order in the sense that the local truncation error is of order one. (We shall return to this concept below.) However, both the Lax–Wendroff and MacCormack methods are of second order. In general, higher-order methods are good for smooth solutions, but also produce solutions that oscillate in the vicinity of discontinuities. On the other hand, lower order methods have "enough diffusion" to prevent oscillations. Therefore, one often uses *hybrid methods*. These methods usually consist of a linear combination of a lower- and a higher-order method. The numerical flux is then given by

$$F(U;j) = \theta(U;j)F_L(U;j) + (1-\theta(U;j))F_H(U;j), \qquad (3.14)$$

where F_L denotes a lower-order numerical flux, and F_H a higher-order numerical flux. The function $\theta(U;j)$ is close to zero, where U is smooth and close to one near discontinuities. Needless to say, choosing appropriate θ's is a discipline in its own right. We have implemented a method (called *fluxlim* in Figure 3.1) that is a combination of the (second-order) MacCormack method and the (first-order) Lax–Friedrichs scheme, and this scheme is compared with the "pure" methods in this figure. We, somewhat arbitrarily used

$$\theta(U;j) = 1 - \frac{1}{1+|\Delta_{j,\Delta x}U|},$$

where $\Delta_{j,\Delta x}U$ is an approximation to the second derivative of U with respect to x,

$$\Delta_{j,\Delta x}U = \frac{U_{j+1} - 2U_j + U_{j-1}}{\Delta x^2}.$$

Another approach is to try to generalize Godunov's method by replacing the piecewise constant data U^n by a smoother function. The simplest such replacement is by a piecewise linear function. To obtain a proper generalization one should then solve a "Riemann problem" with linear initial data to the left and right. While this is difficult to do exactly, one can use approximations instead. One such approximation leads to the following method:

$$F(U^n;j) = \frac{1}{2}(g_j + g_{j+1}) - \frac{1}{2\lambda}\Delta U_j^n.$$

Here $\Delta U_j^n = U_{j+1}^n - U_j^n$, and

$$g_j = f(u_j^{n+1/2}) + \frac{1}{2\lambda}u_j',$$

where

$$u_j' = \text{MinMod}\left(\Delta U_{j-1}^n, \Delta U_j^n\right),$$
$$u_j^{n+1/2} = U_j^n - \frac{\lambda}{2}f'(U_j^n)u_j',$$

and

$$\text{MinMod}(a,b) := \frac{1}{2}\left(\text{sign}(a) + \text{sign}(b)\right)\min(|a|,|b|).$$

This method is labeled *slopelim* in the figures. Now we show how these methods perform on two test examples. In both examples the flux function is given by

$$f(u) = \frac{u^2}{u^2 + (1-u)^2}. \tag{3.15}$$

The example is motivated by applications in oil recovery, where one often encounters flux functions that have a shape similar to that of f; that is, $f' \geq 0$ and $f''(u) = 0$ at a single point u. The model is called the *Buckley–Leverett* equation. The first example uses initial data

$$u_0(x) = \begin{cases} 1 & \text{for } x \leq 0, \\ 0 & \text{for } x > 0. \end{cases} \tag{3.16}$$

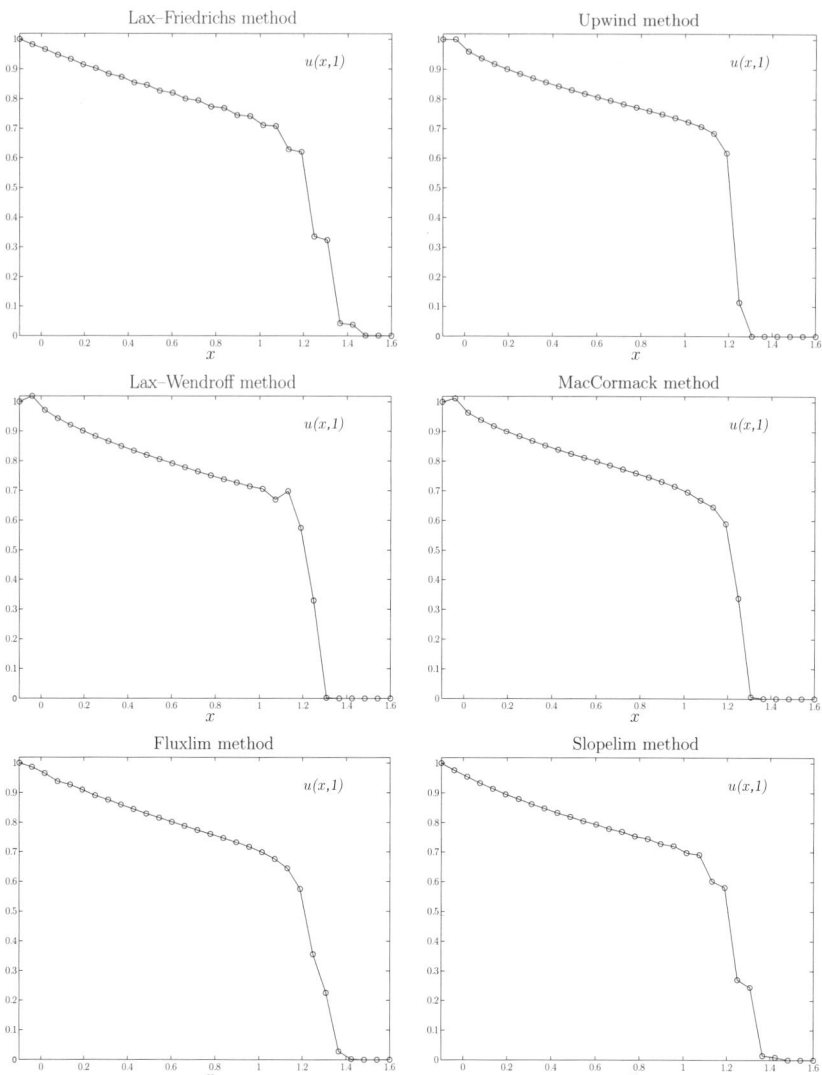

Figure 3.1. Computed solutions at time $t = 1$ for flux function (3.15) and initial data (3.16).

In Figure 3.1 we show the computed solution at time $t = 1$ for all methods, using 30 grid points in the interval $[-0.1, 1.6]$, and $\Delta x = 1.7/29$, $\Delta t = 0.5\Delta x$. The second example uses initial data

$$u_0(x) = \begin{cases} 1 & \text{for } x \in [0, 1], \\ 0 & \text{otherwise,} \end{cases} \qquad (3.17)$$

3.1. Conservative Methods

and 30 grid points in the interval $[-0.1, 2.6]$, $\Delta x = 2.7/29$, $\Delta t = 0.5\Delta x$. In Figure 3.2 we also show a reference solution computed by the upwind method using 500 grid points. The most notable feature of the plots in Figure 3.2 is the solutions computed by the second-order methods. We shall show that if a sequence of solutions produced by a consistent, conservative method converges, then the limit is a weak solution. The exact solution to both these problems can be calculated by the method of characteristics. ◇

The *local truncation error* of a numerical method $L_{\Delta t}$ is defined (formally) as

$$L_{\Delta t}(x) = \frac{1}{\Delta t}\left(S(\Delta t)u - S_N(\Delta t)u\right)(x), \qquad (3.18)$$

where $S(t)$ is the solution operator associated with (3.1); that is, $u = S(t)u_0$ denotes the solution at time t, and $S_N(t)$ is the formal solution operator associated with the numerical method, i.e.,

$$S_N(\Delta t)u(x) = u(x) - \lambda\left(F(u;j) - F(u;j-1)\right).$$

To make matters more concrete, assume that we are studying the upwind method. Then

$$S_N(\Delta t)u(x) = u(x) - \frac{\Delta t}{\Delta x}\left(f(u(x)) - f(u(x - \Delta x))\right).$$

We say that the method is of kth order if for all smooth solutions $u(x,t)$,

$$|L_{\Delta t}(x)| = \mathcal{O}\left(\Delta t^k\right)$$

as $\Delta t \to 0$. That a method is of high order, $k \geq 2$, usually implies that it is "good" for computing smooth solutions.

◇ **Example 3.3 (Local truncation error).**

We verify that the upwind method is of first order:

$$L_{\Delta t}(x) = \frac{1}{\Delta t}\left(u(x, t + \Delta t) - u(x) + \frac{\Delta t}{\Delta x}(f(u(x)) - f(u(x - \Delta x)))\right)$$

$$= \frac{1}{\Delta t}\left(u + \Delta t\, u_t + \frac{(\Delta t)^2}{2}u_{tt} + \cdots - u\right.$$

$$\left. - \lambda\left(f'(u)\left(-u_x\Delta x + \frac{(\Delta x)^2}{2}u_{xx} + \cdots\right)\right.\right.$$

$$\left.\left. + f''(u)\frac{1}{2}(-u_x\Delta x + \cdots)^2\right)\right)$$

$$= \frac{1}{\Delta t}\left(\Delta t\,(u_t + f(u)_x) + \frac{(\Delta t)^2}{2}u_{tt}\right.$$

$$\left. - \frac{\Delta t \Delta x}{2}\left(u_{xx}f'(u) + f''(u)u_x^2\right) + \cdots\right)$$

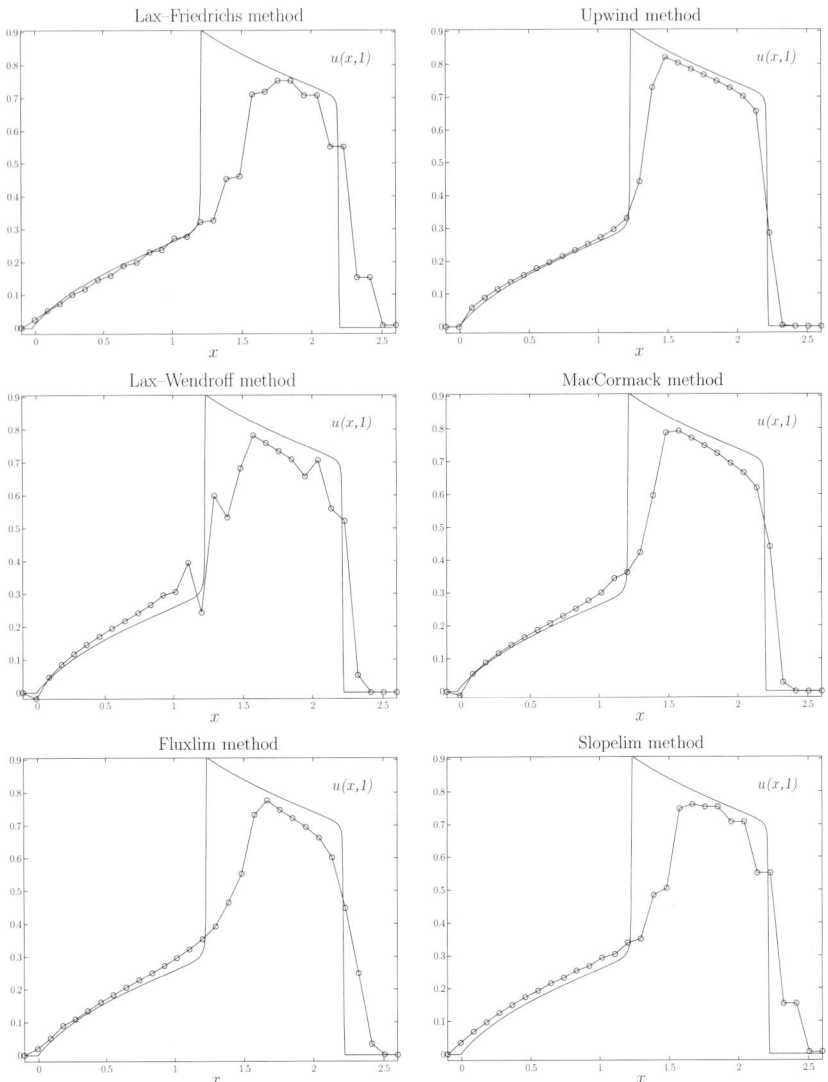

Figure 3.2. Computed solutions at time $t = 1$ for flux function (3.15) and initial data (3.17).

$$= u_t + f(u)_x + \frac{1}{2}\left(\Delta t\, u_{tt} - \Delta x\left(f'(u)u_x\right)_x\right) + \mathcal{O}\left((\Delta t)^2\right)$$
$$= u_t + f(u)_x + \frac{\Delta x}{2}\left(\lambda u_{tt} - \left(f'(u)u_x\right)_x\right) + \mathcal{O}\left((\Delta t)^2\right).$$

Assuming that u is a smooth solution of (3.1), we find that

$$u_{tt} = \left((f'(u))^2 u_x\right)_x,$$

and inserting this into the previous equation we obtain

$$L_{\Delta t} = \frac{\Delta t}{2\lambda} \frac{\partial}{\partial x} \left(f'(u) \left(\lambda f'(u) - 1 \right) u_x \right) + \mathcal{O}\left((\Delta t)^2\right). \quad (3.19)$$

Hence, the upwind method is of first order. The above computations were purely formal, assuming sufficient smoothness for the Taylor expansion to be valid. This means that Godunov's scheme is also of first order. Similarly, computations based on the Lax–Friedrichs scheme yield

$$L_{\Delta t} = \frac{\Delta t}{2\lambda^2} \frac{\partial}{\partial x} \left(\left((\lambda f'(u))^2 - 1 \right) u_x \right) + \mathcal{O}\left(\Delta t^2\right). \quad (3.20)$$

Consequently, the Lax–Friedrichs scheme is also of first order. From the above computations it also emerges that the Lax–Friedrichs scheme is *second-order* accurate on the equation

$$u_t + f(u)_x = \frac{\Delta t}{2\lambda^2} \left(\left(1 - (\lambda f'(u))^2 \right) u_x \right)_x. \quad (3.21)$$

This is called the *model equation* for the Lax–Friedrichs scheme. In order for this to be well posed we must have that the coefficient of u_{xx} on the right-hand side is nonnegative. Hence

$$|\lambda f'(u)| \leq 1. \quad (3.22)$$

This is a stability restriction on λ, and is the Courant–Friedrichs–Lewy (CFL) condition that we encountered in (3.13). The model equation for the upwind method is

$$u_t + f(u)_x = \frac{\Delta t}{2\lambda} \left(f'(u) \left(1 - \lambda f'(u) \right) u_x \right)_x. \quad (3.23)$$

In order for this equation to be well posed, we must have $f'(u) \geq 0$ and $\lambda f'(u) < 1$. $\diamond$

From the above examples, we see that first-order methods have model equations with a diffusive term. Similarly, one finds that second-order methods have model equations with a dispersive right-hand side. Therefore, the oscillations observed in the computations were to be expected.

From now on we let the function $u_{\Delta t}$ be defined by

$$u_{\Delta t}(x,t) = U_j^n, \quad (x,t) \in [j\Delta x, (j+1)\Delta x) \times [n\Delta t, (n+1)\Delta t). \quad (3.24)$$

Observe that

$$\int_{\mathbb{R}} u_{\Delta t}(x,t) \, dx = \Delta x \sum_j U_j^n, \quad \text{for } n\Delta t \leq t < (n+1)\Delta t.$$

We briefly mentioned in Example 3.2 the fact that if $u_{\Delta t}$ converges, then the limit is a weak solution. Precisely, we have the well-known Lax–Wendroff theorem.

Theorem 3.4 (Lax–Wendroff theorem). *Let $u_{\Delta t}$ be computed from a conservative and consistent method. Assume that $\text{T.V.}_x (u_{\Delta t})$ is uniformly*

72   3. A Short Course in Difference Methods

bounded in Δt. Consider a subsequence $u_{\Delta t_k}$ such that $\Delta t_k \to 0$, and assume that $u_{\Delta t_k}$ converges in L^1_{loc} as $\Delta t_k \to 0$. Then the limit is a weak solution to (3.1).

Proof. The proof uses summation by parts. Let $\varphi(x,t)$ be a test function. By the definition of U_j^{n+1},

$$\sum_{n=0}^{N} \sum_{j=-\infty}^{\infty} \varphi(x_j, t_n) \left(U_j^{n+1} - U_j^n\right)$$

$$= -\frac{\Delta t}{\Delta x} \sum_{n=0}^{N} \sum_{j=-\infty}^{\infty} \varphi(x_j, t_n) \left(F(U^n; j) - F(U^n; j-1)\right),$$

where $x_j = j\Delta x$ and $t_n = n\Delta t$, and we choose $T = N\Delta t$ such that $\varphi = 0$ for $t \geq T$. After a summation by parts we get

$$-\sum_{j=-\infty}^{\infty} \varphi(x_j, 0) U_j^0 - \sum_{j=-\infty}^{\infty} \sum_{n=1}^{N} \left(\varphi(x_j, t_n) - \varphi(x_j, t_{n-1})\right) U_j^n$$

$$-\frac{\Delta t}{\Delta x} \sum_{n=0}^{N} \sum_{j=-\infty}^{\infty} \left(\varphi(x_{j+1}, t_n) - \varphi(x_j, t_n)\right) F(U^n; j) = 0.$$

Rearranging, we find that

$$\Delta t \Delta x \sum_{n=1}^{N} \sum_{j=-\infty}^{\infty} \left[\left(\frac{\varphi(x_j, t_n) - \varphi(x_j, t_{n-1})}{\Delta t}\right) U_j^n\right.$$

$$\left. + \left(\frac{\varphi(x_{j+1}, t_n) - \varphi(x_j, t_n)}{\Delta x}\right) F(U^n; j)\right]$$

$$= -\Delta x \sum_{j=-\infty}^{\infty} \varphi(x_j, 0) U_j^0. \quad (3.25)$$

This almost looks like a Riemann sum for the weak formulation of (3.1), were it not for F. To conclude that the limit is a weak solution we must show that

$$\Delta t \Delta x \sum_{n=1}^{N} \sum_{j=-\infty}^{\infty} \left|F(U^n; j) - f(U_j^n)\right| \quad (3.26)$$

tends to zero as $\Delta t \to 0$. Using consistency, we find that (3.26) equals

$$\Delta t \Delta x \sum_{n=1}^{N} \sum_{j=-\infty}^{\infty} \left|F\left(U_{j-p}^n, \ldots, U_{j+q}^n\right) - F\left(U_j^n, \ldots, U_j^n\right)\right|,$$

which by the Lipschitz continuity of F is less than

$$\Delta t \Delta x M \sum_{n=1}^{N} \sum_{j=-\infty}^{\infty} \sum_{k=-p}^{q} |U_{j+k}^n - U_j^n|$$

$$\leq \frac{1}{2}(q(q-1) + p(p-1)) \Delta t \, \Delta x \, M \sum_{n=1}^{N} \sum_{j=-\infty}^{\infty} |U_{j+1}^n - U_j^n|$$

$$\leq (q^2 + p^2) \Delta x \, M \, \text{T.V.} \, (u_{\Delta t}) \, T,$$

where M is larger than the Lipschitz constant of F. Therefore, (3.26) is small for small Δx, and the limit is a weak solution. □

We proved in Theorem 2.14 that the solution of a scalar conservation law in one dimension possesses several properties. The corresponding properties for conservative and consistent numerical schemes read as follows:

Definition 3.5. *Let $u_{\Delta t}$ be computed from a conservative and consistent method.*

- *A method is said to be total variation stable[1] if the total variation of U^n is uniformly bounded, independently of Δx and Δt.*

- *We say that a numerical method is total variation diminishing (TVD) if $\text{T.V.}(U^{n+1}) \leq \text{T.V.}(U^n)$ for all $n \in \mathbb{N}_0$.*

- *A method is called monotonicity preserving if the initial data is monotone implies that so is U^n for all $n \in \mathbb{N}$.*

- *A numerical method is called L^1-contractive if it is L^1-contractive [sic!], i.e., $\|u_{\Delta t}(t) - v_{\Delta t}(t)\|_1 \leq \|u_{\Delta t}(0) - v_{\Delta t}(0)\|_1$ for all $t \geq 0$. Here $v_{\Delta t}$ is another solution with initial data v_0. Alternatively, we can of course write this as*

$$\sum_j |U_j^{n+1} - V_j^{n+1}| \leq \sum_j |U_j^n - V_j^n|, \quad n \in \mathbb{N}_0.$$

- *A method is said to be monotone if for initial data U^0 and V^0, we have*

$$U_j^0 \leq V_j^0, \quad j \in \mathbb{Z} \quad \Rightarrow \quad U_j^n \leq V_j^n, \quad j \in \mathbb{Z}, n \in \mathbb{N}.$$

The above notions are strongly interrelated, as the next theorem shows.

Theorem 3.6. *For conservative and consistent methods the following hold:*

(i) *Any monotone method is L^1-contractive, assuming $u_{\Delta t}(0) - v_{\Delta t}(0) \in L^1(\mathbb{R})$.*

(ii) *Any L^1-contractive method is TVD, assuming that $\text{T.V.}(u_0)$ is finite.*

(iii) *Any TVD method is monotonicity preserving.*

---

[1] This definition is slightly different from the standard definition of T.V. stable methods.

74   3. A Short Course in Difference Methods

Proof. (i) We apply the Crandall–Tartar lemma, Lemma 2.12, with $\Omega = \mathbb{R}$, and D equal to the set of all functions in L^1 that are piecewise constant on the grid $\Delta x \mathbb{Z}$, and finally we let $T(U^0) = U^n$. Since the method is conservative (cf. (3.6)), we have that

$$\sum_j U_j^n = \sum_j U_j^0, \text{ or } \int T(U^0) = \int U^n = \int U^0.$$

Lemma 2.12 immediately implies that

$$\|u_{\Delta t} - v_{\Delta t}\|_1 = \Delta x \sum_j |U_j^n - V_j^n| \leq \Delta x \sum_j |U_j^0 - V_j^0|$$
$$= \|u_{\Delta t}(0) - v_{\Delta t}(0)\|_1.$$

(ii) Assume now that the method is L^1-contractive, i.e.,

$$\sum_j |U_j^{n+1} - V_j^{n+1}| \leq \sum_j |U_j^n - V_j^n|.$$

Let V^n be the numerical solution with initial data

$$V_i^0 = U_{i+1}^0.$$

Then by the translation invariance induced by (3.5), $V_i^n = U_{i+1}^n$ for all n. Furthermore,

$$\text{T.V.}\left(U^{n+1}\right) = \sum_{j=-\infty}^{\infty} |U_{j+1}^{n+1} - U_j^{n+1}| = \sum_j |U_j^{n+1} - V_j^{n+1}|$$
$$\leq \sum_j |U_j^n - V_j^n| = \text{T.V.}\left(U^n\right).$$

(iii) Consider now a TVD method, and assume that we have monotone initial data. Since $\text{T.V.}\left(U^0\right)$ is finite, the limits

$$U_L = \lim_{j \to -\infty} U_j^0 \text{ and } U_R = \lim_{j \to \infty} U_j^0$$

exist. Then $\text{T.V.}\left(U^0\right) = |U_R - U_L|$. If U^1 were not monotone, then $\text{T.V.}\left(U^1\right) > |U_R - U_L| = \text{T.V.}\left(U^0\right)$, which is a contradiction. □

We can summarize the above theorem as follows:

monotone $\Rightarrow$ L^1-contractive $\Rightarrow$ TVD $\Rightarrow$ monotonicity preserving.

Monotonicity is relatively easy to check for explicit methods, e.g., by calculating the partial derivatives $\partial G/\partial U^i$ in (3.3).

◇ **Example 3.7 (Lax–Friedrichs scheme).**

Recall from Example 3.2 that the Lax–Friedrichs scheme is given by

$$U_j^{n+1} = \frac{1}{2}\left(U_{j+1}^n + U_{j-1}^n\right) - \frac{1}{2}\lambda\left(f\left(U_{j+1}^n\right) - f\left(U_{j-1}^n\right)\right).$$

Computing partial derivatives we obtain

$$\frac{\partial U_j^{n+1}}{\partial U_k^n} = \begin{cases} (1 - \lambda f'(U_k^n))/2 & \text{for } k = j+1, \\ (1 + \lambda f'(U_k^n))/2 & \text{for } k = j-1, \\ 0 & \text{otherwise,} \end{cases}$$

and hence we see that the Lax–Friedrichs scheme is monotone as long as the CFL condition

$$\lambda |f'(u)| < 1$$

is fulfilled. $\diamond$

Theorem 3.8. *Let $u_0 \in L^1(\mathbb{R})$ have bounded variation. Assume that $u_{\Delta t}$ is computed with a method that is conservative, consistent, total variation stable, and uniformly bounded; that is,*

$$\text{T.V.}(u_{\Delta t}) \leq M \quad \text{and} \quad \|u_{\Delta t}\|_\infty \leq M,$$

where M is independent of Δx and Δt.

Let $T > 0$. Then $\{u_{\Delta t}(t)\}$ has a subsequence that converges for all $t \in [0, T]$ to a weak solution $u(t)$ in $L^1_{\text{loc}}(\mathbb{R})$. Furthermore, the limit is in $C\left([0, T]; L^1_{\text{loc}}(\mathbb{R})\right)$.

Proof. We intend to apply Theorem A.8. It remains to show that

$$\int_a^b |u_{\Delta t}(x, t) - u_{\Delta t}(x, s)| \, dx \leq C |t - s| + o(1), \quad \text{as } \Delta t \to 0, \quad s, t \in [0, T].$$

Consistency of the scheme implies, for any fixed Δt,

$$\begin{aligned} |U_j^{n+1} - U_j^n| &= \lambda \left| F(U_j^n; j) - F(U_j^n; j-1) \right| \\ &= \lambda \left| F(U_{j-p}^n, \ldots, U_{j+q}^n) - F(U_{j-p-1}^n, \ldots, U_{j+q-1}^n) \right| \\ &\leq \lambda L \left(|U_{j-p}^n - U_{j-p-1}^n| + \cdots + |U_{j+q}^n - U_{j+q-1}^n| \right), \end{aligned}$$

from which we conclude that

$$\begin{aligned} \|u_{\Delta t}(\cdot, t_{n+1}) - u_{\Delta t}(\cdot, t_n)\|_1 &= \sum_{j=-\infty}^\infty |U_j^{n+1} - U_j^n| \Delta x \\ &\leq L(p + q + 1) \text{T.V.}(U^n) \Delta t \\ &\leq L(p + q + 1) M \Delta t, \end{aligned}$$

where L is the Lipschitz constant of F. More generally,

$$\|u_{\Delta t}(\cdot, t_m) - u_{\Delta t}(\cdot, t_n)\|_1 \leq L(p + q + 1) M |n - m| \Delta t.$$

Now let $\tau_1, \tau_2 \in [0, T]$, and choose $\tilde{t}_1, \tilde{t}_2 \in \{n\Delta t \mid 0 \leq n \leq T/\Delta t\}$ such that

$$0 \leq \tau_j - \tilde{t}_j < \Delta t \text{ for } j = 1, 2.$$

By construction $u_{\Delta t}(\tau_j) = u_{\Delta t}(\tilde{t}_j)$, and hence

$$\|u_{\Delta t}(\,\cdot\,,\tau_1) - u_{\Delta t}(\,\cdot\,,\tau_2)\|_1$$
$$\leq \|u_{\Delta t}(\,\cdot\,,\tau_1) - u_{\Delta t}(\,\cdot\,,\tilde{t}_1)\|_1 + \|u_{\Delta t}(\,\cdot\,,\tilde{t}_1) - u_{\Delta t}(\,\cdot\,,\tilde{t}_2)\|_1$$
$$+ \|u_{\Delta t}(\,\cdot\,,\tilde{t}_2) - u_{\Delta t}(\,\cdot\,,\tau_2)\|_1$$
$$\leq (p+q+1)LM\left|\tilde{t}_1 - \tilde{t}_2\right| \leq (p+q+1)LM\left|\tau_1 - \tau_2\right| + \mathcal{O}\left(\Delta t\right).$$

Observe that this estimate is uniform in $\tau_1, \tau_2 \in [0,T]$. We conclude that

$$u_{\Delta t} \to u \text{ in } C([0,T];L^1([a,b]))$$

for a sequence $\Delta t \to 0$. The Lax–Wendroff theorem then says that this limit is a weak solution. □

At this point it is convenient to introduce the concept of *entropy pairs* or *entropy/entropy flux pairs*.[2] Recall that a pair of functions $(\eta(u), q(u))$ with η convex is called an entropy pair if

$$q'(u) = f'(u)\eta'(u). \tag{3.27}$$

The reason for introducing this concept is that the entropy condition can now be reformulated using (η, q). To see this, assume that u is a solution of the viscous conservation law

$$u_t + f(u)_x = \varepsilon u_{xx}. \tag{3.28}$$

Assume, or consult Appendix B, that this equation has a unique twice-differentiable solution. Hence, multiplying by $\eta'(u)$ yields (cf. (2.10))

$$\eta(u)_t + q(u)_x = \varepsilon \eta'(u) u_{xx} = \varepsilon \left(\eta'(u) u_x\right)_x - \varepsilon \eta''(u) \left(u_x\right)^2.$$

If η' is bounded, and $\eta'' > 0$, then the first term on the right of the above equation tends to zero as a distribution as $\varepsilon \to 0$, while the second term is nonpositive. Consequently, if the solution of (3.1) is to be the limit of the solutions of (3.28) as $\varepsilon \to 0$, the solution of (3.1) must satisfy (cf. (2.12))

$$\eta(u)_t + q(u)_x \leq 0 \tag{3.29}$$

as a distribution. Choosing $\eta(u) = |u-k|$ we recover the Kružkov entropy condition; see (2.17). We have demonstrated that that if a function satisfies (2.46) for all k, then it satisfies (3.29) for all convex η and vice versa; see Remark 2.1. Hence, the Kružkov entropy condition is equivalent to demanding (3.29) for all convex η.

The analogue of an entropy pair for difference schemes reads as follows. Write

$$a \vee b = \max(a,b) \quad \text{and} \quad a \wedge b = \min(a,b),$$

---

[2] We have already encountered an entropy pair with $\eta(u) = |u-k|$ and $q(u) = \text{sign}(u-k)(f(u) - f(k))$ when we introduced the Kružkov entropy condition in Chapter 2.

and observe the trivial identity
$$|a-b| = a \vee b - a \wedge b.$$
Then we define the *numerical entropy flux* Q by
$$Q(U;j) = F(U \vee k;j) - F(U \wedge k;j),$$
or explicitly,
$$Q(U_{j-p},\ldots,U_{j+p'})$$
$$= F(U_{j-p} \vee k,\ldots,U_{j+p'} \vee k) - F(U_{j-p} \wedge k,\ldots,U_{j+p'} \wedge k).$$
We have that Q is consistent with the usual entropy flux, i.e.,
$$Q(u,\ldots,u) = \text{sign}(u-k)(f(u)-f(k)).$$
Returning to monotone difference schemes, we have the following result.

Theorem 3.9. *Under the assumptions of Theorem 3.8, the approximate solutions computed by a conservative, consistent, and monotone difference method converge to the entropy solution as $\Delta t \to 0$.*

Proof. Theorem 3.8 allows us to conclude that $u_{\Delta t}$ has a subsequence that converges in $C([0,T];L^1([a,b]))$ to a weak solution. It remains to show that the limit satisfies a discrete Kružkov form. By a direct calculation we find that
$$\left|U_j^n - k\right| - \lambda\left(Q(U^n;j) - Q(U^n;j-1)\right) = G(U^n \vee k;j) - G(U^n \wedge k;j).$$
Using that $U_j^{n+1} = G(U^n;j)$ and that $k = G(k;j)$, the monotonicity of the scheme implies that
$$G(U^n \vee k;j) \geq G(U^n;j) \vee G(k;j) = G(U^n;j) \vee k,$$
$$-G(U^n \wedge k;j) \geq -G(U^n;j) \wedge G(k;j) = -G(U^n;j) \wedge k.$$
Therefore,
$$\left|U_j^{n+1} - k\right| - \left|U_j^n - k\right| + \lambda\left(Q(U^n;j) - Q(U^n;j-1)\right) \leq 0. \tag{3.30}$$
Applying the technique used in proving the Lax–Wendroff theorem to (3.30) gives that the limit u satisfies
$$\iint \left(|u-k|\varphi_t + \text{sign}(u-k)(f(u)-f(k))\varphi_x\right) dx\, dt \geq 0.$$
$\square$

Note that we can also use the above theorem to conclude the existence of weak entropy solutions to scalar conservation laws.

Now we shall examine the local truncation error of a general conservative, consistent, and monotone method. Since this can be written
$$U_j^{n+1} = G(U^n;j) = G\left(U_{j-p-1}^n,\ldots,U_{j+q}^n\right)$$
$$= U_j^n - \lambda\left(F\left(U_{j+q}^n,\ldots,U_{j-p}^n\right) - F\left(U_{j-p-1}^n,\ldots,U_{j+q-1}^n\right)\right),$$

we write
$$G = G(u_1, \ldots, u_{p+q+1}) \quad \text{and} \quad F = F(u_1, \ldots, u_{p+q}).$$

We assume that F, and hence G, is three times continuously differentiable with respect to all arguments, and write the derivatives with respect to the ith argument as
$$\partial_i G(u_1, \ldots, u_{p+q+1}) \quad \text{and} \quad \partial_i F(u_1, \ldots, u_{p+q}).$$

We set $\partial_i F = 0$ if $i = 0$ or $i = p+q+1$. Throughout this calculation, we assume that the jth slot of G contains U_j^n, so that $G(u_1, \ldots, u_{p+q+1}) = u_j - \lambda(\ldots)$. By consistency we have that
$$G(u, \ldots, u) = u \quad \text{and} \quad F(u, \ldots, u) = f(u).$$

Using this we find that
$$\sum_{i=1}^{p+q} \partial_i F(u, \ldots, u) = f'(u), \tag{3.31}$$

$$\partial_i G = \delta_{i,j} - \lambda \left(\partial_{i-1} F - \partial_i F \right), \tag{3.32}$$

and
$$\partial^2_{i,k} G = -\lambda \left(\partial^2_{i-1,k-1} F - \partial^2_{i,k} F \right). \tag{3.33}$$

Therefore,
$$\sum_{i=1}^{p+q+1} \partial_i G(u, \ldots, u) = \sum_{i=1}^{p+q+1} \delta_{i,j} = 1. \tag{3.34}$$

Furthermore,
$$\sum_{i=1}^{p+q+1} (i-j) \partial_i G(u, \ldots, u) = \sum_{i=1}^{p+q+1} (i-j) \delta_{i,j}$$
$$- \lambda (i-j) \left(\partial_{i-1} F(u, \ldots, u) - \partial_i F(u, \ldots, u) \right)$$
$$= -\lambda \sum_{i=1}^{p+q} ((i+1) - i)) \partial_i F(u, \ldots, u)$$
$$= -\lambda f'(u). \tag{3.35}$$

We also find that
$$\sum_{i,k=1}^{p+q+1} (i-k)^2 \partial^2_{i,k} G(u, \ldots, u)$$
$$= -\lambda \sum_{i,k=1}^{p+q+1} (i-k)^2 \left(\partial^2_{i-1,k-1} F(u, \ldots, u) - \partial^2_{i,k} F(u, \ldots, u) \right)$$

$$= -\lambda \sum_{i,k=1}^{p+q} \left(((i+1)-(k+1))^2 - (i-k)^2\right) \partial_{i,k}^2 F(u,\ldots,u)$$
$$= 0. \tag{3.36}$$

Having established this, we now let $u = u(x,t)$ be a smooth solution of the conservation law (3.1). We are interested in applying G to $u(x,t)$, i.e., to calculate

$$G(u(x-p\Delta x, t)\ldots, u(x,t),\ldots, u(x+q\Delta x, t)).$$

Set $u_i = u(x+(i-j)\Delta x, t)$ for $i = 1,\ldots,p+q+1$. Then we find that

$$G(u_1,\ldots,u_{p+q+1})$$
$$= G(u_j,\ldots,u_j) + \sum_{i=1}^{p+q+1} \partial_i G(u_j,\ldots,u_j)(u_i - u_j)$$
$$+ \frac{1}{2} \sum_{i,k=1}^{p+q+1} \partial_{i,k}^2 G(u_j,\ldots,u_j)(u_i - u_j)(u_k - u_j) + \mathcal{O}\left(\Delta x^3\right)$$
$$= u(x,t) + u_x(x,t)\Delta x \sum_{i=1}^{p+q+1}(i-j)\partial_i G(u_j,\ldots,u_j)$$
$$+ \frac{1}{2}u_{xx}(x,t)\Delta x^2 \sum_{i=1}^{p+q+1}(i-j)^2 \partial_i G(u_j,\ldots,u_j)$$
$$+ \frac{1}{2}u_x^2(x,t)\Delta x^2 \sum_{i,k=1}^{p+q+1}(i-j)(k-j)\partial_{i,k}^2 G(u_j,\ldots,u_j) + \mathcal{O}\left(\Delta x^3\right)$$
$$= u(x,t) + u_x(x,t)\Delta x \sum_{i=1}^{p+q+1}(i-j)\partial_i G(u_j,\ldots,u_j)$$
$$+ \frac{1}{2}\Delta x^2 \sum_{i=1}^{p+q+1}(i-j)^2 [\partial_i G(u_j,\ldots,u_j)u_x(x,t)]_x$$
$$- \frac{1}{2}\Delta x^2 u_x^2(x,t) \sum_{i,k}^{p+q+1}\left((i-j)^2 - (i-j)(k-j)\right)\partial_{i,k}^2 G(u_j,\ldots,u_j)$$
$$+ \mathcal{O}\left(\Delta x^3\right).$$

Next we observe, since $\partial_{i,k}^2 G = \partial_{k,i}^2 G$ and using (3.36), that

3. A Short Course in Difference Methods

$$0 = \sum_{i,k}(i-k)^2 \partial_{i,k}^2 G = \sum_{i,k}((i-j)-(k-j))^2 \partial_{i,k}^2 G$$

$$= \sum_{i,k}((i-j)^2 - 2(i-j)(k-j))\partial_{i,k}^2 G + \sum_{i,k}(k-j)^2 \partial_{k,i}^2 G$$

$$= 2\sum_{i,k}((i-j)^2 - (i-j)(k-j))\partial_{i,k}^2 G.$$

Consequently, the penultimate term in the Taylor expansion of G above is zero, and we have that

$$G(u(x-p\Delta x, t), \ldots, u(x+q\Delta x, t)) = u(x,t) - \Delta t f(u(x,t))_x$$
$$+ \frac{\Delta x^2}{2} \sum_i (i-j)^2 \left[\partial_i G(u(x,t), \ldots, u(x,t))u_x\right]_x + \mathcal{O}\left(\Delta x^3\right). \quad (3.37)$$

Since u is a smooth solution of (3.1), we have already established that

$$u(x, t+\Delta t) = u(x,t) - \Delta t f(u)_x + \frac{\Delta t^2}{2} \left[\partial\left(f'(u)\right)^2 u_x\right]_x + \mathcal{O}\left(\Delta t^3\right).$$

Hence, we compute the local truncation error as

$$L_{\Delta t} = \frac{\Delta t}{2\lambda^2} \left[\left(\sum_{i=1}^{p+q+1} (i-j)^2 \partial_i G(u, \ldots, u) - \lambda^2 (f'(u))^2\right) u_x\right]_x$$

$$=: \frac{\Delta t}{2\lambda^2} \left[\beta(u) u_x\right]_x + \mathcal{O}\left(\Delta t^2\right). \quad (3.38)$$

Thus if $\beta > 0$, then the method is of first order. What we have done so far is valid for any conservative and consistent method where the numerical flux function is three times continuously differentiable. Next, we utilize that $\partial_i G \geq 0$, so that $\sqrt{\partial_i G}$ is well-defined. This means that

$$-\lambda f'(u) = \sum_{i=1}^{p+q+1} (i-j)\partial_i G(u, \ldots, u)$$

$$= \sum_{i=1}^{p+q+1} (i-j)\sqrt{\partial_i G(u, \ldots, u)}\sqrt{\partial_i G(u, \ldots, u)}.$$

Using the Cauchy–Schwarz inequality and (3.34) we find that

$$\lambda^2 \left(f'(u)\right)^2 \leq \sum_{i=1}^{p+q+1} (i-j)^2 \partial_i G(u, \ldots, u) \sum_{i=1}^{p+q+1} \partial_i G(u, \ldots, u)$$

$$= \sum_{i=1}^{p+q+1} (i-j)^2 \partial_i G(u, \ldots, u).$$

Thus, $\beta(u) \geq 0$. Furthermore, the inequality is strict if more than one term in the right-hand sum is different from zero. If $\partial_i G(u, \ldots, u) = 0$ except for

$i = k$ for some k, then $G(u_1, \ldots, u_{p+q+1}) = u_k$ by (3.34). Hence the scheme is a linear translation, and by consistency $f(u) = cu$, where $c = (j-k)\lambda$. Therefore, monotone methods for nonlinear conservation laws are at most first-order accurate. This is indeed their main drawback. To recapitulate, we have proved the following theorem:

Theorem 3.10. *Assume that the numerical flux F is three times continuously differentiable, and that the corresponding scheme is monotone. Then the method is at most first-order accurate.*

3.2 Error Estimates

> Let others bring order to chaos.
> I would bring chaos to order instead.
>
> *Kurt Vonnegut, Breakfast of Champions (1973)*

The concept of local error estimates is based on formal computations, and indicates how the method performs in regions where the solution is smooth. Since the convergence of the methods discussed was in L^1, it is reasonable to ask how far the approximated solution is from the true solution in this space.

In this section we will consider functions u that are maps $t \mapsto u(t)$ from $[0,\infty)$ to $L^1_{\text{loc}} \cap BV(\mathbb{R})$ such that the one-sided limits $u(t\pm)$ exist in L^1_{loc}, and for definiteness we assume that this map is right continuous. Furthermore, we assume that

$$\|u(t)\|_\infty \leq \|u(0)\|_\infty, \quad \text{T.V.}(u(t)) \leq \text{T.V.}(u(0)).$$

We denote this class of functions by $\mathcal{K}$. From Theorem 2.14 we know that solutions of scalar conservation laws are in the class $\mathcal{K}$.

It is convenient to introduce *moduli of continuity in time* (see Appendix A)

$$\nu_t(u, \sigma) = \sup_{|\tau| \leq \sigma} \|u(t+\tau) - u(t)\|_1, \quad \sigma > 0,$$

$$\nu(u, \sigma) = \sup_{0 \leq t \leq T} \nu_t(u, \sigma).$$

From Theorem 2.14 we have that

$$\nu(u, \sigma) \leq |\sigma| \, \|f\|_{\text{Lip}} \text{T.V.}(u_0) \qquad (3.39)$$

for weak solutions of conservation laws.

Now let $u(x,t)$ be any function in $\mathcal{K}$, not necessarily a solution of (3.1). In order to measure how far u is from being a solution of (3.1) we insert u

in the Kružkov form (cf. (2.19))

$$\Lambda_T(u,\phi,k) = \int_0^T \int \left(|u-k|\phi_t + q(u,k)\phi_x\right) dx\,ds \qquad (3.40)$$
$$- \int |u(x,T) - k|\phi(x,T)\,dx + \int |u_0(x) - k|\phi(x,0)\,dx.$$

If u is a solution, then $\Lambda_T \geq 0$ for all constants k and all nonnegative test functions ϕ. We shall now use the special test function

$$\Omega(x,x',s,s') = \omega_{\varepsilon_0}(s-s')\omega_\varepsilon(x-x'),$$

where

$$\omega_\varepsilon(x) = \frac{1}{\varepsilon}\omega\left(\frac{x}{\varepsilon}\right)$$

and $\omega(x)$ is an even C^∞ function satisfying

$$0 \leq \omega \leq 1, \quad \omega(x) = 0 \quad \text{for } |x| > 1, \quad \int \omega(x)\,dx = 1.$$

Let $v(x',s')$ be the unique weak solution of (3.1), and define

$$\Lambda_{\varepsilon,\varepsilon_0}(u,v) = \int_0^T \int \Lambda_T\left(u, \Omega(\,\cdot\,, x', \,\cdot\,, s'), v(x',s')\right) dx'\,ds'.$$

The comparison result reads as follows.

Theorem 3.11 (Kuznetsov's lemma). *Let $u(\,\cdot\,,t)$ be a function in $\mathcal{K}$, and v be a solution of (3.1). If $0 < \varepsilon_0 < T$ and $\varepsilon > 0$, then*

$$\|u(\,\cdot\,,T-) - v(\,\cdot\,,T)\|_1 \leq \|u_0 - v_0\|_1 + \text{T.V.}(v_0)\left(2\varepsilon + \varepsilon_0\|f\|_{\text{Lip}}\right)$$
$$+ \nu(u,\varepsilon_0) - \Lambda_{\varepsilon,\varepsilon_0}(u,v), \qquad (3.41)$$

where $u_0 = u(\,\cdot\,,0)$ and $v_0 = v(\,\cdot\,,0)$.

Proof. We use special properties of the test function Ω, namely that

$$\Omega(x,x',s,s') = \Omega(x',x,s,s') = \Omega(x,x',s',s) = \Omega(x',x,s',s) \qquad (3.42)$$

and

$$\Omega_x = -\Omega_{x'}, \quad \text{and} \quad \Omega_s = -\Omega_{s'}. \qquad (3.43)$$

Using (3.42) and (3.43), we find that

$$\Lambda_{\varepsilon,\varepsilon_0}(u,v) = -\Lambda_{\varepsilon,\varepsilon_0}(v,u) - \int_0^T \iint \Omega(x,x',s,T)\big(|u(x,T) - v(x',s)|$$
$$+ |v(x',T) - u(x,s)|\big)\,dx\,dx'\,ds$$
$$+ \int_0^T \iint \Omega(x,x',s,0)\big(|v_0(x') - u(x,s)|$$
$$+ |u_0(x) - v(x',s)|\big)\,dx\,dx'\,ds$$

3.2. Error Estimates

$$:= -\Lambda_{\varepsilon,\varepsilon_0}(v, u) - A + B.$$

Since v is a weak solution, $\Lambda_{\varepsilon,\varepsilon_0}(v, u) \geq 0$, and hence

$$A \leq B - \Lambda_{\varepsilon,\varepsilon_0}(u, v).$$

Therefore, we would like to obtain a lower bound on A and an upper bound on B, the lower bound on A involving $\|u(T) - v(T)\|_1$ and the upper bound on B involving $\|u_0 - v_0\|_1$. We start with the lower bound on A.

Let ρ_ε be defined by

$$\rho_\varepsilon(u, v) = \iint \omega_\varepsilon(x - x') |u(x) - v(x')| \, dx \, dx'. \tag{3.44}$$

Then

$$A = \int_0^T \omega_{\varepsilon_0}(T - s) \left(\rho_\varepsilon(u(T), v(s)) + \rho_\varepsilon(u(s), v(T)) \right) ds.$$

Now by a use of the triangle inequality,

$$\begin{aligned}
\|u(x, T) - v(x', s)\| + |u(x, s) - v(x', T)| \\
\geq |u(x, T) - v(x, T)| + |u(x, T) - v(x, T)| \\
- |v(x, T) - v(x', T)| - |u(x, T) - u(x, s)| \\
- |v(x', T) - v(x', s)| - |v(x, T) - v(x', T)|.
\end{aligned}$$

Hence

$$\begin{aligned}
\rho_\varepsilon(u(T), v(s)) + \rho_\varepsilon(u(s), v(T)) \geq 2\|u(T) - v(T)\|_1 - 2\rho_\varepsilon(v(T), v(T)) \\
- \|u(T) - u(s)\|_1 - \|v(T) - v(s)\|_1.
\end{aligned}$$

Regarding the upper estimate on B, we similarly have that

$$B = \int_0^T \omega_{\varepsilon_0}(s) \left[\rho_\varepsilon(u_0, v(s)) + \rho_\varepsilon(u(s), v_0) \right] ds,$$

and we also obtain

$$\begin{aligned}
\rho_\varepsilon(u_0, v(s)) + \rho_\varepsilon(u(s), v_0) \leq 2\|u_0 - v_0\|_1 + 2\rho_\varepsilon(v_0, v_0) \\
+ \|u_0 - u(s)\|_1 + \|v_0 - v(s)\|_1.
\end{aligned}$$

Since v is a solution, it satisfies the TVD property, and hence

$$\begin{aligned}
\rho_\varepsilon(v(T), v(T)) &= \int \int_{-\varepsilon}^{\varepsilon} \omega_\varepsilon(z) |v(x+z, T) - v(x, T)| \, dz \, dx \\
&\leq \int_{-\varepsilon}^{\varepsilon} \omega_\varepsilon(z) \sup_{|z| \leq \varepsilon} \left(\int |v(x+z, T) - v(x, T)| \, dx \right) dz \\
&= |\varepsilon| \int_{-\varepsilon}^{\varepsilon} \omega_\varepsilon(z) \text{T.V.} (v(T)) \, dz \leq |\varepsilon| \, \text{T.V.} (v_0),
\end{aligned}$$

using (A.10). By the properties of ω,
$$\int_0^T \omega_\varepsilon(T-s)\,ds = \int_0^T \omega_\varepsilon(s)\,ds = \frac{1}{2}.$$

Applying (3.39) we obtain (recall that $\varepsilon_0 < T$)
$$\int_0^T \omega_{\varepsilon_0}(T-s)\|v(T)-v(s)\|_1\,ds$$
$$\leq \int_0^T \omega_{\varepsilon_0}(T-s)\,(T-s)\|f\|_{\text{Lip}}\text{T.V.}(v_0)\,ds$$
$$\leq \frac{1}{2}\varepsilon_0\|f\|_{\text{Lip}}\text{T.V.}(v_0)$$

and
$$\int_0^T \omega_{\varepsilon_0}(s)\|v_0-v(s)\|_1\,ds \leq \frac{1}{2}\varepsilon_0\|f\|_{\text{Lip}}\text{T.V.}(v_0).$$

Similarly,
$$\int_0^T \omega_{\varepsilon_0}(T-s)\|u(T)-u(s)\|_1\,ds \leq \frac{1}{2}\nu(u,\varepsilon_0)$$

and
$$\int_0^T \omega_{\varepsilon_0}(s)\|u_0-u(s)\|_1\,ds \leq \frac{1}{2}\nu(u,\varepsilon_0).$$

If we collect all the above bounds, we should obtain the statement of the theorem. $\square$

Observe that in the special case where u is a solution of the conservation law (3.1), we know that $\Lambda_{\varepsilon,\varepsilon_0}(u,v) \geq 0$, and hence we obtain, when $\varepsilon,\varepsilon_0 \to 0$, the familiar stability result
$$\|u(\,\cdot\,,T)-v(\,\cdot\,,T)\|_1 \leq \|u_0-v_0\|_1.$$

We shall now show in three cases how Kuznetsov's lemma can be used to give estimates on how fast a method converges to the entropy solution of (3.1).

$\diamond$ **Example 3.12 (The smoothing method).**

While not a proper numerical method, the smoothing method provides an example of how the result of Kuznetsov may be used. The smoothing method is a (semi)numerical method approximating the solution of (3.1) as follows: Let $\omega_\delta(x)$ be a standard mollifier with support in $[-\delta,\delta]$, and let $t_n = n\Delta t$. Set $u^0 = u_0 * \omega_\delta$. For $0 \leq t < \Delta t$ define u^1 to be the solution of (3.1) with initial data u^0. If Δt is small enough, u^1 remains differentiable for $t < \Delta t$. In the interval $[(n-1)\Delta t, n\Delta t)$, we define u^n to be the solution of (3.1), with $u^n(x,(n-1)\Delta t) = u^{n-1}(\,\cdot\,,t_n-) * \omega_\delta$.

3.2. Error Estimates

The advantage of doing this is that u^n will remain differentiable in x for all times, and the solution in the strips $[t_n, t_{n+1})$ can be found by, e.g., the method of characteristics. To show that u^n is differentiable, we calculate

$$|u_x^n(x, t_{n-1})| = \left| \int u_x^{n-1}(y, t_{n-1}) \omega_\delta(x - y) \, dy \right|$$

$$\leq \frac{1}{\delta} \text{T.V.} \left(u^{n-1}(t_{n-1})\right) \leq \frac{\text{T.V.}(u_0)}{\delta}.$$

Let $\mu(t) = \max_x |u_x(x, t)|$. Using that u is a classical solution of (3.1), we find by differentiating (3.1) with respect to x that

$$u_{xt} + f'(u) u_{xx} + f''(u) u_x^2 = 0.$$

Write

$$\mu(t) = u_x(x_0(t), t),$$

where $x_0(t)$ is the location of the maximum of $|u_x|$. Then

$$\mu'(t) = u_{xx}(x_0(t), t) x_0'(t) + u_{xt}(x_0(t), t)$$
$$\leq u_{xt}(x_0(t), t) = -f''(u) \big(u_x(x_0(t), t)\big)^2$$
$$\leq c\mu(t)^2,$$

since $u_{xx} = 0$ at an extremum of u_x. Thus

$$\mu'(t) \leq c\mu^2(t), \tag{3.45}$$

where $c = \|f''\|_\infty$. The idea is now that (3.45) has a blowup at some finite time, and we choose Δt less than this time. We shall be needing a precise relation between the Δt and δ and must therefore investigate (3.45) further. Solving (3.45) we obtain

$$\mu(t) \leq \frac{\mu(t_n)}{1 - c\mu(t_n)(t - t_n)} \leq \frac{\text{T.V.}(u_0)}{\delta - c\text{T.V.}(u_0) \Delta t}.$$

So if

$$\Delta t < \frac{\delta}{c\text{T.V.}(u_0)}, \tag{3.46}$$

the method is well-defined. Choosing $\Delta t = \delta/(2c\text{T.V.}(u_0))$ will do. Since u is an exact solution in the strips $[t_n, t_{n+1})$, we have

$$\int_{t_n}^{t_{n+1}} \int \left(|u - k|\phi_t + q(u, k)\phi_x\right) dx \, dt$$
$$+ \int \Big(|u(x, t_n+) - k|\phi(x, t_n) - |u(x, t_{n+1}-) - k|\phi(x, t_{n+1})\Big) dx \geq 0.$$

Summing these inequalities, and setting $k = v(y, s)$ where v is an exact solution of (3.1), we obtain

$$\Lambda_T(u, \Omega, v(y,s)) \geq -\sum_{n=0}^{N-1} \int \Omega(x, y, t_n, s) \Big(|u(x, t_n+) - v(y,s)| \\ - |u(x, t_n-) - v(y,s)| \Big) dx,$$

where we use the test function $\Omega(x, y, t, s) = \omega_{\varepsilon_0}(t - s)\omega_\varepsilon(x - y)$. Integrating this over y and s, and letting ε_0 tend to zero, we get

$$\liminf_{\varepsilon_0 \to 0} \Lambda_{\varepsilon,\varepsilon_0}(u, v) \geq -\sum_{n=0}^{N-1} \big(\rho_\varepsilon(u(t_n+), v(t_n)) - \rho_\varepsilon(u(t_n-), v(t_n))\big).$$

Using this in Kuznetsov's lemma, and letting $\varepsilon_0 \to 0$, we obtain

$$\|u(T) - v(T)\|_1 \leq \|u_0 - u^0\|_1 + 2\varepsilon \text{ T.V.}(u_0) \quad (3.47)$$
$$+ \sum_{n=0}^{N-1} \big(\rho_\varepsilon(u(t_n+), v(t_n)) - \rho_\varepsilon(u(t_n-), v(t_n))\big),$$

where we have used that $\lim_{\varepsilon_0 \to 0} \nu_t(u, \varepsilon_0) = 0$, which holds because u is a solution of the conservation law in each strip $[t_n, t_{n+1})$.

To obtain a more explicit bound on the difference of u and v, we investigate $\rho_\varepsilon(\omega_\delta * u, v) - \rho_\varepsilon(u, v)$, where ρ_ε is defined by (3.44),

$$\rho_\varepsilon(u * \omega_\delta, v) - \rho_\varepsilon(u, v) \leq \iiint_{|z| \leq 1} \omega_\varepsilon(x - y)\omega(z)\Big(|u(x + \delta z) - v(y)| \\ - |u(x) - v(y)|\Big) dx\, dy\, dz$$
$$= \frac{1}{2} \iiint_{|z| \leq 1} (\omega_\varepsilon(x - y) - \omega_\varepsilon(x + \delta z - y))\omega(z) \\ \times (|u(x + \delta z) - v(y)| - |u(x) - v(y)|)\, dx\, dy\, dz,$$

which follows after writing $\iiint = \frac{1}{2}\iiint + \frac{1}{2}\iiint$ and making the substitution $x \to x - \delta z$, $z \to -z$ in one of these integrals. Therefore,

$$\rho_\varepsilon(u * \omega_\delta, v) - \rho_\varepsilon(u, v) \leq \frac{1}{2} \iiint_{|z| \leq 1} |\omega_\varepsilon(y + \delta z) - \omega_\varepsilon(y)| \\ \times \omega(z)|u(x + \delta z) - u(x)|\, dx\, dy\, dz$$
$$\leq \frac{1}{2} \text{T.V.}(\omega_\varepsilon) \text{ T.V.}(u) \delta^2$$
$$\leq \text{T.V.}(u) \frac{\delta^2}{\varepsilon},$$

by the triangle inequality and a further substitution $y \mapsto x - y$. Since $N = T/\Delta t$, the last term in (3.47) is less than

$$N \operatorname{T.V.}(u_0) \frac{\delta^2}{\varepsilon} \leq (\operatorname{T.V.}(u_0))^2 \, 2cT \frac{\delta}{\varepsilon},$$

using (3.46). Furthermore, we have that

$$\|u^0 - u_0\|_1 \leq \delta \operatorname{T.V.}(u_0).$$

Letting $K = \operatorname{T.V.}(u_0) c$, we find that

$$\|u(T) - v(T)\|_1 \leq 2 \operatorname{T.V.}(u_0) \left[\delta + \varepsilon + \frac{KT\delta}{\varepsilon}\right],$$

using (3.47). Minimizing with respect to ε, we find that

$$\|u(T) - v(T)\|_1 \leq 2 \operatorname{T.V.}(u_0) \left(\delta + 2\sqrt{KT\delta}\right). \tag{3.48}$$

So, we have shown that the smoothing method is of order $\frac{1}{2}$ in the smoothing coefficient δ. $\diamond$

$\diamond$ **Example 3.13 (The method of vanishing viscosity).**

Another (semi)numerical method for (3.1) is the method of vanishing viscosity. Here we approximate the solution of (3.1) by the solution of

$$u_t + f(u)_x = \delta u_{xx}, \quad \delta > 0, \tag{3.49}$$

using the same initial data. Let u^δ denote the solution of (3.49). Due to the dissipative term on the right-hand side, the solution of (3.49) remains a classical (twice differentiable) solution for all $t > 0$. Furthermore, the solution operator for (3.49) is TVD. Hence a numerical method for (3.49) will (presumably) not experience the same difficulties as a numerical method for (3.1). If (η, q) is a convex entropy pair, we have, using the differentiability of the solution, that

$$\eta(u)_t + q(u)_x = \delta \eta'(u) u_{xx} = \delta \left(\eta(u)_{xx} - \eta''(u) u_x^2\right).$$

Multiplying by a nonnegative test function φ and integrating by parts, we get

$$\iint \left(\eta(u)\varphi_t + q(u)\varphi_x\right) dx\, dt \geq \delta \iint \eta(u)_x \varphi_x\, dx\, dt,$$

where we have used the convexity of η. Applying this with $\eta = |u^\delta - u|$ and $q = F(u^\delta, u)$ we can bound $\lim_{\varepsilon_0 \to 0} \Lambda_{\varepsilon,\varepsilon_0}(u^\delta, u)$ as follows:

$$-\lim_{\varepsilon_0 \to 0} \Lambda_{\varepsilon,\varepsilon_0}(u^\delta, u)$$

$$\leq \delta \int_0^T \iint \left|\frac{\partial \omega_\varepsilon(x-y)}{\partial x}\right| \frac{\partial |u^\delta(x,t) - u(y,t)|}{\partial x} \, dx \, dy \, dt$$

$$\leq \delta \int_0^T \iint \left|\frac{\partial \omega_\varepsilon(x-y)}{\partial x}\right| \left|\frac{\partial u^\delta(x,t)}{\partial x}\right| \, dx \, dy \, dt$$

$$\leq 2\text{T.V.}(u^\delta) \, T \frac{\delta}{\varepsilon}$$

$$\leq 2T \, \text{T.V.}(u_0) \, \frac{\delta}{\varepsilon}.$$

Now letting $\varepsilon_0 \to 0$ in (3.41) we obtain

$$\|u^\delta(T) - u(T)\|_1 \leq \min_\varepsilon \left(2\varepsilon + \frac{2T\delta}{\varepsilon}\right) \text{T.V.}(u_0) = 2\text{T.V.}(u_0) \sqrt{T\delta}.$$

So the method of vanishing viscosity also has order $\frac{1}{2}$. ◇

◇ **Example 3.14 (Monotone schemes).**

We will here show that monotone schemes converge in L^1 to the solution of (3.1) at a rate of $(\Delta t)^{1/2}$. In particular, this applies to the Lax–Friedrichs scheme.

Let $u_{\Delta t}$ be defined by (3.24), where U_j^n is defined by (3.5), that is,

$$U_j^{n+1} = U_j^n - \lambda\Big(F\left(U_{j-p}^n, \ldots, U_{j+p'}^n\right) - F\left(U_{j-1-p}^n, \ldots, U_{j-1+p'}^n\right)\Big), \tag{3.50}$$

for a scheme that is assumed to be monotone; cf. Definition 3.5. In the following we use the notation

$$\eta_j^n = \left|U_j^n - k\right|, \quad q_j^n = f\left(U_j^n \vee k\right) - f\left(U_j^n \wedge k\right).$$

We find that

$$-\Lambda_T(u_{\Delta t}, \phi, k)$$

$$= -\sum_j \sum_{n=0}^{N-1} \int_{x_j}^{x_{j+1}} \int_{t_n}^{t_{n+1}} \left(\eta_j^n \phi_t(x,s) + q_j^n \phi_x(x,s)\right) ds \, dx$$

$$- \sum_j \int_{x_j}^{x_{j+1}} \eta_j^0 \phi(x,0) \, dx + \sum_j \int_{x_j}^{x_{j+1}} \eta_j^N \phi(x,T) \, dx$$

$$= -\sum_j \left[\sum_{n=0}^{N-1} \int_{x_j}^{x_{j+1}} \eta_j^n \left(\phi(x, t_{n+1}) - \phi(x, t_n) \right) dx \right.$$
$$+ \int_{x_j}^{x_{j+1}} \eta_j^0 \phi(x, 0) \, dx - \int_{x_j}^{x_{j+1}} \eta_j^N \phi(x, T) \, dx$$
$$\left. + \sum_{n=0}^{N-1} \int_{t_n}^{t_{n+1}} q_j^n \left(\phi(x_{j+1}, s) - \phi(x_j, s) \right) ds \right]$$
$$= \sum_j \sum_{n=0}^{N-1} \left((\eta_j^{n+1} - \eta_j^n) \int_{x_j}^{x_{j+1}} \phi(x, t_{n+1}) \, dx \right.$$
$$\left. + (q_j^n - q_{j-1}^n) \int_{t_n}^{t_{n+1}} \phi(x_j, s) \, ds \right)$$

by a summation by parts. Recall that we define the numerical entropy flux by

$$Q_j^n = Q(U^n; j) = F(U^n \vee k; j) - F(U^n \wedge k; j).$$

Monotonicity of the scheme implies, cf. (3.30), that

$$\eta_j^{n+1} - \eta_j^n + \lambda(Q_j^n - Q_{j-1}^n) \leq 0.$$

For a nonnegative test function ϕ we obtain

$$-\Lambda_T(u_{\Delta t}, \phi, k)$$
$$\leq \sum_j \sum_{n=0}^{N-1} \left(-\lambda(Q_j^n - Q_{j-1}^n) \int_{x_j}^{x_{j+1}} \phi(x, t_{n+1}) \, dx \right.$$
$$\left. + (q_j^n - q_{j-1}^n) \int_{t_n}^{t_{n+1}} \phi(x_j, s) \, ds \right)$$
$$= \sum_j \sum_{n=0}^{N-1} \left[\lambda(Q_j^n - q_j^n) \left(\int_{x_{j+1}}^{x_{j+2}} \phi(x, t_{n+1}) \, dx - \int_{x_j}^{x_{j+1}} \phi(x, t_{n+1}) \, dx \right) \right.$$
$$\left. + (q_j^n - q_{j-1}^n) \left(\int_{t_n}^{t_{n+1}} \phi(x_j, s) \, ds - \lambda \int_{x_j}^{x_{j+1}} \phi(x, t_{n+1}) \, dx \right) \right].$$

We also have that

$$|Q_j^n - q_j^n| \leq 2\|f\|_{\text{Lip}} \sum_{m=-p}^{p'} |U_{j+m}^n - U_j^n|,$$

and

$$|q_j^n - q_{j-1}^n| \leq 2\|f\|_{\text{Lip}} |U_j^n - U_{j-1}^n|,$$

which implies that

$$
\begin{aligned}
-&\Lambda_T(u_{\Delta t}, \phi, k) \\
&\leq 2\|f\|_{\text{Lip}} \sum_j \sum_{n=0}^{N-1} \Bigg[\Big(\sum_{m=-p}^{p'} |U_{j+m}^n - U_j^n| \Big) \\
&\qquad \times \int_{x_j}^{x_{j+1}} |\phi(x+\Delta x, t_{n+1}) - \phi(x, t_{n+1})|\, dx \\
&\qquad + |U_j^n - U_{j-1}^n| \\
&\qquad \times \Big| \int_{t_n}^{t_{n+1}} \phi(x_j, s)\, ds - \lambda \int_{x_j}^{x_{j+1}} \phi(x, t_{n+1})\, dx \Big| \Bigg].
\end{aligned}
$$

Next, we subtract $\phi(x_j, t_n)$ from the integrand in each of the latter two integrals. Since $\Delta t = \lambda \Delta x$, the extra terms cancel, and we obtain

$$
-\Lambda_T(u_{\Delta t}, \phi, k) \tag{3.51}
$$

$$
\begin{aligned}
&\leq 2\|f\|_{\text{Lip}} \sum_j \sum_{n=0}^{N-1} \Bigg[\Big(\sum_{m=-p}^{p'} |U_{j+m}^n - U_j^n| \Big) \\
&\qquad \times \int_{x_j}^{x_{j+1}} |\phi(x+\Delta x, t_{n+1}) - \phi(x, t_{n+1})|\, dx \\
&\qquad + |U_j^n - U_{j-1}^n| \Big(\int_{t_n}^{t_{n+1}} |\phi(x_j, s) - \phi(x_j, t_n)|\, ds \\
&\qquad + \lambda \int_{x_j}^{x_{j+1}} |\phi(x, t_{n+1}) - \phi(x_j, t_n)|\, dx \Big) \Bigg].
\end{aligned}
$$

Let $v = v(x, t)$ denote the unique weak solution of (3.1), and let $k = v(x', s')$. Choose the test function as $\phi(x, s) = \omega_\varepsilon(x - x')\omega_{\varepsilon_0}(s - s')$, and observe that

$$
\int_0^T \int_{\mathbb{R}} |\omega_\varepsilon(x + \Delta x - x') - \omega_\varepsilon(x - x')| \omega_{\varepsilon_0}(t_n - s')\, dx'\, ds'
$$

$$
\leq \Delta x \, \text{T.V.}(\omega_\varepsilon) \leq 2\frac{\Delta x}{\varepsilon}.
$$

Similarly,

$$
\int_0^T \int_{\mathbb{R}} \omega_\varepsilon(x_j - x') |\omega_{\varepsilon_0}(s - s') - \omega_{\varepsilon_0}(t_n - s')|\, dx'\, ds' \leq 2\frac{\Delta t}{\varepsilon_0},
$$

whenever $|s - t_n| \leq \Delta t$, and

$$
\int_0^T \int_{\mathbb{R}} |\omega_\varepsilon(x - x')\omega_{\varepsilon_0}(t_{n+1} - s')
$$

$$-\omega_\varepsilon(x_j - x')\omega_{\varepsilon_0}(t_n - s')\Big| \, dx' \, ds' \leq 2\Big(\frac{\Delta t}{\varepsilon_0} + \frac{\Delta x}{\varepsilon}\Big).$$

Integrating (3.51) over (x', s') with $0 \leq s' \leq T$ we obtain

$$-\Lambda_{\varepsilon,\varepsilon_0}(u_{\Delta t}, v)$$

$$\leq 4\|f\|_{\text{Lip}} \sum_{n=0}^{N-1} \Bigg[\sum_j \sum_{m=-p}^{p'} |U_{j+m}^n - U_j^n| \frac{\Delta x}{\varepsilon} \Delta x$$

$$+ \sum_j |U_j^n - U_{j-1}^n| \left(\frac{\Delta t}{\varepsilon_0}\Delta t + \lambda(\frac{\Delta x}{\varepsilon} + \frac{\Delta t}{\varepsilon_0})\Delta x\right) \Bigg]$$

$$\leq 4\|f\|_{\text{Lip}} \text{T.V.}(u_{\Delta t}(0))$$

$$\times \sum_{n=0}^{N-1} \left[\frac{1}{2}(p + p' + 1)^2 \frac{(\Delta x)^2}{\varepsilon} + \frac{(\Delta t)^2}{\varepsilon_0} + \lambda(\frac{(\Delta x)^2}{\varepsilon} + \frac{\Delta x \Delta t}{\varepsilon_0}) \right]$$

$$\leq K \, T \, \text{T.V.}(u_{\Delta t}(0)) \left(\frac{1}{\varepsilon} + \frac{1}{\varepsilon_0}\right) \Delta t$$

for some constant K, by using the estimate

$$\sum_j \sum_{m=-p}^{p'} |U_{j+m}^n - U_j^n| \leq \frac{1}{2}(p + p' + 1)^2 \text{T.V.}(U^n).$$

Recalling Kuznetsov's lemma

$$\|u_{\Delta t}(T) - v(T)\|_1 \leq \|u_{\Delta t}(0) - v_0\|_1 + \text{T.V.}(v_0)(2\varepsilon + \varepsilon_0 \|f\|_{\text{Lip}})$$

$$+ \frac{1}{2}\left(\nu_T(u_{\Delta t}, \varepsilon_0) + \nu_0(u_{\Delta t}, \varepsilon_0)\right) - \Lambda_{\varepsilon,\varepsilon_0}.$$

We have that

$$\text{T.V.}(u_{\Delta t}(\cdot, t)) \leq \text{T.V.}(u_{\Delta t}(\cdot, 0))$$

and

$$\nu_t(u_{\Delta t}, \varepsilon) \leq K_1 (\Delta t + \varepsilon) \, \text{T.V.}(u_{\Delta t}(\cdot, 0)).$$

Choose the initial approximation such that

$$\|u_{\Delta t}(0) - v_0\|_1 \leq \Delta x \, \text{T.V.}(v_0).$$

This implies

$$\|u_{\Delta t}(T) - v(T)\|_1$$
$$\leq \text{T.V.}(v_0)(\Delta x + 2\varepsilon + \varepsilon_0 \|f\|_{\text{Lip}})$$
$$+ \text{T.V.}(u_{\Delta t}(\cdot, 0))\left(K_1(\Delta t + \varepsilon_0) + KT\Delta t\left(\frac{1}{\varepsilon_0} + \frac{1}{\varepsilon}\right)\right)$$
$$\leq K_2 \text{T.V.}(v_0)\left[\Delta t + \varepsilon + \frac{T\Delta t}{\varepsilon} + \varepsilon_0 + \frac{T\Delta t}{\varepsilon_0}\right].$$

Minimizing with respect to ε_0 and ε, we obtain the final bound

$$\|u_{\Delta t}(T) - v(T)\|_1 \leq K_4 \text{T.V.}(v_0)\left(\Delta t + 4\sqrt{T\Delta t}\right). \tag{3.52}$$

Thus, as promised, we have shown that monotone schemes are of order $(\Delta t)^{1/2}$. $\diamond$

If one uses Kuznetsov's lemma to estimate the error of a scheme, one must estimate the modulus of continuity $\tilde{\nu}_t(u, \varepsilon_0)$ and the term $\Lambda_{\varepsilon,\varepsilon_0}(u, v)$. In other words, one must obtain regularity estimates on the *approximation* u. Therefore, this approach gives a posteriori error estimates, and perhaps the proper use for this approach should be in adaptive methods, in which it would provide error control and govern mesh refinement. However, despite this weakness, Kuznetsov's theory is still actively used.

3.3 A Priori Error Estimates

We shall now describe an application of a variation of Kuznetsov's approach, in which we obtain an error estimate for the method of vanishing viscosity, without using the regularity properties of the viscous approximation. Of course, this application only motivates the approach, since regularity of the solutions of parabolic equations is not difficult to obtain elsewhere. Nevertheless, it is interesting in its own right, since many difference methods have (3.53) as their model equation. We first state the result.

Theorem 3.15. *Let $v(x,t)$ be a solution of (3.1) with initial value v_0, and let u solve the equation*

$$u_t + f(u)_x = (\delta(u)u_x)_x, \qquad u(x,0) = u_0(x), \tag{3.53}$$

in the classical sense, with $\delta(u) > 0$. Then

$$\|u(T) - v(T)\|_1 \leq 2\|u_0 - v_0\|_1 + 4\text{T.V.}(v_0)\sqrt{8T\|\delta\|_v},$$

where

$$\|\delta\|_v = \sup_{\substack{t\in[0,T] \\ x\in\mathbb{R}}} \tilde{\delta}\left(v(x-,t), v(x+,t)\right)$$

and

$$\tilde{\delta}(a,b) = \frac{1}{b-a}\int_a^b \delta(c)\,dc.$$

This result is not surprising, and in some sense is weaker than the corresponding result found by using Kuznetsov's lemma. The new element here is that the proof does *not* rely on any smoothness properties of the function

u, and is therefore also considerably more complicated than the proof using Kuznetsov's lemma.

Proof. The proof consists in choosing new Λ's, and using a special form of the test function φ. Let w^∞ be defined as

$$w^\infty(x) = \begin{cases} \frac{1}{2} & \text{for } |x| \leq 1, \\ 0 & \text{otherwise.} \end{cases}$$

We will consider a family of smooth functions w such that $w \to w^\infty$. To keep the notation simple we will not add another parameter to the functions w, but rather write $w \to w^\infty$ when we approach the limit. Let

$$\varphi(x, y, t, s) = w_\varepsilon(x - y) w_{\varepsilon_0}(t - s)$$

with $w_\alpha(x) = (1/\alpha) w(x/\alpha)$ as usual. In this notation

$$w_\varepsilon^\infty(x) = \begin{cases} 1/(2\varepsilon) & \text{for } |x| \leq \varepsilon, \\ 0 & \text{otherwise.} \end{cases}$$

In the following we will use the entropy pair

$$\eta(u, k) = |u - k| \quad \text{and} \quad q(u, k) = \operatorname{sign}(u - k)(f(u) - f(k)),$$

and except where explicitly stated, we always let $u = u(y, s)$ and $v = v(x, t)$. Let $\eta_\sigma(u, k)$ and $q_\sigma(u, k)$ be smooth approximations to η and q such that

$$\eta_\sigma(u) \to \eta(u) \quad \text{as } \sigma \to 0, \qquad q_\sigma(u, k) = \int \eta_\sigma'(z - k)(f(z) - f(k))\, dz.$$

For a test function φ define

$$\Lambda_T^\sigma(u, k) = \int_0^T \int \eta_\sigma'(u - k) \left(u_s + f(u)_y - (\delta(u) u_y)_y \right) \varphi\, dy\, ds$$

(which is clearly zero because of (3.53)) and

$$\Lambda_{\varepsilon, \varepsilon_0}^\sigma(u, v) = \int_0^T \int \Lambda_T^\sigma(u, v(x, t))\, dx\, dt.$$

Note that since u satisfies (3.53), $\Lambda_{\varepsilon, \varepsilon_0}^\sigma = 0$ for every v. We now split $\Lambda_{\varepsilon, \varepsilon_0}^\sigma$ into two parts. Writing (cf. (2.10))

$$(u_s + f(u)_x - (\delta(u) u_y)_y) \eta_\sigma'(u - k)$$
$$= \eta(u - k)_s + ((f(u) - f(k))' \eta_\sigma'(u - k) u_y - (\delta(u) u_y)_y \eta_\sigma'(u - k)$$
$$= \eta_\sigma(u - k)_s + q_\sigma(u, k)_u u_y - (\delta(u) u_y)_y \eta_\sigma'(u - k)$$
$$= \eta_\sigma(u - k)_s + q_\sigma(u, k)_y - (\delta(u) \eta_\sigma(u - k)_y)_y + \eta_\sigma''(u - k) \delta(u)(u_y)^2$$
$$= \eta_\sigma(u - k)_s + (q_\sigma(u, k) - \delta(u) \eta_\sigma(u - k)_y)_y + \eta''(u - k) \delta(u)(u_y)^2,$$

we may introduce

$$\Lambda_1^\sigma(u,v) = \int_0^T \int \int_0^T \int \eta_\sigma''(u-v)\delta(u)\,(u_y)^2\,\varphi\,dy\,ds\,dx\,dt,$$

$$\Lambda_2^\sigma(u,v)$$
$$= \int_0^T \int \int_0^T \int \Big(\eta_\sigma(u-v)_s + (q_\sigma(u,v) - \delta(u)\eta_\sigma(u-v)_y)_y\Big)\varphi\,dy\,ds\,dx\,dt,$$

such that $\Lambda_{\varepsilon,\varepsilon_0}^\sigma = \Lambda_1^\sigma + \Lambda_2^\sigma$. Note that if $\delta(u) > 0$, we always have $\Lambda_1^\sigma \geq 0$, and hence $\Lambda_2^\sigma \leq 0$. Then we have that

$$\Lambda_2 := \limsup_{\sigma \to 0} \Lambda_2^\sigma \leq 0.$$

To estimate Λ_2, we integrate by parts:

$$\Lambda_2(u,v)$$
$$= \int_0^T \int \int_0^T \int \left(-\eta(u-v)\varphi_s - q(u,v)\varphi_y + V(u,v)\varphi_{yy}\right)dy\,ds\,dx\,dt$$
$$+ \int_0^T \int\!\!\int \eta(u(T)-v)\varphi|_{s=T}\,dy\,dx\,dt - \int_0^T \int\!\!\int \eta(u_0-v)\varphi|_{s=0}\,dy\,dx\,dt$$
$$= \int_0^T \int \int_0^T \int \left(\eta(u-v)\varphi_t + F(u,v)\varphi_x - V(u,v)\varphi_{xy}\right)dy\,ds\,dx\,dt$$
$$+ \int_0^T \int\!\!\int \eta(u(T)-v)\varphi|_{s=T}\,dy\,dx\,dt - \int_0^T \int\!\!\int \eta(u_0-v)\varphi|_{s=0}\,dy\,dx\,dt,$$

where

$$V(u,v) = \int_u^v \delta(s)\eta'(s-v)\,ds.$$

Now define (the "dual of Λ_2")

$$\Lambda_2^* := -\int_0^T \int \int_0^T \int \left(\eta(u-v)\varphi_t + q(u,v)\varphi_x - V(u,v)\varphi_{xy}\right)dy\,ds\,dx\,dt$$
$$- \int_0^T \int\!\!\int \eta(u-v(T))\varphi \Big|_{t=0}^{t=T} dx\,dy\,ds.$$

Then we can write

$$\Lambda_2 = -\Lambda_2^*$$
$$+ \underbrace{\int_0^T \int\!\!\int (\eta(u(T)-v)\varphi)|_{s=T}\,dy\,dx\,dt}_{\Phi_1}$$
$$\underbrace{- \int_0^T \int\!\!\int (\eta(u_0-v)\varphi)|_{s=0}\,dy\,dx\,dt}_{\Phi_2}$$

3.3. A Priori Error Estimates

$$+ \underbrace{\int_0^T \iint (\eta(u - v(T))\varphi)|_{t=T}\, dx\, dy\, ds}_{\Phi_3}$$

$$- \underbrace{\int_0^T \iint (\eta(u_0 - v_0)\varphi)|_{t=0}\, dx\, dy\, ds}_{\Phi_4}$$

$$=: -\Lambda_2^* + \Phi.$$

We will need later that

$$\Phi = \Lambda_2^* + \Lambda_2 \leq \Lambda_2^*. \tag{3.54}$$

Let

$$\Omega_{\varepsilon_0}(t) = \int_0^t \omega_{\varepsilon_0}(s)\, ds$$

and

$$e(t) = \|u(t) - v(t)\|_1 = \int \eta(u(x,t) - v(x,t))\, dx.$$

To continue estimating, we need the following proposition.

Proposition 3.16.

$$\Phi \geq \Omega_{\varepsilon_0}(T)e(T) - \Omega_{\varepsilon_0}(T)e(0) + \int_0^T \omega_{\varepsilon_0}(T-t)e(t)\, dt - \int_0^T \omega_{\varepsilon_0}(t)e(t)\, dt$$
$$- 4\Omega_{\varepsilon_0}(T)\left(\varepsilon_0\|f\|_{\mathrm{Lip}} + \varepsilon\right) \mathrm{T.V.}(v_0).$$

Proof (of Proposition 3.16). We start by estimating Φ_1. First note that

$$\eta(u(y,T) - v(x,t)) = |u(y,T) - v(x,t)|$$
$$\geq |u(y,T) - v(y,T)|$$
$$\quad - |v(y,T) - v(y,t)| - |v(y,t) - v(x,t)|$$
$$= \eta(u(y,T) - v(y,T))$$
$$\quad - |v(y,T) - v(y,t)| - |v(y,t) - v(x,t)|.$$

Thus

$$\Phi_1 \geq \int_0^T \iint \eta(u(y,T) - v(y,T))\varphi|_{s=T}\, dy\, dx\, dt$$
$$\quad - \int_0^T \iint |v(y,T) - v(y,t)|\, \varphi|_{s=T}\, dy\, dx\, dt$$
$$\quad - \int_0^T \iint |v(y,t) - v(x,t)|\, \varphi|_{s=T}\, dy\, dx\, dt$$
$$\geq \Omega_{\varepsilon_0}(T)e(T) - \Omega_{\varepsilon_0}(T)\left(\varepsilon_0\|f\|_{\mathrm{Lip}} + \varepsilon\right) \mathrm{T.V.}(v_0).$$

96   3. A Short Course in Difference Methods

Here we have used that v is an exact solution. The estimate for Φ_2 is similar, yielding
$$\Phi_2 \geq -\Omega_{\varepsilon_0}(T)e(0) - \Omega_{\varepsilon_0}(T)\left(\varepsilon_0\|f\|_{\text{Lip}} + \varepsilon\right)\text{T.V.}(v_0).$$

To estimate Φ_3 we proceed in the same manner:
$$\eta(u(y,s) - v(x,T)) \geq \eta(u(y,s) - v(y,s)) - |v(y,s) - v(x,s)|$$
$$- |v(x,s) - v(x,T)|.$$

This gives
$$\Phi_3 \geq \int_0^T w_{\varepsilon_0}(T-t)e(t)\,dt - \Omega_{\varepsilon_0}(T)\left(\varepsilon_0\|f\|_{\text{Lip}} + \varepsilon\right)\text{T.V.}(v_0),$$

while by the same reasoning, the estimate for Φ_4 reads
$$\Phi_4 \geq -\int_0^T w_{\varepsilon_0}(t)e(t)\,dt - \Omega_{\varepsilon_0}(T)\left(\|f\|_{\text{Lip}}\varepsilon_0 + \varepsilon\right)\text{T.V.}(v_0).$$

The proof of Proposition 3.16 is complete. □

To proceed further, we shall need the following Gronwall-type lemma:

Lemma 3.17. *Let θ be a nonnegative function that satisfies*
$$\Omega_{\varepsilon_0}^\infty(\tau)\theta(\tau) + \int_0^\tau w_{\varepsilon_0}^\infty(\tau-t)\theta(t)\,dt \leq C\Omega_{\varepsilon_0}^\infty(\tau) + \int_0^\tau w_{\varepsilon_0}^\infty(t)\theta(t)\,dt, \quad (3.55)$$
for all $\tau \in [0,T]$ and some constant C. Then
$$\theta(\tau) \leq 2C.$$

Proof (of Lemma 3.17). If $\tau \leq \varepsilon_0$, then for $t \in [0,\tau]$, $w_{\varepsilon_0}^\infty(t) = w_{\varepsilon_0}^\infty(\tau-t) = 1/(2\varepsilon_0)$. In this case (3.55) immediately simplifies to $\theta(t) \leq C$.

For $\tau > \varepsilon_0$, we can write (3.55) as
$$\theta(\tau) \leq C + \frac{1}{\Omega_{\varepsilon_0}^\infty(\tau)}\int_0^{\varepsilon_0}\left(w_{\varepsilon_0}^\infty(t) - w_{\varepsilon_0}^\infty(\tau-t)\right)\theta(t)\,dt.$$

For $t \in [0,\varepsilon_0]$ we have $\theta(t) \leq C$, and this implies
$$\theta(\tau) \leq C\left(1 + \frac{1}{\Omega_{\varepsilon_0}^\infty(\tau)}\int_0^{\varepsilon_0}\left(w_{\varepsilon_0}^\infty(t) - w_{\varepsilon_0}^\infty(\tau-t)\right)dt\right) \leq 2C.$$

This concludes the proof of the lemma. □

Now we can continue the estimate of $e(T)$.

Proposition 3.18. *We have that*
$$e(T) \leq 2e(0) + 8\left(\varepsilon + \varepsilon_0\|f\|_{\text{Lip}}\right)\text{T.V.}(v_0) + 2\lim_{\omega\to\omega^\infty}\sup_{t\in[0,T]}\frac{\Lambda_2^*(u,v)}{\Omega_{\varepsilon_0}^\infty(t)}.$$

3.3. A Priori Error Estimates

Proof (of Proposition 3.18). Starting with the inequality (3.54), using the estimate for Φ from Proposition 3.16, we have, after passing to the limit $\omega \to \omega^\infty$, that

$$\Omega^\infty_{\varepsilon_0}(T)e(T) + \int_0^T \omega^\infty_{\varepsilon_0}(T-t)e(t)\,dt \leq \Omega^\infty_{\varepsilon_0}(t)e(0) + \int_0^T \omega^\infty_{\varepsilon_0}(t)e(t)\,dt$$
$$+ 4\Omega^\infty_{\varepsilon_0}(t)\left(\varepsilon + \varepsilon_0 \|f\|_{\mathrm{Lip}}\right) \mathrm{T.V.}(v_0)$$
$$+ \Omega^\infty_{\varepsilon_0}(T) \lim_{\omega \to \omega^\infty} \sup_{t \in [0,T]} \frac{\Lambda_2^*(u,v)}{\Omega^\infty_{\varepsilon_0}(t)}.$$

We apply Lemma 3.17 with

$$C = 4\left(\varepsilon + \varepsilon_0 \|f\|_{\mathrm{Lip}}\right) \mathrm{T.V.}(v_0) + \lim_{\omega \to \omega^\infty} \sup_{t \in [0,T]} \frac{\Lambda_2^*(u,v)}{\Omega^\infty_{\varepsilon_0}(t)} + e(0)$$

to complete the proof. □

To finish the proof of the theorem, it remains only to estimate

$$\lim_{\omega \to \omega^\infty} \sup_{t \in [0,T]} \frac{\Lambda_2^*(u,v)}{\Omega(t)}.$$

We will use the following inequality:

$$\left|\frac{V(u,v^+) - V(u,v^-)}{v^+ - v^-}\right| \leq \frac{1}{v^+ - v^-} \int_{v^-}^{v^+} \delta(s)\,ds. \tag{3.56}$$

Since v is an entropy solution to (3.1), we have that

$$\Lambda_2^* \leq -\int_0^T \int_0^T \int \int V(u,v)\varphi_{xy}\,dy\,ds\,dx\,dt. \tag{3.57}$$

Since v is of bounded variation, it suffices to study the case where v is differentiable except on a countable number of curves $x = x(t)$. We shall bound Λ_2^* in the case that we have one such curve; the generalization to more than one is straightforward. Integrating (3.57) by parts, we obtain

$$\Lambda_2^* \leq \int_0^T \int \Psi(y,s)\,dy\,ds, \tag{3.58}$$

where Ψ is given by

$$\Psi(y,s) = \int_0^T \left(\int_{-\infty}^{x(t)} V(u,v)_v\,v_x\varphi_y\,dx \right.$$
$$\left. + \frac{[V]}{[v]}[v]\varphi_y|_{x=x(t)} + \int_{x(t)}^\infty V(u,v)_v\,v_x\varphi_y\,dx\right)dt.$$

As before, $[\![a]\!]$ denotes the jump in a, i.e., $[\![a]\!] = a(x(t)+,t) - a(x(t)-,t)$. Using (3.56), we obtain

$$|\Psi(y,s)| \leq \|\delta\|_v \int_0^T \Big(\int_{-\infty}^{x(t)} |v_x| |\varphi_y| \, dx \\ + |[\![v]\!]| |\varphi_y|_{x=x(t)}| + \int_{x(t)}^{\infty} |v_x| |\varphi_y| \, dx \Big) dt. \tag{3.59}$$

Let D be given by

$$D(x,t) = \int_0^T \int |\varphi_y| \, dy \, ds.$$

A simple calculation shows that

$$D(x,t) = \frac{1}{\varepsilon} \int_0^T \omega_{\varepsilon_0}(t-s) \, ds \int |\omega'(y)| \, dy \leq \frac{1}{\varepsilon} \int_0^T \omega_{\varepsilon_0}(t-s) \, ds.$$

Consequently,

$$\int_0^T \sup_x D(x,t) \, dt \leq \frac{1}{\varepsilon} \int_0^T \int_0^T \omega_{\varepsilon_0}(t-s) \, ds \, dt$$

$$= \frac{2}{\varepsilon} \int_0^T (T-t) \omega_{\varepsilon_0}(t) \, dt$$

$$\leq \frac{2T\Omega(T)}{\varepsilon}.$$

Inserting this in (3.59), and the result in (3.58), we find that

$$\Lambda_2^*(u,v,T) \leq \frac{2}{\varepsilon} T \, \mathrm{T.V.}(v_0) \|\delta\|_v \Omega(T).$$

Summing up, we have now shown that

$$e(T) \leq 2e(0) + 8\left(\varepsilon + \varepsilon_0 \|f\|_{\mathrm{Lip}}\right) \mathrm{T.V.}(v_0) + \frac{4}{\varepsilon} T \, \mathrm{T.V.}(v_0) \|\delta\|_v.$$

We can set ε_0 to zero, and minimize over ε, obtaining

$$\|u(T) - v(T)\|_1 \leq 2\|u_0 - v_0\|_1 + 4\mathrm{T.V.}(v_0) \sqrt{8T \|\delta\|_v}.$$

The theorem is proved. □

The main idea behind this approach to getting a priori error estimates, is to choose the "Kuznetsov-type" form $\Lambda_{\varepsilon,\varepsilon_0}$ such that

$$\Lambda_{\varepsilon,\varepsilon_0}(u,v) = 0$$

for every function v, and then writing $\Lambda_{\varepsilon,\varepsilon_0}$ as a sum of a nonnegative and a nonpositive part. Given a numerical scheme, the task is then to prove a discrete analogue of the previous theorem.

3.4 Measure-Valued Solutions

> You try so hard, but you don't understand ...
>
> Bob Dylan, Ballad of a Thin Man (1965)

Monotone methods are at most first-order accurate. Consequently, one must work harder to show that higher-order methods converge to the entropy solution. While this is possible in one space dimension, i.e., in the above setting, it is much more difficult in several space dimensions. One useful tool to aid the analysis of higher-order methods is the concept of *measure-valued solutions*. This is a rather complicated concept, which requires a background from analysis beyond this book. Therefore, the presentation in this section is brief, and is intended to give the reader a first flavor, and an idea of what this method can accomplish.

Consider the case where a numerical scheme gives a sequence U_j^n that is uniformly bounded in $L^\infty(\mathbb{R} \times [0, \infty))$, and with the L^1-norm Lipschitz continuous in time, but such that there is no bound on the total variation. We can still infer the existence of a weak limit
$$u_{\Delta t} \overset{*}{\rightharpoonup} u,$$
but the problem is to show that
$$f(u_{\Delta t}) \overset{*}{\rightharpoonup} f(u).$$
Here, we have introduced the concept of weak-$*$ L^∞ convergence. A sequence $\{u_n\}$ that is bounded in L^∞ is said to converge weakly-$*$ to u if for all $v \in L^1$,
$$\int u_n v \, dx \to \int u v \, dx, \quad \text{as } n \to \infty.$$
Since $u_{\Delta t}$ is bounded, $f(u_{\Delta t})$ is also bounded and converges weakly, and thus
$$f(u_{\Delta t}) \overset{*}{\rightharpoonup} \bar{f},$$
but $\bar{f}$ is in general not equal to $f(u)$. We provide a simple example of the problem.

◇ **Example 3.19.**

Let $u_n = \sin(nx)$ and $f(u) = u^2$. Then
$$\left| \int \sin(nx) \varphi(x) \, dx \right| \le \frac{1}{n} \left| \int \cos(nx) \varphi'(x) \, dx \right| \le \frac{C}{n} \to 0 \text{ as } n \to \infty.$$
On the other hand, $f(u_n) = \sin^2(nx) = (1 - \cos(2nx))/2$, and hence a similar estimate shows that
$$\left| \int (f(u_n) - \frac{1}{2}) \varphi(x) \, dx \right| \le \frac{C}{n} \to 0 \text{ as } n \to \infty.$$

Thus we conclude that
$$u_n \overset{*}{\rightharpoonup} 0, \quad f(u_n) \overset{*}{\rightharpoonup} \frac{1}{2} \neq 0 = f(0).$$

$\diamond$

To be able to treat this situation, we will further weaken the requirements to solutions of conservation laws.

Let $\mathcal{M}(\mathbb{R})$ denote the space of bounded Radon measures on $\mathbb{R}$ and
$$\mathcal{C}_0(\mathbb{R}) = \left\{ g \in C(\mathbb{R}) \mid \lim_{|\lambda| \to \infty} g(\lambda) = 0 \right\}.$$
If $\mu \in \mathcal{M}(\mathbb{R})$, then we write
$$\langle \mu, g \rangle = \int_{\mathbb{R}} g(\lambda) \, d\mu(\lambda), \quad \text{for all} \quad g \in \mathcal{C}_0(\mathbb{R}).$$
Now we have that $\mu \in \mathcal{M}(\mathbb{R})$ if and only if $|\langle \mu, g \rangle| \leq C \|g\|_{L^\infty(\mathbb{R})}$ for all $g \in \mathcal{C}_0(\mathbb{R}^n)$. We can then define a norm on $\mathcal{M}(\mathbb{R})$ by
$$\|\mu\|_{\mathcal{M}(\mathbb{R})} = \sup \left\{ |\langle \mu, \psi \rangle| \mid \psi \in \mathcal{D}(\mathbb{R}), \|\psi\|_{L^\infty(\mathbb{R})} \leq 1 \right\}.$$
The space $\mathcal{M}(\mathbb{R})$ equipped with the norm $\|\cdot\|_{\mathcal{M}(\mathbb{R})}$ is a Banach space, and it is isometrically isomorphic to the dual space of $\mathcal{C}_0(\mathbb{R})$ with the norm $\|\cdot\|_{L^\infty(\mathbb{R})}$; see, e.g., [107, p. 149]. Next we define the space of probability measures $\text{Prob}(\mathbb{R})$ as
$$\text{Prob}(\mathbb{R}) = \left\{ \mu \in \mathcal{M}(\mathbb{R}) \mid \mu \text{ is nonnegative and } \|\mu\|_{\mathcal{M}(\mathbb{R})} = 1 \right\}.$$
Then we can state the fundamental theorem in the theory of compensated compactness.

Theorem 3.20 (Young's theorem). *Let $K \subset \mathbb{R}$ be a bounded open set and $\{u_\varepsilon\}$ a sequence of functions $u_\varepsilon \colon \mathbb{R} \times [0, T] \to K$. Then there exists a family of probability measures $\{\nu_{(x,t)}(\lambda) \in \text{Prob}(\mathbb{R})\}_{(x,t) \in \mathbb{R} \times [0,T]}$ (depending weak-$*$ measurably on (x, t)) such that for any continuous function $g \colon K \to \mathbb{R}$, we have a subsequence*
$$g(u_\varepsilon) \overset{*}{\rightharpoonup} \bar{g} \text{ in } L^\infty(\mathbb{R} \times [0, T]) \text{ as } \varepsilon \to 0,$$
where (the exceptional set depends possibly on g)
$$\bar{g}(x, t) := \langle \nu_{(x,t)}, g \rangle = \int_{\mathbb{R}} g(\lambda) \, d\nu_{(x,t)}(\lambda) \text{ for a.e. } (x, t) \in \mathbb{R} \times [0, T].$$
Furthermore,
$$\operatorname{supp} \nu_{(x,t)} \subset \bar{K} \text{ for a.e. } (x, t) \in \mathbb{R} \times [0, T].$$

This theorem is indeed the main reason why measure-valued solutions are easier to obtain than weak solutions, since for any bounded sequence of approximations to a solution of a conservation law we can associate (at least) one probability measure $\nu_{(x,t)}$ representing the weak-star limits of the sequence. Thus we avoid having to show that the method is TVD stable and use Helly's theorem to be able to work with the limit of the sequence. The measures associated with weakly convergent sequences are frequently called Young measures.

Intuitively, when we are in the situation that we have no knowledge of eventual oscillations in u_ε as $\varepsilon \to 0$, the Young measure $\nu_{(x,t)}(E)$ can be thought of as the probability that the "limit" at the point (x,t) takes a value in the set E. To be a bit more precise, define

$$\nu_{(x,t)}^{\varepsilon,r}(E) = \frac{1}{r^2} \operatorname{meas}\Big\{ (y,s) \mid |x-y|, |t-s| \le r \quad \text{and} \quad u_\varepsilon(y,s) \in E \Big\}.$$

Then for small r, $\nu_{(x,t)}^{\varepsilon,r}(E)$ is the probability that u^ε takes values in E near x. It can be shown that

$$\nu_{(x,t)} = \lim_{r\to 0} \lim_{\varepsilon \to 0} \nu_{(x,t)}^{\varepsilon,r};$$

see [6].

We also have the following corollary of Young's theorem.

Corollary 3.21. *The sequence u_ε converges strongly to u if and only if the measure $\nu_{(x,t)}$ reduces to a Dirac measure located at $u(x,t)$, i.e., $\nu_{(x,t)} = \delta_{u(x,t)}$.*

Now we can define measure-valued solutions. A probability measure $\nu_{(x,t)}$ is a measure-valued solution to (3.1) if

$$\langle \nu_{(x,t)}, \operatorname{Id} \rangle_t + \langle \nu_{(x,t)}, f \rangle_x = 0$$

in the distributional sense, where Id is the identity map, $\operatorname{Id}(\lambda) = \lambda$. As with weak solutions, we call a measure-valued solution compatible with the entropy pair (η, q) if

$$\langle \nu_{(x,t)}, \eta \rangle_t + \langle \nu_{(x,t)}, q \rangle_x \le 0 \tag{3.60}$$

in the distributional sense. If (3.60) holds for *all* convex η, we call $\nu_{(x,t)}$ a measure-valued entropy solution. Clearly, weak solutions are also measure-valued solutions, as we can see by setting

$$\nu_{(x,t)} = \delta_{u(x,t)}$$

for a weak entropy solution u. But measure-valued solutions are more general than weak solutions, since for any two measure-valued solutions $\nu_{(x,t)}$ and $\mu_{(x,t)}$, and for any $\theta \in [0,1]$ the convex combination

$$\theta \nu_{(x,t)} + (1-\theta) \mu_{(x,t)} \tag{3.61}$$

is also a measure-valued solution. It is not clear, however, what are the initial data satisfied by the measure-valued solution defined by (3.61). We would like our measure-valued solutions initially to be Dirac masses, i.e., $\nu_{(x,0)} = \delta_{u_0(x)}$. Concretely we shall assume the following:

$$\lim_{T \downarrow 0} \frac{1}{T} \int_0^T \int_{-A}^A \langle \nu_{(x,t)}, |\mathrm{Id} - u_0(x)| \rangle \, dx \, dt = 0 \tag{3.62}$$

for every A. For any Young measure $\nu_{(x,t)}$ we have the following lemma.

Lemma 3.22. *Let $\nu_{(x,t)}$ be a Young measure (with support in $[-K, K]$), and let ω_ε be a standard mollifier in x and t. Then:*

(i) *there exists a Young measure $\nu^\varepsilon_{(x,t)}$ defined by*

$$\begin{aligned}\left\langle \nu^\varepsilon_{(x,t)}, g \right\rangle &= \langle \nu_{(x,t)}, g \rangle * \omega_\varepsilon \\ &= \iint \omega_\varepsilon(x-y) \omega_\varepsilon(t-s) \langle \nu_{(y,s)}, g \rangle \, dy \, ds.\end{aligned} \tag{3.63}$$

(ii) *For all $(x,t) \in \mathbb{R} \times [0,T]$ there exist bounded measures $\partial_x \nu^\varepsilon_{(x,t)}$ and $\partial_t \nu^\varepsilon_{(x,t)}$, defined by*

$$\begin{aligned}\left\langle \partial_t \nu^\varepsilon_{(x,t)}, g \right\rangle &= \partial_t \left\langle \nu^\varepsilon_{(x,t)}, g \right\rangle, \\ \left\langle \partial_x \nu^\varepsilon_{(x,t)}, g \right\rangle &= \partial_x \left\langle \nu^\varepsilon_{(x,t)}, g \right\rangle.\end{aligned} \tag{3.64}$$

Proof. Clearly, the right-hand side of (3.63) is a bounded linear functional on $C_0(\mathbb{R})$, and hence the Riesz representation theorem guarantees the existence of $\nu^\varepsilon_{(x,t)}$. To show that $\|\nu^\varepsilon_{(x,t)}\|_{\mathcal{M}(\mathbb{R})} = 1$, let $\{\psi_n\}$ be a sequence of test functions such that

$$\langle \nu_{(x,t)}, \psi_n \rangle \to 1, \quad \text{as } n \to \infty.$$

Then for any $1 > \kappa > 0$ we can find an N such that

$$\langle \nu_{(x,t)}, \psi_n \rangle > 1 - \kappa,$$

for $n \geq N$. Thus, for such n

$$\left\langle \nu^\varepsilon_{(x,t)}, \psi_n \right\rangle \geq 1 - \kappa,$$

and therefore $\|\nu^\varepsilon_{(x,t)}\|_{\mathcal{M}(\mathbb{R})} \geq 1$. The opposite inequality is immediate, since

$$\left| \left\langle \nu^\varepsilon_{(x,t)}, \psi \right\rangle \right| \leq \left| \langle \nu_{(x,t)}, \psi \rangle \right|$$

for all test functions ψ. Therefore, $\nu^\varepsilon_{(x,t)}$ is a probability measure. Similarly, the existence of $\partial_x \nu^\varepsilon_{(x,t)}$ and $\partial_t \nu^\varepsilon_{(x,t)}$ follows by the Riesz representation theorem. Since $\nu_{(x,t)}$ is bounded, the boundedness of $\partial_x \nu^\varepsilon_{(x,t)}$ and $\partial_t \nu^\varepsilon_{(x,t)}$ follows for each fixed $\varepsilon > 0$. □

Now that we have established the existence of the "smooth approximation" to a Young measure, we can use this to prove the following lemma.

3.4. Measure-Valued Solutions

Lemma 3.23. *Assume that f is a Lipschitz continuous function and that $\nu_{(x,t)}(\lambda)$ and $\sigma_{(x,t)}(\mu)$ are measure-valued solutions with support in $[-K, K]$. Then*

$$\partial_t \langle \nu_{(x,t)} \otimes \sigma_{(x,t)}, |\lambda - \mu| \rangle + \partial_x \langle \nu_{(x,t)} \otimes \sigma_{(x,t)}, q(\lambda, \mu) \rangle \leq 0, \qquad (3.65)$$

in the distributional sense, where

$$q(\lambda, \mu) = \operatorname{sign}(\lambda - \mu)(f(\lambda) - f(\mu)),$$

and $\nu_{(x,t)} \otimes \sigma_{(x,t)}$ denotes the product measure $d\nu_{(x,t)} d\sigma_{(x,t)}$ on $\mathbb{R} \times \mathbb{R}$.

Proof. If $\nu_{(x,t)}^\varepsilon$ and $\sigma_{(x,t)}^\varepsilon$ are defined by (3.63), and $\varphi \in C_0^\infty(\mathbb{R} \times [0, T])$, then we have that

$$\iint_{\mathbb{R} \times [0,T]} \langle \nu_{(x,t)}, g \rangle \partial_t (\varphi * \omega_\varepsilon) \, dx \, dt = \iint_{\mathbb{R} \times [0,T]} \langle \nu_{(x,t)}^\varepsilon, g \rangle \partial_t \varphi \, dx \, dt$$

$$= -\iint_{\mathbb{R} \times [0,T]} \langle \partial_t \nu_{(x,t)}^\varepsilon, g \rangle \varphi \, dx \, dt,$$

and similarly,

$$\iint_{\mathbb{R} \times [0,T]} \langle \nu_{(x,t)}, g \rangle \partial_x (\varphi * \omega_\varepsilon) \, dx \, dt = -\iint_{\mathbb{R} \times [0,T]} \langle \partial_x \nu_{(x,t)}^\varepsilon, g \rangle \varphi \, dx \, dt,$$

and analogous identities also hold for $\sigma_{(x,t)}$. Therefore,

$$\langle \partial_t \nu_{(x,t)}^\varepsilon, |\lambda - \mu| \rangle + \langle \partial_x \nu_{(x,t)}^\varepsilon, q(\lambda, \mu) \rangle \leq 0, \qquad (3.66)$$

$$\langle \partial_t \sigma_{(x,t)}^\varepsilon, |\lambda - \mu| \rangle + \langle \partial_x \sigma_{(x,t)}^\varepsilon, q(\lambda, \mu) \rangle \leq 0. \qquad (3.67)$$

Next, we observe that for any continuous function g,

$$\partial_t \langle \nu_{(x,t)}^\varepsilon \otimes \sigma_{(x,t)}^\varepsilon, g(\lambda, \mu) \rangle = \int_{\mathbb{R}} \partial_t \left(\int_{\mathbb{R}} g(\lambda, \mu) \, d\nu_{(x,t)}^\varepsilon(\lambda) \right) d\sigma_{(x,t)}^\varepsilon(\mu)$$

$$+ \int_{\mathbb{R}} \partial_t \left(\int_{\mathbb{R}} g(\lambda, \mu) \, d\sigma_{(x,t)}^\varepsilon(\mu) \right) d\nu_{(x,t)}^\varepsilon(\lambda)$$

$$= \int_{\mathbb{R}} \langle \partial_t \nu_{(x,t)}^\varepsilon, g(\lambda, \mu) \rangle \, d\sigma_{(x,t)}^\varepsilon(\mu)$$

$$+ \int_{\mathbb{R}} \langle \partial_t \sigma_{(x,t)}^\varepsilon, g(\lambda, \mu) \rangle \, d\nu_{(x,t)}^\varepsilon(\lambda),$$

and an analogous equality holds for

$$\partial_x \langle \nu_{(x,t)}^\varepsilon \otimes \sigma_{(x,t)}^\varepsilon, g(\lambda, \mu) \rangle.$$

Therefore, we find that

$$\iint_{\mathbb{R} \times [0,T]} \langle \nu_{(x,t)}^{\varepsilon_1} \otimes \sigma_{(x,t)}^{\varepsilon_2}, |\lambda - \mu| \rangle \varphi_t + \langle \nu_{(x,t)}^{\varepsilon_1} \otimes \sigma_{(x,t)}^{\varepsilon_2}, q(\lambda, \mu) \rangle \varphi_x(x,t) \, dx \, dt$$

$$= -\iint_{\mathbb{R}\times[0,T]} \left(\int_{\mathbb{R}} \left\langle \partial_t \nu^{\varepsilon_1}_{(x,t)}, |\lambda - \mu| \right\rangle \right.$$
$$\left. + \left\langle \partial_x \nu^{\varepsilon_1}_{(x,t)}, q(\lambda,\mu) \right\rangle d\sigma^{\varepsilon_2}_{(x,t)}(\mu) \right) \varphi \, dx \, dt$$
$$- \iint_{\mathbb{R}\times[0,T]} \left(\int_{\mathbb{R}} \left\langle \partial_t \sigma^{\varepsilon_2}_{(x,t)}, |\lambda - \mu| \right\rangle \right.$$
$$\left. + \left\langle \partial_x \sigma^{\varepsilon_2}_{(x,t)}, q(\lambda,\mu) \right\rangle d\nu^{\varepsilon_1}_{(x,t)}(\lambda) \right) \varphi \, dx \, dt$$
$$\geq 0,$$

for any nonnegative test function φ. Now we would like to conclude the proof by sending ε_1 and ε_2 to zero. Consider the second term

$$I^{\varepsilon_1,\varepsilon_2} = \iint_{\mathbb{R}\times[0,T]} \left\langle \nu^{\varepsilon_1}_{(x,t)} \otimes \sigma^{\varepsilon_2}_{(x,t)}, q(\lambda,\mu) \right\rangle \varphi_x(x,t) \, dx \, dt$$
$$= \iint_{\mathbb{R}\times[0,T]} \iint \int \left\langle \sigma^{\varepsilon_2}_{(x,t)}, q(\lambda,\mu) \right\rangle d\nu_{(y,s)}$$
$$\times \omega_{\varepsilon_1}(x-y)\omega_{\varepsilon_1}(t-s)\varphi_x(x,t) \, dy \, ds \, dx \, dt.$$

Since

$$\iint \int \left\langle \sigma^{\varepsilon_2}_{(x,t)}, q(\lambda,\mu) \right\rangle d\nu_{(y,s)} \omega_{\varepsilon_1}(x-y)\omega_{\varepsilon_1}(t-s)\varphi_x(x,t) \, dy \, ds$$
$$\to \int \left\langle \sigma^{\varepsilon_2}_{(x,t)}, q(\lambda,\mu) \right\rangle d\nu_{(x,t)} \varphi_x(x,t) < \infty$$

for almost all (x,t) as $\varepsilon_1 \to 0$, we can use the Lebesgue bounded convergence theorem to conclude that

$$\lim_{\varepsilon_1 \to 0} I^{\varepsilon_1,\varepsilon_2} = \iint_{\mathbb{R}\times[0,T]} \left\langle \nu_{(x,t)} \otimes \sigma^{\varepsilon_2}_{(x,t)}, q(\lambda,\mu) \right\rangle \varphi_x(x,t) \, dx \, dt.$$

We can apply this argument once more for ε_2, obtaining

$$\lim_{\varepsilon_2 \to 0} \lim_{\varepsilon_1 \to 0} I^{\varepsilon_1,\varepsilon_2} = \iint_{\mathbb{R}\times[0,T]} \left\langle \nu_{(x,t)} \otimes \sigma_{(x,t)}, q(\lambda,\mu) \right\rangle \varphi_x(x,t) \, dx \, dt. \quad (3.68)$$

Similarly, we obtain

$$\lim_{\varepsilon_2 \to 0} \lim_{\varepsilon_1 \to 0} \iint_{\mathbb{R}\times[0,T]} \left\langle \nu^{\varepsilon_1}_{(x,t)} \otimes \sigma^{\varepsilon_2}_{(x,t)}, |\lambda-\mu| \right\rangle \varphi_t(x,t) \, dx \, dt$$
$$= \iint_{\mathbb{R}\times[0,T]} \left\langle \nu_{(x,t)} \otimes \sigma_{(x,t)}, |\lambda-\mu| \right\rangle \varphi_t(x,t) \, dx \, dt.$$
(3.69)

This concludes the proof of the lemma. $\square$

Let $\{u_\varepsilon\}$ and $\{v_\varepsilon\}$ be the sequences associated with $\nu_{(x,t)}$ and $\sigma_{(x,t)}$, and assume that for $t \leq T$, the support of $u_\varepsilon(\cdot,t)$ and $v_\varepsilon(\cdot,t)$ is contained in a finite interval I. Then both $u_\varepsilon(\cdot,t)$ and $v_\varepsilon(\cdot,t)$ are in $L^1(\mathbb{R})$ uniformly

in ε. This means that both
$$\langle \nu_{(x,t)}, |\lambda| \rangle \quad \text{and} \quad \langle \sigma_{(x,t)}, |\lambda| \rangle$$
are in $L^1(\mathbb{R})$ for almost all t. Using this observation, and the preceding lemma, Lemma 3.23, we can continue. Define
$$\phi_\varepsilon(t) = \int_0^t \left(\omega_\varepsilon(s-t_1) - \omega_\varepsilon(s-t_2)\right) ds,$$
where $t_2 > t_1 > 0$ and ω_ε is the usual mollifier. Also define
$$\psi_n(x) = \begin{cases} 1 & \text{for } |x| \leq n, \\ 2(1 - x/(2n)) & \text{for } n < |x| \leq 2n, \\ 0 & \text{otherwise,} \end{cases}$$
and set $\psi_{\varepsilon,n} = \psi_n * \omega_\varepsilon(x)$. Hence
$$\varphi(x,t) = \phi_\varepsilon(t)\psi_{\varepsilon,n}(x)$$
is an admissible test function. Furthermore, $|\psi'_{\varepsilon,n}| \leq 1/n$, and $\phi_\varepsilon(t)$ tends to the characteristic function of the interval $[t_1, t_2]$ as $\varepsilon \to 0$. Therefore,
$$-\lim_{\varepsilon \to 0} \iint_{\mathbb{R} \times [0,T]} \Big[\langle \nu_{(x,t)} \otimes \sigma_{(x,t)}, |\lambda - \mu|\rangle \varphi_t$$
$$+ \langle \nu_{(x,t)} \otimes \sigma_{(x,t)}, q(\lambda, \mu)\rangle \varphi_x\Big] dx\, dt \leq 0.$$

Set
$$A_n(t) = \int_\mathbb{R} \langle \nu_{(x,t)} \otimes \sigma_{(x,t)}, |\lambda - \mu|\rangle \psi_n(x)\, dx.$$

Using this definition we find that
$$A_n(t_2) - A_n(t_1) \leq \int_{t_1}^{t_2} \int_\mathbb{R} \langle \nu_{(x,t)} \otimes \sigma_{(x,t)}, |\lambda - \mu|\rangle |\psi'_n(x)|\, dx\, dt. \quad (3.70)$$

The right-hand side of this is bounded by
$$\|f\|_{\text{Lip}} \frac{1}{n} \left(\|\langle \nu_{(x,t)}, |\lambda|\rangle\|_{L^1(\mathbb{R})} + \|\langle \sigma_{(x,t)}, |\mu|\rangle\|_{L^1(\mathbb{R})}\right) \to 0$$
as $n \to \infty$. Since $\nu_{(x,t)}$ and $\sigma_{(x,t)}$ are probability measures, for almost all t, the set
$$\{x \mid \langle \nu_{(x,t)}, 1\rangle \neq 1 \quad \text{and} \quad \langle \sigma_{(x,t)}, 1\rangle \neq 1\}$$
has zero Lebesgue measure. Therefore, for almost all t,
$$A_n(t) \leq \int_\mathbb{R} \langle \nu_{(x,t)} \otimes \sigma_{(x,t)}, |\lambda - u_0(x)| + |\mu - u_0(x)|\rangle\, dx$$
$$= \int_\mathbb{R} \langle \nu_{(x,t)}, |\lambda - u_0(x)|\rangle\, dx + \int_\mathbb{R} \langle \sigma_{(x,t)}, |\mu - u_0(x)|\rangle\, dx.$$

Integrating (3.70) with respect to t_1 from 0 to T, then dividing by T and sending T to 0, using (3.62), and finally sending $n \to \infty$, we find that

$$\iint_{\mathbb{R}\times\mathbb{R}} |\lambda - \mu|\, d\nu_{(x,t)}\, d\sigma_{(x,t)} = 0, \quad \text{for } (x,t) \notin E, \tag{3.71}$$

where the Lebesgue measure of the (exceptional) set E is zero. Suppose now that for $(x,t) \notin E$ there is a $\bar\lambda$ in the support of $\nu_{(x,t)}$ and a $\bar\mu$ in the support of $\sigma_{(x,t)}$ and $\bar\lambda \neq \bar\mu$. Then we can find positive functions g and h such that

$$0 \leq g \leq 1, \quad 0 \leq h \leq 1,$$

and $\operatorname{supp}(g) \cap \operatorname{supp}(h) = \emptyset$. Furthermore,

$$\langle \nu_{(x,t)}, g\rangle > 0 \quad \text{and} \quad \langle \sigma_{(x,t)}, h\rangle > 0.$$

Thus

$$0 < \iint_{\mathbb{R}\times\mathbb{R}} g(\lambda)h(\mu)\, d\nu_{(x,t)}\, d\sigma_{(x,t)}$$

$$\leq \sup_{\lambda,\mu} \left|\frac{g(\lambda)h(\mu)}{\lambda - \mu}\right| \iint_{\mathbb{R}\times\mathbb{R}} |\lambda - \mu|\, d\nu_{(x,t)}\, d\sigma_{(x,t)} = 0.$$

This contradiction shows that both $\nu_{(x,t)}$ and $\sigma_{(x,t)}$ are unit point measures with support at a common point. Precisely, we have proved the following theorem:

Theorem 3.24. *Suppose that $\nu_{(x,t)}$ and $\sigma_{(x,t)}$ are measure-valued entropy solutions to the conservation law*

$$u_t + f(u)_x = 0,$$

and that both $\nu_{(x,t)}$ and $\sigma_{(x,t)}$ satisfy the initial condition (3.62), and that $\langle \nu_{(x,t)}, |\lambda|\rangle$ and $\langle \sigma_{(x,t)}, |\mu|\rangle$ are in $L^\infty([0,T]; L^1(\mathbb{R}))$. Then there exists a function $u \in L^\infty([0,T]; L^1(\mathbb{R})) \cap L^\infty(\mathbb{R} \times [0,T])$ such that

$$\nu_{(x,t)} = \sigma_{(x,t)} = \delta_{u(x,t)}, \quad \text{for almost all } (x,t).$$

In order to avoid checking (3.62) directly, we can use the following lemma.

Lemma 3.25. *Let $\nu_{(x,t)}$ be a probability measure, and assume that for all test functions $\varphi(x)$ we have*

$$\lim_{\tau\to 0^+} \frac{1}{\tau} \int_0^\tau \int \langle \nu_{(x,t)}, \operatorname{Id}\rangle \varphi(x)\, dx\, dt = \int u_0(x)\varphi(x)\, dx, \tag{3.72}$$

and that for all nonnegative $\varphi(x)$ and for at least one strictly convex continuous function η,

$$\limsup_{\tau\to 0^+} \frac{1}{\tau} \int_0^\tau \int \langle \nu_{(x,t)}, \eta\rangle \varphi(x)\, dx\, dt \leq \int \eta(u_0(x))\varphi(x)\, dx. \tag{3.73}$$

Then (3.62) holds.

3.4. Measure-Valued Solutions

Proof. We shall prove

$$\lim_{\tau \to 0+} \frac{1}{\tau} \int_0^\tau \int_{-A}^A \langle \nu_{(x,t)}, (\mathrm{Id} - u_0(x))^+ \rangle \, dx \, dt = 0, \tag{3.74}$$

from which the desired result will follow from (3.72) and the identity

$$|\lambda - u_0(x)| = 2(\lambda - u_0(x))^+ - (\lambda - u_0(x)),$$

where $a^+ = \max\{a, 0\}$ denotes the positive part of a. To get started, we write η'_+ for the right-hand derivative of η. It exists by virtue of the convexity of η; moreover,

$$\eta(\lambda) \geq \eta(y) + \eta'_+(y)(\lambda - y)$$

for all λ. Whenever $\varepsilon > 0$, write

$$\zeta(y, \varepsilon) = \frac{\eta(y + \varepsilon) - \eta(y)}{\varepsilon} - \eta'_+(y).$$

Since η is *strictly* convex, $\zeta(y, \varepsilon) > 0$, and this quantity is an increasing function of ε. In particular, if $\lambda > y + \varepsilon$, then $\zeta(y, \lambda - y) > \zeta(y, \varepsilon)$, or

$$\eta(\lambda) > \eta(y) + \eta'_+(y)(\lambda - y) + \zeta(y, \varepsilon)(\lambda - y).$$

In *every* case, then,

$$\eta(\lambda) > \eta(y) + \eta'_+(y)(\lambda - y) + \zeta(y, \varepsilon)\big((\lambda - y)^+ - \varepsilon\big). \tag{3.75}$$

On the other hand, whenever $y < \lambda < y + \varepsilon$, then $\zeta(y, \lambda - y) > \zeta(y, \varepsilon)$, so

$$\eta(\lambda) < \eta(y) + \eta'_+(y)(\lambda - y) + \varepsilon \zeta(y, \varepsilon) \qquad (y \leq \lambda < y + \varepsilon). \tag{3.76}$$

Let us now assume that $\varphi \geq 0$ is such that

$$\varphi(x) \neq 0 \Rightarrow y \leq u_0(x) < y + \varepsilon. \tag{3.77}$$

We use (3.75) on the left-hand side and (3.76) on the right-hand side of (3.73), and get

$$\limsup_{\tau \to 0+} \frac{1}{\tau} \int_0^\tau \int_{\mathbb{R}} \langle \nu_{(x,t)}, [\eta(y) + \eta'_+(y)(\mathrm{Id} - y) \\ + \zeta(y, \varepsilon)\big((\mathrm{Id} - y)^+ - \varepsilon\big)] \rangle \varphi(x) \, dx \, dt \\ \leq \int_{\mathbb{R}} [\eta(y) + \eta'_+(y)(u(x_0) - y) + \varepsilon \zeta(y, \varepsilon)] \varphi(x) \, dx.$$

Here, thanks to (3.72) and the fact that $\nu_{(x,t)}$ is a probability measure, all the terms not involving $\zeta(y, \varepsilon)$ cancel, and then we can divide by $\zeta(y, \varepsilon) \neq 0$ to arrive at

$$\limsup_{\tau \to 0+} \frac{1}{\tau} \int_0^\tau \int_{\mathbb{R}} \langle \nu_{(x,t)}, (\mathrm{Id} - y)^+ \rangle \varphi(x) \, dx \, dt \leq 2\varepsilon \int_{\mathbb{R}} \varphi(x) \, dx.$$

Now, remembering (3.77) we see that whenever $\varphi(x) \neq 0$ we have $(\lambda - y)^+ \leq (\lambda - u_0(x))^+ + \varepsilon$, so the above implies

$$\limsup_{\tau \to 0+} \frac{1}{\tau} \int_0^\tau \int_{\mathbb{R}} \langle \nu_{(x,t)}, (\mathrm{Id} - u_0(x))^+ \rangle \varphi(x) \, dx \, dt \leq 3\varepsilon \int_{\mathbb{R}} \varphi(x) \, dx$$

whenever (3.77) holds.

It remains only to divide up the common support $[-M, M]$ of all the measures $\nu_{(x,t)}$, writing $y_i = -M + i\varepsilon$ for $i = 0, 1, \ldots, N-1$, where $\varepsilon = 2M/N$; let φ_i be the characteristic function of $[-A, A] \cap u_0^{-1}([y_i, y_i + \varepsilon))$; and add together the above inequality for each i to arrive at

$$\limsup_{\tau \to 0+} \frac{1}{\tau} \int_0^\tau \int_{-A}^A \langle \nu_{(x,t)}, (\mathrm{Id} - u_0(x))^+ \rangle \varphi(x) \, dx \, dt \leq 3\varepsilon \int_{-A}^A \varphi(x) \, dx.$$

Since ε can be made arbitrarily small, (3.74) follows, and the proof is complete. $\square$

Remark 3.26. We *cannot* conclude that[3]

$$\lim_{\tau \to 0+} \frac{1}{\tau} \int_0^\tau \int_{\mathbb{R}} \langle \nu_{(x,t)}, |\mathrm{Id} - u_0(x)| \rangle \, dx \, dt = 0 \tag{3.78}$$

from the present assumptions. Here is an example to show this.

Let $\nu_{(x,t)} = \mu_{\gamma(x,t)}$, where $\mu_\beta = \frac{1}{2}(\delta_{-\beta} + \delta_\beta)$ and γ is a continuous, nonnegative function with $\gamma(x, 0) = 0$. Let $u_0(x) = 0$ and $\eta(y) = y^2$. Then (3.72) holds trivially, and (3.73) becomes

$$\limsup_{\tau \to 0+} \frac{1}{\tau} \int_0^\tau \int_{\mathbb{R}} \gamma(x,t)^2 \varphi(x) \, dx \, dt = 0,$$

which is also true due to the stated assumptions on γ.

The desired conclusion (3.78), however, is now

$$\limsup_{\tau \to 0+} \frac{1}{\tau} \int_0^\tau \int_{\mathbb{R}} \gamma(x,t) \, dx \, dt = 0.$$

But the simple choice

$$\gamma(x,t) = te^{-(xt)^2}$$

yields

$$\limsup_{\tau \to 0+} \frac{1}{\tau} \int_0^\tau \int_{\mathbb{R}} \gamma(x,t) \, dx \, dt = \sqrt{\pi}.$$

We shall now describe a framework that allows one to prove convergence of a sequence of approximations without proving that the method is TV stable. Unfortunately, the application of this method to concrete examples,

---

[3] Where the integral over the compact interval $[-A, A]$ in (3.62) has been replaced by an integral over the entire real line.

while not very difficult, involves quite large calculations, and will be omitted here. Readers are encouraged to try their hands at it themselves.

We give one application of these concepts. The setting is as follows. Let U^n be computed from a conservative and consistent scheme, and assume uniform boundedness of U^n. Young's theorem states that there exists a family of probability measures $\nu_{(x,t)}$ such that $g(U^n) \overset{*}{\rightharpoonup} \langle \nu_{(x,t)}, g \rangle$ for Lipschitz continuous functions g. We assume that the CFL condition, $\lambda \sup_u |f'(u)| \leq 1$, is satisfied. The next theorem states conditions, strictly weaker than TVD, for which we prove that the limit measure $\nu_{(x,t)}$ is a measure-valued solution of the scalar conservation law.

Theorem 3.27. *Assume that the sequence $\{U^n\}$ is the result of a conservative, consistent method, and define $u_{\Delta t}$ as in (3.24). Assume that $u_{\Delta t}$ is uniformly bounded in $L^\infty(\mathbb{R} \times [0,T])$, $T = n\Delta t$. Let $\nu_{(x,t)}$ be the Young measure associated with $u_{\Delta t}$, and assume that U_j^n satisfies the estimate*

$$(\Delta x)^\beta \sum_{n=0}^N \sum_j |U_{j+1}^n - U_j^n| \Delta t \leq C(T), \qquad (3.79)$$

for some $\beta \in [0,1)$ and some constant $C(T)$. Then $\nu_{(x,t)}$ is a measure-valued solution to (3.1).

Furthermore, let (η, q) be a strictly convex entropy pair, and let Q be a numerical entropy flux consistent with q. Write $\eta_j^n = \eta(U_j^n)$ and $Q_j^n = Q(U_j^n)$. Assume that

$$\frac{1}{\Delta t} \left(\eta_j^{n+1} - \eta_j^n \right) + \frac{1}{\Delta x} \left(Q_j^n - Q_{j-1}^n \right) \leq R_j^n \qquad (3.80)$$

for all n and j, where R_j^n satisfies

$$\lim_{\Delta t \to 0} \sum_{n=0}^N \sum_j \varphi(j\Delta x, n\Delta t) R_j^n \Delta x \Delta t = 0 \qquad (3.81)$$

for all nonnegative $\varphi \in C_0^1$. Then $\nu_{(x,t)}$ is a measure-valued solution compatible with (η, q), and the initial data is assumed in the sense of (3.72), (3.73). If (3.80) and (3.81) hold for all entropy pairs (η, q), $\nu_{(x,t)}$ is a measure-valued entropy solution to (3.1).

Remark 3.28. For $\beta = 0$, (3.79) is the standard TV estimate, while for $\beta > 0$, (3.79) is genuinely weaker than a TV estimate.

Proof. We start by proving the first statement in the theorem, assuming (3.79). As before, we obtain (3.25) by rearranging. For simplicity, we now write $F_j^n = F(U^n; j)$, $f_j^n = f(U_j^n)$, and observe that $F_j^n = f_j^n + (F_j^n - f_j^n)$, getting

$$\iint \left(u_{\Delta t} \Delta_t^{n,j} \varphi + f(u_{\Delta t}) \Delta_x^{n,j} \varphi \right) dx \, dt = \sum_{j,n} \Delta_x^{n,j} \varphi \left(F_j^n - f_j^n \right) \Delta t \Delta x. \qquad (3.82)$$

Here we use the notation

$$u_{\Delta t} = U_j^n \quad \text{for} \quad (x,t) \in [j\Delta x, (j+1)\Delta x) \times [n\Delta t, (n+1)\Delta t),$$

and

$$\Delta_t^{n,j}\varphi = \frac{1}{\Delta t}\left(\varphi(j\Delta x, (n+1)\Delta t) - \varphi(j\Delta x, n\Delta t)\right),$$

$$\Delta_x^{n,j}\varphi = \frac{1}{\Delta x}\left(\varphi((j+1)\Delta x, n\Delta t) - \varphi(j\Delta x, n\Delta t)\right).$$

The first term on the left-hand side in (3.82) reads

$$\iint u_{\Delta t}\Delta_t^{n,j}\varphi \, dx \, dt$$

$$= \iint \langle \nu_{(x,t)}, \mathrm{Id}\rangle \varphi_t \, dx \, dt + \iint \left(u_{\Delta t} - \langle \nu_{(x,t)}, \mathrm{Id}\rangle\right) \varphi_t \, dx \, dt$$

$$+ \iint u_{\Delta t}\left(\Delta_t^{n,j}\varphi - \varphi_t\right) dx \, dt. \tag{3.83}$$

The third term on the right-hand side of (3.83) clearly tends to zero as Δt goes to zero. Furthermore, by definition of the Young measure $\nu_{(x,t)}$, the second term tends to zero as well. Thus the left-hand side of (3.83) approaches $\iint \langle \nu_{(x,t)}, \mathrm{Id}\rangle \varphi_t \, dx \, dt$.

One can use a similar argument for the second term on the left-hand side of (3.82) to show that the (whole) left-hand side of (3.82) tends to

$$\iint \left(\langle \nu_{(x,t)}, \mathrm{Id}\rangle \varphi_t + \langle \nu_{(x,t)}, f\rangle \varphi_x\right) dx \, dt \tag{3.84}$$

as $\Delta t \to 0$. We now study the right-hand side of (3.82). Mimicking the proof of the Lax–Wendroff theorem, we have

$$\left|F_j^n - f_j^n\right| \leq C \sum_{k=-p}^{q} \left|U_{j+k}^n - U_j^n\right|.$$

Therefore,

$$\left|\sum_{j,n} \Delta_x^{n,j}\varphi \left(F_j^n - f_j^n\right) \Delta t \Delta x\right|$$

$$\leq C\|\varphi\|_{\mathrm{Lip}}(p+q+1)\sum_{n=0}^{N}\sum_{j}\left|U_{j+1}^n - U_j^n\right|\Delta t \Delta x$$

$$\leq C\|\varphi\|_{\mathrm{Lip}}(p+q+1)(\Delta x)^{1-\beta}, \tag{3.85}$$

using the assumption (3.79). Thus the right-hand side of (3.85), and hence also of (3.82), tends to zero. Since the left-hand side of (3.82) tends to (3.84), we conclude that $\nu_{(x,t)}$ is a measure-valued solution. Using similar calculations, and (3.81), one shows that $\nu_{(x,t)}$ is also an entropy measure-valued solution.

It remains to show consistency with the initial condition, i.e., (3.72) and (3.73). Let $\varphi(x)$ be a test function, and we use the notation $\varphi(j\Delta x) = \varphi_j$. From the definition of U_j^{n+1}, after a summation by parts, we have that

$$\sum_j \varphi_j \left(U_j^{n+1} - U_j^n\right) \Delta x = \Delta t \sum_j F_j^n \Delta_x^{n,j} \varphi_{j+1} \Delta x \le \mathcal{O}(1)\Delta t,$$

since U_j^n is bounded. Remembering that $\varphi = \varphi(x)$, we get

$$\left| \sum_j \varphi_j \left(U_j^n - U_j^0\right) \Delta x \right| \le \mathcal{O}(1) n\Delta t. \qquad (3.86)$$

Let $t_1 = n_1 \Delta t$ and $t_2 = n_2 \Delta t$. Then (3.86) yields

$$\left| \frac{1}{(n_2 + 1 - n_1)\Delta t} \sum_{n=n_1}^{n_2} \sum_j \varphi_j \left(U_j^n - U_j^0\right) \Delta x \Delta t \right| \le \mathcal{O}(1) t_2,$$

which implies that the Young measure $\nu_{(x,t)}$ satisfies

$$\left| \frac{1}{t_2 - t_1} \int_{t_1}^{t_2} \varphi(x) \langle \nu_{(x,t)}, \mathrm{Id} \rangle \, dx \, dt - \int \varphi(x) u_0(x) \, dx \right| \le \mathcal{O}(1) t_2. \qquad (3.87)$$

We let $t_1 \to 0$ and set $t_2 = \tau$ in (3.87), obtaining

$$\left| \frac{1}{\tau} \int_0^\tau \int \varphi(x) \langle \nu_{(x,t)}, \mathrm{Id} \rangle \, dx \, dt - \int \varphi(x) u_0(x) \, dx \right| \le \mathcal{O}(1) \tau, \qquad (3.88)$$

which proves (3.72). Now for (3.73). We have that there exists a strictly convex entropy η for which (3.80) holds. Now let $\varphi(x)$ be a nonnegative test function. Using (3.80), and proceeding as before, we obtain

$$\left| \sum_j \left(\eta_j^n - \eta_j^0\right) \varphi_j \Delta x \right| \le \mathcal{O}(1) n\Delta t + \sum_{\ell=0}^n \sum_j R_j^\ell \varphi_j \Delta t \Delta x.$$

By using this estimate and the assumption on R_j^ℓ, (3.81), we can use the same arguments as in proving (3.88) to prove (3.73). The proof of the theorem is complete. $\qquad \square$

A trivial application of this approach is found by considering monotone schemes. Here we have seen that (3.79) holds for $\beta = 0$, and (3.80) for $R_j^n = 0$. The theorem then gives the convergence of these schemes without using Helly's theorem. However, in this case the application does not give the existence of a solution, since we must have this in order to use DiPerna's theorem. The main usefulness of the method is for schemes in several space dimensions, where TV bounds are more difficult to obtain.

3.5 Notes

The Lax–Friedrichs scheme was introduced by Lax in 1954; see [94]. Godunov discussed what has later become the Godunov scheme in 1959 as a method to study gas dynamics; see [60]. The Lax–Wendroff theorem, Theorem 3.4, was first proved in [97]. Theorem 3.8 was proved by Oleĭnik in her fundamental paper [110]; see also [130]. Several of the fundamental results concerning monotone schemes are due to Crandall and Majda [39], [38]. Theorem 3.10 is due to Harten, Hyman, and Lax; see [62].

The error analysis is based on the fundamental analysis by Kuznetsov, [89], where one also can find a short discussion of the examples we have analyzed, namely the smoothing method, the method of vanishing viscosity, as well as monotone schemes. Our presentation of the a priori estimates follows the approach due to Cockburn and Gremaud; see [31] and [32], where also applications to numerical methods are given.

The concept of measure-valued solutions is due to DiPerna, and the key result, Corollary 3.21, can be found in [45], while Lemma 3.25 is to be found in [44]. The proof of Lemma 3.25 and Remark 3.26 are due to H. Hanche-Olsen. Our presentation of the uniqueness of measure-valued solutions, Theorem 3.24, is taken mainly from Szepessy, [134]. Theorem 3.27 is due to Coquel and LeFloch, [35]; see also [36], where several extensions are discussed.

Exercises

3.1 Show that the Lax–Wendroff and the MacCormack methods are of second order.

3.2 The Engquist–Osher (or generalized upwind) method, see [46], is a conservative difference scheme with a numerical flux defined as follows:
$$F(U;j) = f^{\text{EO}}(U_j, U_{j+1}), \quad \text{where}$$
$$f^{\text{EO}}(u, v) = \int_0^u \max(f'(s), 0)\, ds + \int_0^v \min(f'(s), 0)\, ds + f(0).$$

a. Show that this method is consistent and monotone.
b. Find the order of the scheme.
c. Show that the Engquist–Osher flux f^{EO} can be written
$$f^{\text{EO}}(u, v) = \frac{1}{2}\left(f(u) + f(v) - \int_u^v |f'(s)|\, ds\right).$$
d. If $f(u) = u^2/2$, show that the numerical flux can be written
$$f^{\text{EO}}(u, v) = \frac{1}{2}\left(\max(u, 0)^2 + \min(v, 0)^2\right).$$

Generalize this simple expression to the case where $f''(u) \neq 0$ and $\lim_{|u| \to \infty} |f(u)| = \infty$.

3.3 Why does the method
$$U_j^{n+1} = U_j^n - \frac{\Delta t}{2\Delta x} \left(f\left(U_{j+1}^n\right) - f\left(U_{j-1}^n\right) \right)$$
not give a viable difference scheme?

3.4 We study a nonconservative method for Burgers' equation. Assume that $U_j^0 \in [0, 1]$ for all j. Then the characteristic speed is nonnegative, and we define
$$U_j^{n+1} = U_j^n - \lambda U_j^{n+1} \left(U_j^n - U_{j-1}^n\right), \quad n \geq 0, \quad (3.89)$$
where $\lambda = \Delta t / \Delta x$.

    a. Show that this yields a monotone method, provided that a CFL condition holds.

    b. Show that this method is consistent and determine the truncation error.

3.5 Assume that $f'(u) > 0$ and that $f''(u) \geq 2c > 0$ for all u in the range of u_0. We use the upwind method to generate approximate solutions to
$$u_t + f(u)_x = 0, \quad u(x, 0) = u_0(x); \quad (3.90)$$
i.e., we set
$$U_j^{n+1} = U_j^n - \lambda \left(f(U_j^n) - f(U_{j-1}^n) \right).$$
Set
$$V_j^n = \frac{U_j^n - U_{j-1}^n}{\Delta x}.$$

    a. Show that
$$V_j^{n+1} = \left(1 - \lambda f'(U_{j-1}^n)\right) V_j^n + \lambda f'(U_{j-1}^n) V_{j-1}^n$$
$$- \frac{\Delta t}{2} \left(f''(\eta_{j-1/2}) \left(V_j^n\right)^2 + f''(\eta_{j-3/2}) \left(V_{j-1}^n\right)^2 \right),$$
where $\eta_{j-1/2}$ is between U_j^n and U_{j-1}^n.

    b. Next, assume inductively that
$$V_j^n \leq \frac{1}{(n+2)c\Delta t}, \quad \text{for all } j,$$
and set $\hat{V}^n = \max(\max_j V_j^n, 0)$. Then show that
$$\hat{V}^{n+1} \leq \hat{V}^n - c\Delta t \left(\hat{V}^n\right)^2.$$

114   3. A Short Course in Difference Methods

    c. Use this to show that
$$\hat{V}^n \le \frac{\hat{V}^0}{1+\hat{V}^0 cn\Delta t}.$$

    d. Show that this implies that
$$U_i^n - U_j^n \le \Delta x (i-j) \frac{\hat{V}^0}{1+\hat{V}^0 cn\Delta t},$$
for $i \ge j$.

    e. Let u be the entropy solution of (3.90), and assume that $0 \le \max_x u_0'(x) = M < \infty$, show that for almost every x, y, and t we have that
$$\frac{u(x,t) - u(y,t)}{x - y} \le \frac{M}{1+cMt}. \qquad (3.91)$$
This is the Oleĭnik entropy condition for convex scalar conservation laws.

3.6 Assume that f is as in the previous exercise, and that u_0 is periodic with period p.

    a. Use uniqueness of the entropy solution to (3.90) to show that the entropy solution $u(x,t)$ is also periodic in x with period p.

    b. Then use the Oleĭnik entropy condition (3.91) to deduce that
$$\sup_x u(x,t) - \inf_x u(x,t) \le \frac{Mp}{1+cMt}.$$
Thus $\lim_{t\to\infty} u(x,t) = \bar{u}$ for some constant $\bar{u}$.

    c. Use conservation to show that
$$\bar{u} = \frac{1}{p} \int_0^p u_0(x)\, dx.$$

3.7 Assume that $g(x)$ is a continuously differentiable function with period 2π. Then we have that the Fourier representation
$$g(x) = \frac{a_0}{2} + \sum_{k=1}^{\infty} \big(a_k \cos(kx) + b_k \sin(kx)\big)$$
holds pointwise, where
$$a_0 = \frac{1}{\pi} \int_0^{2\pi} g(x)\, dx \quad \text{and} \quad \begin{cases} a_k = \dfrac{1}{2\pi} \int_0^{2\pi} g(x) \cos(kx)\, dx, \\ b_k = \dfrac{1}{2\pi} \int_0^{2\pi} g(x) \sin(kx)\, dx, \end{cases}$$
for $k \ge 1$.

    a. Use this to show that
$$g(nx) \overset{*}{\rightharpoonup} \frac{a_0}{2}.$$

b. Find a regular measure ν such that for any continuously differentiable h,

$$h(\sin(nx)) \stackrel{*}{\rightharpoonup} \int h(\lambda)\,d\nu(\lambda).$$

Thus we have found an explicit form of the Young measure associated with the sequence $\{\sin(nx)\}$.

3.8 We shall consider a scalar conservation law with a "fractal" function as the initial data. Define the set of piecewise linear functions

$$\mathcal{D} = \{\phi(x) = Ax + B \mid x \in [a,b],\ A, B \in \mathbb{R}\},$$

and the map

$$F(\phi) = \begin{cases} 2D(x-a) + \phi(a) & \text{for } x \in [a, a+L/3], \\ -D(x-a) + \phi(a) & \text{for } x \in [a+L/3, a+2L/3], \\ 2D(x-b) + \phi(b) & \text{for } x \in [a+2L/3, b], \end{cases}$$

$\phi \in \mathcal{D}$, where $L = b - a$ and $D = (\phi(b) - \phi(a))/L$. For a nonnegative integer k introduce $\chi_{j,k}$ as the characteristic function of the interval $I_{j,k} = [j/3^k, (j+1)/3^k]$, $j = 0, \ldots, 3^{k+1} - 1$. We define functions $\{v_k\}$ recursively as follows. Let

$$v_0(x) = \begin{cases} 0 & \text{for } x \leq 0, \\ x & \text{for } 0 \leq x \leq 1, \\ 1 & \text{for } 1 \leq x \leq 2, \\ 3 - x & \text{for } 2 \leq x \leq 3, \\ 0 & \text{for } 3 \leq x. \end{cases}$$

Assume that $v_{j,k}$ is linear on $I_{j,k}$ and let

$$v_k = \sum_{j=-3^k}^{3^k - 1} v_{j,k} \chi_{j,k}, \qquad (3.92)$$

and define the next function v_{k+1} by

$$v_{k+1} = \sum_{j=0}^{3^{k+1}-1} F(v_{j,k}) \chi_{j,k} = \sum_{j=0}^{3^{k+2}-1} v_{j,k+1} \chi_{j,k+1}. \qquad (3.93)$$

In the left part of Figure 3.3 we show the effect of the map F, and on the right we show $v_5(x)$ (which is piecewise linear on $3^6 = 729$ segments).

a. Show that the sequence $\{v_k\}_{k \geq 1}$ is a Cauchy sequence in the supremum norm, and hence we can define a continuous function v by setting

$$v(x) = \lim_{k \to \infty} v_k(x).$$

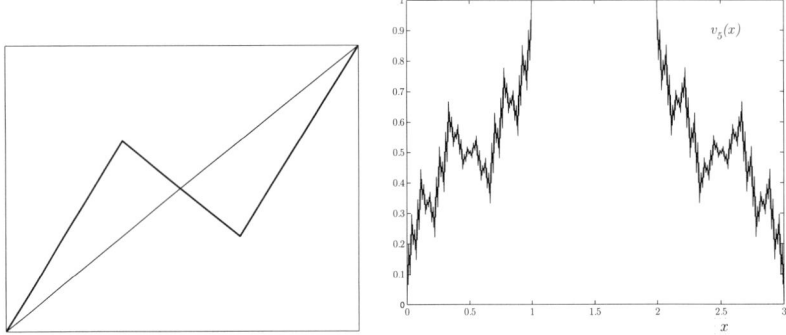

Figure 3.3. Left: the construction of $F(\phi)$ from ϕ. Right: $v_5(x)$.

b. Show that v is not of bounded variation, and determine the total variation of v_k.

c. Show that
$$v(j/3^k) = v_k(j/3^k),$$
for any integers $j = 0, \ldots, 3^{k+1}$, $k \in \mathbb{N}$.

d. Assume that f is a C^1 function on $[0, 1]$ with $0 \leq f'(u) \leq 1$. We are interested in solving the conservation law
$$u_t + f(u)_x = 0, \quad u_0(x) = v(x).$$
To this end we shall use the upwind scheme defined by (3.8), with $\Delta t = \Delta x = 1/3^k$, and
$$U_j^0 = v(j\Delta x).$$
Show that $u_{\Delta t}(x, t)$ converges to an entropy solution of the conservation law above.

4
Multidimensional Scalar Conservation Laws

> Just send me the theorems,
> then I shall find the proofs.[1]
>
> *Chrysippus told Cleanthes, 3th century* B.C.

Our analysis has so far been confined to scalar conservation laws in one dimension. Clearly, the multidimensional case is considerably more important. Luckily enough, the analysis in one dimension can be carried over to higher dimensions by essentially treating each dimension separately. This technique is called *dimensional splitting*. The final results are very much the natural generalizations one would expect.

The same splitting techniques of dividing complicated differential equations into several simpler parts, can in fact be used to handle other problems. These methods are generally denoted *operator splitting methods* or *fractional steps methods*.

4.1 Dimensional Splitting Methods

We will in this section show how one can analyze scalar multidimensional conservation laws by dimensional splitting, which amounts to solving one

---

[1] Lucky guy! Paraphrased from Diogenes Laertius, *Lives of Eminent Philosophers*, c. A.D. 200.

space direction at a time. To be more concrete, let us consider the two-dimensional conservation law

$$u_t + f(u)_x + g(u)_y = 0, \qquad u(x,y,0) = u_0(x,y). \tag{4.1}$$

If we let $S_t^{f,x} u_0$ denote the solution of

$$v_t + f(v)_x = 0, \qquad v(x,y,0) = u_0(x,y)$$

(where y is a passive parameter), and similarly let $S_t^{g,y} u_0$ denote the solution of

$$w_t + g(w)_y = 0, \qquad w(x,y,0) = u_0(x,y)$$

(x is a parameter), then the idea of dimensional splitting is to approximate the solution of (4.1) as follows:

$$u(x,y,n\Delta t) \approx \left[S_{\Delta t}^{g,y} \circ S_{\Delta t}^{f,x} \right]^n u_0. \tag{4.2}$$

◇ **Example 4.1 (A single discontinuity).**

We first show how this works on a concrete example. Let

$$f(u) = g(u) = \frac{1}{2} u^2$$

and

$$u_0(x,y) = \begin{cases} u_l & \text{for } x < y, \\ u_r & \text{for } x \geq y, \end{cases}$$

with $u_r > u_l$. The solution in the x-direction for fixed y gives a rarefaction wave, the left and right part moving with speeds u_l and u_r, respectively. With a quadratic flux, the rarefaction wave is a linear interpolation between the left and right states. Thus

$$u^{1/2} := S_{\Delta t}^{f,x} u_0 = \begin{cases} u_l & \text{for } x < y + u_l \Delta t, \\ (x-y)/\Delta t & \text{for } y + u_l \Delta t < x < y + u_r \Delta t, \\ u_r & \text{for } x > y + u_r \Delta t. \end{cases}$$

The solution in the y-direction for fixed x with initial state $u^{1/2}$ will exhibit a focusing of characteristics. The left state, which now equals u_r, will move with speed given by the derivative of the flux function, in this case u_r, and hence overtake the right state, given by u_l, which moves with smaller speed, namely u_l. The characteristics interact at a time t given by

$$u_r t + x - u_r \Delta t = u_l t + x - u_l \Delta t,$$

or $t = \Delta t$. At that time we are back to the original Riemann problem between states u_l and u_r at the point $x = y$. Thus

$$u^1 := S_{\Delta t}^{g,y} u^{1/2} = u_0.$$

Another application of $S_{\Delta t}^{f,x}$ will of course give
$$u^{3/2} := S_{\Delta t}^{f,x} u^1 = u^{1/2}.$$
So we have that $u^n = u_0$ for all $n \in \mathbb{N}$. Introducing coordinates
$$\xi = \frac{1}{\sqrt{2}}(x+y), \qquad \eta = \frac{1}{\sqrt{2}}(x-y),$$
the equation transforms into
$$u_t + \left(\frac{1}{\sqrt{2}} u^2\right)_\xi = 0, \quad u(\xi, \eta, 0) = \begin{cases} u_l & \text{for } \eta \leq 0, \\ u_r & \text{for } \eta > 0. \end{cases}$$
We see that $u(x, y, t) = u_0(x, y)$, and consequently $\lim_{\Delta t \to 0} u^n = u_0$ (where we keep $n\Delta t = t$ fixed). Thus the dimensional splitting procedure produces approximate solutions converging to the right solution in this case. ◇

We will state all results for the general case of arbitrary dimension, while proofs will be carried out in two dimensions only, to keep the notation simple. We first need to define precisely what is meant by a weak entropy solution of the initial value problem
$$u_t + \sum_{j=1}^m f_j(u)_{x_j} = 0, \quad u(x_1, \ldots, x_m, 0) = u_0(x_1, \ldots, x_m). \tag{4.3}$$

Here we adopt the *Kružkov entropy condition* from Chapter 2, and say that u is a (weak) Kružkov entropy solution of (4.3) for time $[0, T]$ if u is a bounded function that satisfies[2]

$$\int_0^T \int_{\mathbb{R}^m} \left(|u-k|\varphi_t + \text{sign}(u-k) \sum_{j=1}^m (f_j(u) - f_j(k)) \varphi_{x_j} \right) dx_1 \cdots dx_m \, dt$$
$$+ \int_{\mathbb{R}^m} \left(\varphi|_{t=0} |u_0 - k| - (|u-k|\varphi)|_{t=T} \right) dx_1 \cdots dx_m \geq 0, \tag{4.4}$$

for all constants $k \in \mathbb{R}$ and all nonnegative test functions $\varphi \in C_0^\infty(\mathbb{R}^m \times [0, T])$. It certainly follows as in the one-dimensional case that u is a weak solution, i.e.,

$$\int_0^\infty \int_{\mathbb{R}^m} \left(u\varphi_t + \sum_{j=1}^m f_j(u)\varphi_{x_j} \right) dx_1 \cdots dx_m \, dt$$
$$+ \int_{\mathbb{R}^m} \varphi|_{t=0} u_0 \, dx_1 \cdots dx_m = 0, \tag{4.5}$$

for all test functions $\varphi \in C_0^\infty(\mathbb{R}^m \times [0, \infty))$.

---

[2] If we want a solution for all time we disregard the last term in (4.4) and integrate t over $[0, \infty)$.

Our analysis aims at two different goals. We first show that the dimensional splitting indeed produces a sequence of functions that converges to a solution of the multidimensional equation (4.3). Our discussion will here be based on the more or less standard argument using Kolmogorov's compactness theorem. The argument is fairly short. To obtain stability in the multidimensional case in the sense of Theorem 2.13, we show that dimensional splitting preserves this stability. Furthermore, we show how one can use front tracking as our solution operator in one dimension in combination with dimensional splitting. Finally, we determine the appropriate convergence rate of this procedure. This analysis strongly uses Kuznetsov's theory from Section 3.2, but matters are more complicated and technical than in one dimension.

We shall now show that dimensional splitting produces a sequence that converges to the entropy solution u of (4.3); that is, the limit u should satisfy (4.4). As promised, our analysis will be carried out in the two-dimensional case only, i.e., for equation (4.1). Assume that u_0 is a compactly supported function in $L^\infty(\mathbb{R}^2) \cap BV(\mathbb{R}^2)$ (consult Appendix A for a definition of $BV(\mathbb{R}^2)$). Let $t_n = n\Delta t$ and $t_{n+1/2} = \left(n + \frac{1}{2}\right)\Delta t$. Define

$$u^0 = u_0, \qquad u^{n+1/2} = S_{\Delta t}^{f,x} u^n, \qquad u^{n+1} = S_{\Delta t}^{g,y} u^{n+1/2}, \qquad (4.6)$$

for $n \in \mathbb{N}_0$. We shall also be needing an approximate solution for $t \neq t_n$. We want the approximation to be an exact solution to a one-dimensional conservation law in each interval $[t_j, t_{j+1/2}]$, $j = k/2$ and $k \in \mathbb{N}_0$. The way to do this is to make "time go twice as fast" in each such interval; i.e., let $u_{\Delta t}$ be defined by[3]

$$u_{\Delta t}(x,t) = \begin{cases} S_{2(t-t_n)}^{f,x} u^n & \text{for } t_n \leq t \leq t_{n+1/2}, \\ S_{2(t-t_{n+1/2})}^{g,y} u^{n+1/2} & \text{for } t_{n+1/2} \leq t \leq t_{n+1}. \end{cases} \qquad (4.7)$$

We first show that the sequence $\{u_{\Delta t}\}$ is compact. Since both operators $S^{f,x}$ and $S^{g,y}$ take L^∞ into itself, $u_{\Delta t}$ will be uniformly bounded, i.e.,

$$\|u_{\Delta t}\|_\infty \leq M \qquad (4.8)$$

for some constant M determined by the initial data and not depending on Δt.

For the solution constructed from dimensional splitting we have

$$\text{T.V.}_{\cdot x, y}\left(u^{n+1/2}\right)$$
$$= \int \text{T.V.}_{\cdot x}\left(S_{\Delta t}^{f,x} u^n\right) dy + \int \text{T.V.}_{\cdot y}\left(S_{\Delta t}^{f,x} u^n\right) dx$$
$$\leq \int \text{T.V.}_{\cdot x}\left(u^n\right) dy + \int \lim_{h \to 0} \frac{1}{h} \int \left|u^{n+1/2}(x, y+h) - u^{n+1/2}(x,y)\right| dy\, dx$$

---

[3] We will keep the ratio $\lambda = \Delta t / \Delta x$ fixed, and thus we index with only Δt.

$$
\begin{aligned}
&= \int \text{T.V.}_{\cdot x}\left(u^n\right) dy + \lim_{h \to 0} \frac{1}{h} \iint \left|u^{n+1/2}(x, y+h) - u^{n+1/2}(x, y)\right| dx\, dy \\
&\leq \int \text{T.V.}_{\cdot x}\left(u^n\right) dy + \lim_{h \to 0} \frac{1}{h} \iint \left|u^n(x, y+h) - u^n(x, y)\right| dx\, dy \\
&= \int \text{T.V.}_{\cdot x}\left(u^n\right) dy + \int \lim_{h \to 0} \frac{1}{h} \int \left|u^n(x, y+h) - u^n(x, y)\right| dy\, dx \\
&= \text{T.V.}_{\cdot x, y}\left(u^n\right),
\end{aligned}
\tag{4.9}
$$

using first the TVD property of $S^{f,x}$ and Lemma A.1, and subsequently the L^1-stability in the x-direction. The interchange of integrals and limits is justified using Lebesgue's dominated convergence. Similarly, $\text{T.V.}_{\cdot x,y}\left(u^{n+1}\right) \leq \text{T.V.}_{\cdot x,y}\left(u^{n+1/2}\right)$, and thus

$$\text{T.V.}_{\cdot x, y}\left(u^n\right) \leq \text{T.V.}_{\cdot x, y}\left(u_0\right)$$

follows by induction. This extends to

$$\text{T.V.}_{\cdot x, y}\left(u_{\Delta t}\right) \leq \text{T.V.}_{\cdot x, y}\left(u_0\right). \tag{4.10}$$

We now want to establish Lipschitz continuity in time of the L^1-norm, i.e.,

$$\|u_{\Delta t}(t) - u_{\Delta t}(s)\|_1 \leq C\,|t - s| \tag{4.11}$$

for some constant C. By repeated use of the triangle inequality it suffices to show that

$$
\begin{aligned}
\|u_{\Delta t}((n+1)\Delta t) - u_{\Delta t}(n\Delta t)\|_1 &\leq \left\|u^{n+1} - u^{n+1/2}\right\|_1 + \left\|u^{n+1/2} - u^n\right\|_1 \\
&= \left\|S^{f,x}_{\Delta t} u^n - u^n\right\|_1 \\
&\quad + \left\|S^{g,y}_{\Delta t} u^{n+1/2} - u^{n+1/2}\right\|_1 \\
&\leq C \Delta t.
\end{aligned}
\tag{4.12}
$$

To this end, assume that

$$u_0 = 0 \text{ for } |x| \geq N \text{ or } |y| \geq N. \tag{4.13}$$

Then

$$
\begin{aligned}
\left\|S^{f,x}_{\Delta t} u^n - u^n\right\|_1 &= \iint \left|S^{f,x}_{\Delta t} u^n(x, y) - u^n(x, y)\right| dx\, dy \\
&\leq \int_{|y| \leq N + \|g\|_{\text{Lip}} \Delta t} M \Delta t\, dy \\
&\leq C \Delta t,
\end{aligned}
$$

which proves (4.12) and consequently (4.11).

Using Theorem A.8 we conclude the existence of a convergent subsequence, also labeled $\{u_{\Delta t}\}$, and set $u = \lim_{\Delta t \to 0} u_{\Delta t}$.

Let $\phi = \phi(x, y, t)$ be a nonnegative test function, and define φ by $\varphi(x, y, t) = \phi(x, y, t/2 + t_n)$. By defining $\tau = 2(t - n\Delta t)$ we have that for each y, the function $u_{\Delta t}$ is a weak solution in x on the strip $t \in [t_n, t_{n+1/2}]$ satisfying the inequality

$$\int_0^{\Delta t}\int \left(|u_{\Delta t} - k|\varphi_\tau + q^f(u_{\Delta t}, k)\varphi_x\right) dx\, d\tau \qquad (4.14)$$
$$\geq \int \left|u^{n+1/2} - k\right| \varphi(\Delta t)\, dx - \int |u^n - k|\varphi(0)\, dx,$$

for all constants k. Here $q^f(u,k) = \operatorname{sign}(u-k)(f(u)-f(k))$. Changing back to the t variable, we find that

$$2\int_{t_n}^{t_{n+1/2}}\int \left(\frac{1}{2}|u_{\Delta t} - k|\phi_t + q^f(u_{\Delta t}, k)\phi_x\right) dx\, dt$$
$$\geq \int \left|u^{n+1/2} - k\right| \phi(t_{n+1/2})\, dx - \int |u^n - k|\phi(t_n)\, dx. \quad (4.15)$$

Similarly,

$$2\int_{t_{n+1/2}}^{t_{n+1}}\int \left(\frac{1}{2}|u_{\Delta t} - k|\phi_t + q^g(u_{\Delta t}, k)\phi_y\right) dy\, dt$$
$$\geq \int \left|u^{n+1} - k\right|\phi(t_{n+1})\, dy - \int \left|u^{n+1/2} - k\right|\phi(t_{n+1/2})\, dy. \quad (4.16)$$

Here q^g is defined similarly to q^f, using g instead of f. Integrating (4.15) over y and (4.16) over x and adding the two results and summing over n, we obtain

$$2\int_0^T \iint \Bigg(\frac{1}{2}|u_{\Delta t} - k|\phi_t + \sum_n \chi_n q^f(u_{\Delta t}, k)\phi_x$$
$$+ \sum_n \tilde{\chi}_n q^g(u_{\Delta t}, k)\phi_y\Bigg) dx\, dy\, dt$$
$$\geq \iint (|u_{\Delta t} - k|\phi)|_{t=T}\, dx\, dy - \iint |u_0 - k|\phi(0)\, dx\, dy,$$

where χ_n and $\tilde{\chi}_n$ denote the characteristic functions of the strips $t_n \leq t \leq t_{n+1/2}$ and $t_{n+1/2} \leq t \leq t_{n+1}$, respectively. As Δt tends to zero, it follows that

$$\sum_n \chi_n \overset{*}{\rightharpoonup} \frac{1}{2}, \quad \sum_n \tilde{\chi}_n \overset{*}{\rightharpoonup} \frac{1}{2}.$$

Specifically, for continuous functions ψ of compact support we see that

$$\sum_n \int_0^\infty \chi_n \psi \, dt = \sum_n \int_{t_n}^{t_{n+1/2}} \psi \, dt$$
$$= \sum_n \psi(t_n^*) \frac{\Delta t}{2}$$
$$= \frac{1}{2} \sum_n \psi(t_n^*) \Delta t$$
$$\to \frac{1}{2} \int \psi \, dt \text{ as } \Delta t \to 0$$

(where t_n^* is in $[t_n, t_{n+1/2}]$), by definition of the Riemann integral. The general case follows by approximation.

Letting $\Delta t \to 0$, we thus obtain

$$\int_0^T \iint \left(|u - k|\phi_t + q^f(u,k)\phi_x + q^g(u,k)\phi_y \right) dx\, dy\, dt$$
$$+ \iint |u_0 - k|\, \phi|_{t=0}\, dx\, dy \geq \iint (|u(T) - k|\phi)|_{t=T}\, dx\, dy,$$

which proves that $u(x,y,t)$ is a solution to (4.1) satisfying the Kružkov entropy condition.

Next, we want to prove uniqueness of solutions of multidimensional conservation laws. Let u and v be two Kružkov entropy solutions of the conservation law

$$u_t + f(u)_x + g(u)_y = 0 \qquad (4.17)$$

with initial data u_0 and v_0, respectively. The argument in Section 2.4 leads, with no fundamental changes in the multidimensional case, to the same result (2.58), namely,

$$\|u(t) - v(t)\|_1 \leq \|u_0 - v_0\|_1, \qquad (4.18)$$

thereby proving uniqueness. Observe that we do not need compact support of u_0 and v_0, but only that $u_0 - v_0$ is integrable. Using the fact that if every subsequence of a sequence has a further subsequence converging to the same limit, the whole sequence converges to that (unique) limit, we find that the whole sequence $\{u_{\Delta t}\}$ converges, not just a subsequence. We have proved the following result.

Theorem 4.2. *Let f_j be Lipschitz continuous functions, and furthermore, let u_0 be a bounded and compactly supported function in $BV(\mathbb{R}^m)$. Define the sequence of functions $\{u^n\}$ by $u^0 = u_0$ and*

$$u^{n+j/m} = S_{\Delta t}^{f_j, x_j} u^{n+(j-1)/m}, \quad j = 1, \ldots, m, \quad n \in \mathbb{N}_0.$$

Introduce the function (where $t_r = r\Delta t$ for any rational number r) $u_{\Delta t} = u_{\Delta t}(x_1, \ldots, x_m, t)$ by

$$u_{\Delta t}(x_1, \ldots, x_m, t) = S^{f_j, x_j}_{m(t - t_{n+(j-1)/m})} u^{n+(j-1)/m},$$

for $t \in [t_{n+(j-1)/m}, t_{n+j/m}]$. Fix $T > 0$, then for any sequence $\{\Delta t\}$ such that $\Delta t \to 0$, for all $t \in [0, T]$ the function $u_{\Delta t}(t)$ converges to the unique weak solution $u(t)$ of (4.3) satisfying the Kružkov entropy condition (4.4). The limit is in $C([0, T]; L^1_{\mathrm{loc}}(\mathbb{R}^m))$.

To prove stability of the solution with respect to flux functions, we will show that the one-dimensional stability result (2.71) in Section 2.4 remains valid with obvious modifications in several dimensions. Let u and v denote the unique solutions of

$$u_t + f(u)_x + g(u)_y = 0, \quad u|_{t=0} = u_0,$$

and

$$v_t + \tilde{f}(v)_x + \tilde{g}(v)_y = 0, \quad v|_{t=0} = v_0,$$

respectively, that satisfy the Kružkov entropy condition. Here we do *not* assume compact support of u_0 and v_0, but only that $u_0 - v_0$ is integrable. We want to estimate the L^1-norm of the difference between the two solutions. To this end we first consider

$$\left\| u^{n+1/2} - v^{n+1/2} \right\|_1 = \iint \left| u^{n+1/2} - v^{n+1/2} \right| dx\, dy$$
$$\leq \int \left(\int |u^n - v^n|\, dx \right.$$
$$\left. + \Delta t \min\{\mathrm{T.V.}_{\cdot x}(u^n), \mathrm{T.V.}_{\cdot x}(v^n)\} \|f - \tilde{f}\|_{\mathrm{Lip}} \right) dy$$
$$= \|u^n - v^n\|_1$$
$$+ \Delta t \|f - \tilde{f}\|_{\mathrm{Lip}} \int \min\{\mathrm{T.V.}_{\cdot x}(u^n), \mathrm{T.V.}_{\cdot x}(v^n)\}\, dy.$$

Next we employ the trivial, but useful, inequality

$$a \wedge b + c \wedge d \leq (a+c) \wedge (b+d), \quad a, b, c, d \in \mathbb{R}.$$

Thus

$$\left\| u^{n+1} - v^{n+1} \right\|_1 = \iint \left| u^{n+1} - v^{n+1} \right| dx\, dy$$
$$\leq \int \left(\int \left| u^{n+1/2} - v^{n+1/2} \right| dy \right.$$
$$\left. + \Delta t \min\left\{\mathrm{T.V.}_{\cdot y}\left(u^{n+1/2}\right), \mathrm{T.V.}_{\cdot y}\left(v^{n+1/2}\right)\right\} \|g - \tilde{g}\|_{\mathrm{Lip}} \right) dx$$

$$\leq \left\| u^{n+1/2} - v^{n+1/2} \right\|_1$$
$$+ \Delta t \|g - \tilde{g}\|_{\text{Lip}} \int \min \left\{ \text{T.V.}_y \left(u^{n+1/2} \right), \text{T.V.}_y \left(v^{n+1/2} \right) \right\} dx$$
$$\leq \|u^n - v^n\|_1 + \Delta t \max \left\{ \|f - \tilde{f}\|_{\text{Lip}}, \|g - \tilde{g}\|_{\text{Lip}} \right\}$$
$$\times \left(\min \left\{ \int \text{T.V.}_x (u^n) \, dy, \int \text{T.V.}_x (v^n) \, dy \right\} \right.$$
$$\left. + \min \left\{ \int \text{T.V.}_y (u^n) \, dx, \int \text{T.V.}_y (v^n) \, dx \right\} \right)$$
$$\leq \|u^n - v^n\|_1$$
$$+ \Delta t \max\{\|f - \tilde{f}\|_{\text{Lip}}, \|g - \tilde{g}\|_{\text{Lip}}\}$$
$$\times \min \left\{ \begin{array}{l} \int \text{T.V.}_x (u^n) \, dy + \int \text{T.V.}_y (u^n) \, dx, \\ \int \text{T.V.}_x (v^n) \, dy + \int \text{T.V.}_y (v^n) \, dx \end{array} \right\}$$
$$= \|u^n - v^n\|_1$$
$$+ \Delta t \max\{\|f - \tilde{f}\|_{\text{Lip}}, \|g - \tilde{g}\|_{\text{Lip}}\} \min \left\{ \text{T.V.}(u^n), \text{T.V.}(v^n) \right\},$$

which implies

$$\|u^n - v^n\|_1 \leq \|u_0 - v_0\|_1$$
$$+ n \Delta t \max\{\|f - \tilde{f}\|_{\text{Lip}}, \|g - \tilde{g}\|_{\text{Lip}}\} \min\{\text{T.V.}(u_0), \text{T.V.}(v_0)\}. \quad (4.19)$$

Consider next $t \in [t_n, t_{n+1/2})$. Then the continuous interpolants defined by (4.7) satisfy

$$\|u_{\Delta t}(t) - v_{\Delta t}(t)\|_{L^1(\mathbb{R}^2)} = \left\| S^{f,x}_{2(t-t_n)} u^n - S^{\tilde{f},x}_{2(t-t_n)} v^n \right\|_{L^1(\mathbb{R}^2)}$$
$$\leq \int \left[\int |u^n - v^n| \, dx \right.$$
$$\left. + 2(t - t_n) \min\{\text{T.V.}_x(u^n), \text{T.V.}_x(v^n)\} \|f - \tilde{f}\|_{\text{Lip}} \right] dy$$
$$= \|u^n - v^n\|_{L^1(\mathbb{R}^2)} \quad (4.20)$$
$$+ 2(t - t_n) \|f - \tilde{f}\|_{\text{Lip}} \int \min\{\text{T.V.}_x(u^n), \text{T.V.}_x(v^n)\} \, dy$$
$$\leq \|u_0 - v_0\|_1$$
$$+ t_n \max\{\|f - \tilde{f}\|_{\text{Lip}}, \|g - \tilde{g}\|_{\text{Lip}}\} \min\{\text{T.V.}(u_0), \text{T.V.}(v_0)\}$$
$$+ 2(t - t_n) \min\{\text{T.V.}(u_0), \text{T.V.}(v_0)\} \max\{\|f - \tilde{f}\|_{\text{Lip}}, \|g - \tilde{g}\|_{\text{Lip}}\}$$
$$\leq \|u_0 - v_0\|_1 + t \min\{\text{T.V.}(u_0), \text{T.V.}(v_0)\} \max\{\|f - \tilde{f}\|_{\text{Lip}}, \|g - \tilde{g}\|_{\text{Lip}}\}.$$

Observe that the above argument also holds mutatis mutandis in the general case of a scalar conservation law in any dimension. We summarize our results in the following theorem.

126   4. Multidimensional Scalar Conservation Laws

Theorem 4.3. *Let u_0 be in $L^1(\mathbb{R}^m) \cap L^\infty(\mathbb{R}^m) \cap BV(\mathbb{R}^m)$, and let f_j be Lipschitz continuous functions for $j = 1, \ldots, m$. Then there exists a unique solution $u(x_1, \ldots, x_m, t)$ of the initial value problem*

$$u_t + \sum_{j=1}^m f_j(u)_{x_j} = 0, \quad u(x_1, \ldots, x_m, 0) = u_0(x_1, \ldots, x_m), \qquad (4.21)$$

that satisfies the Kružkov entropy condition (4.4). *The solution satisfies*

$$\text{T.V.}(u(t)) \leq \text{T.V.}(u_0). \qquad (4.22)$$

Furthermore, if v_0 and g share the same properties as u_0 and f, respectively, then the unique weak Kružkov entropy solution of

$$v_t + \sum_{j=1}^m g_j(v)_{x_j} = 0, \quad v(x_1, \ldots, x_m, 0) = v_0(x_1, \ldots, x_m), \qquad (4.23)$$

satisfies

$$\|u(\,\cdot\,,t) - v(\,\cdot\,,t)\|_1 \leq \|u_0 - v_0\|_1 \qquad (4.24)$$
$$+ t \min\{\text{T.V.}(u_0), \text{T.V.}(v_0)\} \max_j\{\|f_j - g_j\|_{\text{Lip}}\}.$$

Proof. It remains to consider the case where u_0 no longer is assumed to have compact support. Observe that we only used this assumption to prove (4.11). In particular, the estimate (4.22) carries over with no changes.

Let $u_{0,k}$ be compactly supported functions in $L^1(\mathbb{R}^m) \cap BV(\mathbb{R}^m)$ such that

$$\|u_{0,k} - u_0\|_1 \to 0 \text{ when } k \to \infty.$$

Furthermore, let $u_{\Delta t, k}$ and $u_{\Delta t}$ denote the approximate solution with initial data u_k and u_0, respectively, constructed using dimensional splitting. We know from Theorem 4.2 that $u_{\Delta t, k} \to u_k$, the unique solution with initial data $u_{0,k}$, in $C([0, T]; L^1_{\text{loc}}(\mathbb{R}^m))$ as $\Delta t \to 0$ for each fixed $T > 0$. From (4.18) we know that

$$\sup_t \|u_k(t) - u_\ell(t)\|_1 \leq \|u_{0,k} - u_{0,\ell}\|_1,$$

and hence u_k is a Cauchy sequence in $C([0, T]; L^1_{\text{loc}}(\mathbb{R}^m))$ that converges to some function u as $k \to \infty$. Furthermore, by considering the Kružkov entropy condition applied to u_k and taking the limit $k \to \infty$ we conclude that u is a weak Kružkov entropy solution. But

$$\|u_{\Delta t}(t) - u(t)\|_1 \leq \|u_{\Delta t}(t) - u_{\Delta t, k}(t)\|_1 + \|u_{\Delta t, k}(t) - u_k(t)\|_1$$
$$+ \|u_k(t) - u(t)\|_1$$
$$\leq \|u_0 - u_{0,k}\|_1 + \|u_{\Delta t, k}(t) - u_k(t)\|_1 + \|u_k(t) - u(t)\|_1,$$

where for the first term we have used (4.20). The first and the third terms are small by taking k large, while the middle term is small by taking Δt

small. We conclude that $u_{\Delta t}(t) \to u(t)$ in L^1_{loc}. Finally, by taking the limit $\Delta t \to 0$ in (4.20), we see that (4.24) holds. □

4.2 Dimensional Splitting and Front Tracking

> It doesn't matter if the cat is black or white.
> As long as it catches rats, it's a good cat.
>
> *Deng Xiaoping (1904–97)*

In this section we will study the case where we use front tracking to solve the one-dimensional conservation laws. More precisely, we replace the flux functions f and g (in the two-dimensional case) by piecewise linear continuous interpolations f_δ and g_δ, with the interpolation points spaced a distance δ apart. The aim is to determine the convergence rate towards the solution of the full two-dimensional conservation law as $\delta \to 0$ and $\Delta t \to 0$.

With the front-tracking approximation the one-dimensional solutions will be piecewise constant if the initial condition is piecewise constant. In order to prevent the number of discontinuities from growing without bound, we will project the one-dimensional solution $S^{f_\delta,x}u$ onto a fixed grid in the x, y plane before applying the operator $S^{g_\delta,y}$.

To be more concrete, let the grid spacing in the x and y directions be given by Δx and Δy, respectively, and let I_{ij} denote the grid cell

$$I_{ij} = \{(x,y) \mid i\Delta x \leq x < (i+1)\Delta x,\ j\Delta y \leq y < (j+1)\Delta y\}.$$

The projection operator π is defined by

$$\pi u(x,y) = \frac{1}{\Delta x \Delta y} \iint_{I_{ij}} u\, dx\, dy \text{ for } (x,y) \in I_{ij}.$$

Let the approximate solution at the discrete times t_ℓ be defined as

$$u^{n+1/2} = \pi \circ S^{f_\delta,x}_{\Delta t} u^n \text{ and } u^{n+1} = \pi \circ S^{g_\delta,y}_{\Delta t} u^{n+1/2},$$

for $n = 0, 1, 2, \ldots$, with $u^0 = \pi u_0$. We collect the discretization parameters in $\eta = (\delta, \Delta x, \Delta y, \Delta t)$. In analogy to (4.7), we define u_η as

$$u_\eta(t) = \begin{cases} S^{f_\delta,x}_{2(t-t_n)} u^n & \text{for } t_n \leq t < t_{n+1/2}, \\ u^{n+1/2} & \text{for } t = t_{n+1/2}, \\ S^{g_\delta,y}_{2(t-t_{n+1/2})} u^{n+1/2} & \text{for } t_{n+1/2} \leq t < t_{n+1}, \\ u^{n+1} & \text{for } t = t_{n+1}. \end{cases} \quad (4.25)$$

If Figure 4.1 we illustrate how this works. Starting in the upper left corner, the operator $S^{f_\delta,x}_{\Delta t}$ takes us to the upper right corner; then we apply π and move to the lower right corner. Next, $S^{g_\delta,y}_{\Delta t}$ takes us to the lower left corner, and finally π takes us back to the upper left corner, this time with

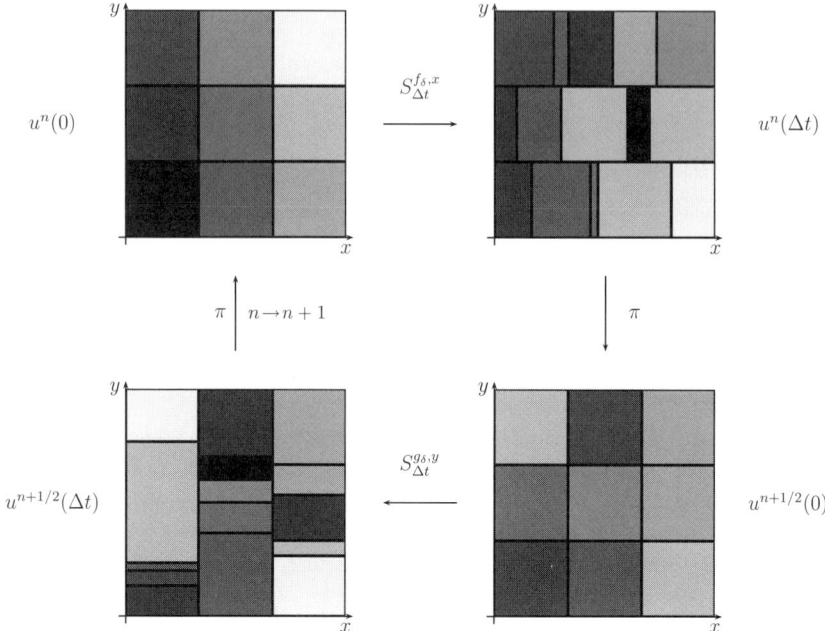

Figure 4.1. Front tracking and dimensional splitting on a 3×3 grid.

n incremented by 1. To prove that u_η converges to the unique solution u as $\eta \to 0$, we essentially mimic the approach we just used to prove Theorem 4.2. First of all we observe that

$$\|u_\eta(t)\|_\infty \leq \|u^0\|_\infty, \tag{4.26}$$

since $S^{f_\delta,x}$, $S^{g_\delta,y}$, and π all obey a maximum principle. On each rectangle I_{ij} the function u_η is constant for $t = \Delta t$. In a desperate attempt to simplify the notation we write

$$u^n_{ij} = u_\eta(x, y, n\Delta t) \text{ for } (x, y) \in I_{ij}.$$

Next we go carefully through one full-time step in this construction, starting with u^n_{ij}. At each step we define a shorthand notation that we will use in the estimates. When we consider u^n_{ij} as a function of x only, we write

$$u^n_j(0) = u^n_{ij} = u_\eta(\,\cdot\,, j\Delta y, n\Delta t).$$

(The argument "0" on the left-hand side indicates the start of the time variable before we advance time an interval Δt using $S^{f_\delta,x}_{\Delta t}$.) Advancing the solution in time by Δt by applying front tracking in the x-variable produces

$$u^n_j(\Delta t) = \left(S^{f_\delta,x}_{\Delta t} u^n_j\right)(x).$$

4.2. Dimensional Splitting and Front Tracking

(The x-dependence is suppressed in the notation on the left-hand side.) We now apply the projection π, which yields

$$u_{ij}^{n+1/2} = \pi u_j^n(\Delta t).$$

After this sweep in the x-variable, it is time to do the y-direction. Considering $u_{ij}^{n+1/2}$ as a function of y we write

$$u_i^{n+1/2}(0) = u_{ij}^{n+1/2} = u_\eta\left(i\Delta x, \cdot, \left(n + \frac{1}{2}\right)\Delta t\right),$$

to which we apply the front-tracking solution operator in the y-direction

$$u_i^{n+1/2}(\Delta t) = \left(S_{\Delta t}^{g_\delta, y} u_i^{n+1/2}\right)(y).$$

(The y-dependence is suppressed in the notation on the left-hand side.) One full time step is completed by a final projection

$$u_{ij}^{n+1} = \pi u_i^{n+1/2}(\Delta t).$$

Using this notation we first want to prove the analogue of Lipschitz continuity in time of the spatial L^1-norm as expressed in (4.11). In this context the result reads

$$\|u_\eta(t_m) - u_\eta(t_n)\|_1 = \sum_{i,j} |u_{ij}^m - u_{ij}^n|\,\Delta x \Delta y$$

$$\leq \Big(\max\{\|f_\delta\|_{\mathrm{Lip}}, \|g_\delta\|_{\mathrm{Lip}}\}\Delta t + (\Delta x + \Delta y)\Big)$$
$$\times \mathrm{T.V.}\left(u^0\right)|m - n|. \qquad (4.27)$$

To prove (4.27) it suffices to show that

$$\sum_{i,j} |u_{ij}^{n+1} - u_{ij}^n|\,\Delta x \Delta y$$

$$\leq \big(\max\{\|f_\delta\|_{\mathrm{Lip}}, \|g_\delta\|_{\mathrm{Lip}}\}\Delta t + (\Delta x + \Delta y)\big)\mathrm{T.V.}\left(u^0\right). \qquad (4.28)$$

We start by writing

$$\begin{aligned}\left|u_{ij}^{n+1} - u_{ij}^n\right| &\leq \left|u_{ij}^{n+1} - u_i^{n+1/2}(\Delta t)\right| + \left|u_{ij}^{n+1/2} - u_j^n(\Delta t)\right|\\ &\quad + \left|u_i^{n+1/2}(\Delta t) - u_i^{n+1/2}(0)\right| + \left|u_j^n(\Delta t) - u_j^n(0)\right|\\ &= \left|\pi u_i^{n+1/2}(\Delta t) - u_i^{n+1/2}(\Delta t)\right| + \left|\pi u_j^n(\Delta t) - u_j^n(\Delta t)\right|\\ &\quad + \left|u_i^{n+1/2}(\Delta t) - u_i^{n+1/2}(0)\right| + \left|u_j^n(\Delta t) - u_j^n(0)\right|.\end{aligned}$$

130   4. Multidimensional Scalar Conservation Laws

Integrating this inequality over $\mathbb{R}^2$ gives

$$\sum_{i,j}\left|u_{ij}^{n+1}-u_{ij}^n\right|\Delta x\Delta y \leq \iint\left|\pi u_i^{n+1/2}(\Delta t)-u_i^{n+1/2}(\Delta t)\right|\,dx\,dy$$
$$+\iint\left|\pi u_j^n(\Delta t)-u_j^n(\Delta t)\right|\,dx\,dy$$
$$+\iint\left|u_i^{n+1/2}(\Delta t)-u_i^{n+1/2}(0)\right|\,dx\,dy$$
$$+\iint\left|u_j^n(\Delta t)-u_j^n(0)\right|\,dx\,dy. \qquad (4.29)$$

We see that two terms involve the projection operator π. For these terms we prove the estimate

$$\iint|\pi\psi-\psi|\,dx\,dy \leq (\Delta x+\Delta y)\,\mathrm{T.V.}(\psi) = \mathcal{O}\left(\max\{\Delta x,\Delta y\}\right). \qquad (4.30)$$

We will prove (4.30) in the one-dimensional case only. Consider (where $I_i = \langle i\Delta x, (i+1)\Delta x\rangle$)

$$\int|\pi\psi-\psi|\,dx = \sum_i \int_{I_i}|\pi\psi(x)-\psi(x)|\,dx$$
$$= \sum_i\int_{I_i}\left|\frac{1}{\Delta x}\int_{I_i}\psi(y)\,dy-\psi(x)\right|\,dx$$
$$= \frac{1}{\Delta x}\sum_i\int_{I_i}\left|\int_{I_i}(\psi(y)-\psi(x))\,dy\right|\,dx$$
$$\leq \frac{1}{\Delta x}\sum_i\int_{I_i}\int_{I_i}|\psi(y)-\psi(x)|\,dy\,dx$$
$$= \frac{1}{\Delta x}\sum_i\int_{I_i}\int_{-x+I_i}|\psi(x+\xi)-\psi(x)|\,d\xi\,dx$$
$$\leq \frac{1}{\Delta x}\sum_i\int_{I_i}\int_{-\Delta x}^{\Delta x}|\psi(x+\xi)-\psi(x)|\,d\xi\,dx$$
$$= \frac{1}{\Delta x}\int_{-\Delta x}^{\Delta x}\int_{\mathbb{R}}|\psi(x+\xi)-\psi(x)|\,dx\,d\xi$$
$$\leq \frac{1}{\Delta x}\int_{-\Delta x}^{\Delta x}|\xi|\,\mathrm{T.V.}(\psi)\,d\xi$$
$$= \Delta x\,\mathrm{T.V.}(\psi). \qquad (4.31)$$

For the two remaining terms in (4.29) we find, using the Lipschitz continuity in time in the L^1 norm in the x-variable (see Theorem 2.14) that

4.2. Dimensional Splitting and Front Tracking

$$\iint |u_j^n(\Delta t) - u_j^n(0)|\,dx\,dy \le \Delta t \,\|f_\delta\|_{\mathrm{Lip}} \int \mathrm{T.V.}_{\cdot x}\left(u_j^n(0)\right) dy$$

$$\le \Delta t \,\|f_\delta\|_{\mathrm{Lip}} \mathrm{T.V.}\left(u^n\right). \tag{4.32}$$

Combining this result with (4.30) we conclude that (4.28), and hence also (4.27), holds.

Finally, we want to show that the total variation is bounded in the sense that

$$\mathrm{T.V.}\left(u^n\right) \le \mathrm{T.V.}\left(u_0\right). \tag{4.33}$$

We will show that

$$\mathrm{T.V.}\left(u^{n+1/2}\right) \le \mathrm{T.V.}\left(u^n\right); \tag{4.34}$$

an analogous argument gives $\mathrm{T.V.}\left(u^{n+1}\right) \le \mathrm{T.V.}\left(u^{n+1/2}\right)$, from which we conclude that

$$\mathrm{T.V.}\left(u^{n+1}\right) \le \mathrm{T.V.}\left(u^n\right),$$

and (4.33) follows by induction. By definition

$$\mathrm{T.V.}\left(u^{n+1/2}\right) = \sum_{i,j} \left(\left|u_{i+1,j}^{n+1/2} - u_{i,j}^{n+1/2}\right|\Delta y + \left|u_{i,j+1}^{n+1/2} - u_{i,j}^{n+1/2}\right|\Delta x\right), \tag{4.35}$$

while

$$\mathrm{T.V.}\left(u^n\right) = \sum_{i,j} \left(\left|u_{i+1,j}^n - u_{i,j}^n\right|\Delta y + \left|u_{i,j+1}^n - u_{i,j}^n\right|\Delta x\right). \tag{4.36}$$

We first consider

$$\sum_i \left|u_{i+1,j}^{n+1/2} - u_{i,j}^{n+1/2}\right| = \mathrm{T.V.}_{\cdot x}\left(\pi u_j^n(\Delta t)\right)$$

$$\le \mathrm{T.V.}_{\cdot x}\left(u_j^n(\Delta t)\right) \le \mathrm{T.V.}_{\cdot x}\left(u_j^n(0)\right)$$

$$= \sum_i \left|u_{i+1,j}^n - u_{i,j}^n\right|, \tag{4.37}$$

where we first used that $\mathrm{T.V.}(\pi\phi) \le \mathrm{T.V.}(\phi)$ for step functions ϕ, and then that $\mathrm{T.V.}(v) \le \mathrm{T.V.}(v_0)$ for solutions v of one-dimensional conservation laws with initial data v_0. For the second term in the definition of $\mathrm{T.V.}\left(u^{n+1/2}\right)$ we obtain (cf. (4.9))

$$\sum_{i,j} \left|u_{i,j+1}^{n+1/2} - u_{i,j}^{n+1/2}\right|\Delta x \Delta y = \sum_{i,j} \int_{I_{ij}} \left|u_{i,j+1}^{n+1/2} - u_{i,j}^{n+1/2}\right| dx\,dy$$

$$= \sum_{i,j} \int_{I_{ij}} \left|\pi\left(u_{j+1}^n(\Delta t) - u_j^n(\Delta t)\right)\right| dx\,dy$$

$$\le \sum_{i,j} \int_{I_{ij}} \pi\left(\left|u_{j+1}^n(\Delta t) - u_j^n(\Delta t)\right|\right) dx\,dy$$

$$= \sum_{i,j} \int_{I_{ij}} \left| u_{j+1}^n(\Delta t) - u_j^n(\Delta t) \right| dx\, dy$$

$$= \sum_{i,j} \Delta y \int_{i\Delta x}^{(i+1)\Delta x} \left| u_{j+1}^n(\Delta t) - u_j^n(\Delta t) \right| dx$$

$$= \sum_j \Delta y \int_{\mathbb{R}} \left| u_{j+1}^n(x, \Delta t) - u_j^n(x, \Delta t) \right| dx$$

$$\leq \sum_j \Delta y \int_{\mathbb{R}} \left| u_{j+1}^n(x, 0) - u_j^n(x, 0) \right| dx$$

$$= \sum_{i,j} \left| u_{i,j+1}^n - u_{i,j}^n \right| \Delta x \Delta y. \tag{4.38}$$

The first inequality follows from $|\pi\phi| \leq \pi|\phi|$; thereafter, we use $\int_{I_{ij}} \pi\phi = \int_{I_{ij}} \phi$, and finally we use the L^1-contractivity, $\|v - w\|_1 \leq \|v_0 - w_0\|_1$, of solutions of one-dimensional conservation laws. Multiplying (4.37) by Δy, summing over j, dividing (4.38) by Δx, and finally adding the results, gives (4.34).

So far we have obtained the following estimates:

(i) Uniform boundedness,

$$\|u_\eta(t)\|_\infty \leq \|u^0\|_\infty.$$

(ii) Uniform bound on the total variation,

$$\text{T.V.}(u^n) \leq \text{T.V.}(u_0).$$

(iii) Lipschitz continuity in time,

$$\|u_\eta(t_m) - u_\eta(t_n)\|_1 \leq \left(C + \mathcal{O}\left(\frac{1}{\Delta t} \max\{\Delta x, \Delta y\} \right) \right) \text{T.V.}(u_0) |t_m - t_n|. \tag{4.39}$$

From Theorem A.8 we conclude that the sequence $\{u_\eta\}$ has a convergent subsequence as $\eta \to 0$, provided that the ratio $\max\{\Delta x, \Delta y\}/\Delta t$ remains bounded. We let u denote its limit. Furthermore, this sequence converges in $C([0,T]; L^1_{\text{loc}}(\mathbb{R}^2))$ for any positive T.

It remains to prove that the limit is indeed an entropy solution of the full two-dimensional conservation law. We first use that $u_j^n(x,t)$ (suppressing the y-dependence) is a solution of the one-dimensional conservation law in the time interval $[t_n, t_{n+1/2}]$. Hence we know that

$$\int_{\mathbb{R}} \int_{t_n}^{t_{n+1/2}} \left(\frac{1}{2} |u_j^n(x,t) - k| \phi_t + q^{f_\delta}(u_j^n(x,t), k)\phi_x \right) dt\, dx$$
$$- \frac{1}{2} \int_{\mathbb{R}} \left| u_j^n(x, t_{n+1/2}-) - k \right| \phi(x, t_{n+1/2})\, dx$$

$$+ \frac{1}{2} \int_{\mathbb{R}} \left| u_j^n(x, t_n) - k \right| \phi(x, t_n) \, dx \geq 0.$$

Similarly, we obtain for the y-direction

$$\int_{\mathbb{R}} \int_{t_{n+1/2}}^{t_{n+1}} \left(\frac{1}{2} \left| u_i^{n+1/2}(y, t) - k \right| \phi_t + q^{G_\delta}(u_i^{n+1/2}(y, t), k) \phi_y \right) dt \, dy$$

$$- \frac{1}{2} \int_{\mathbb{R}} \left| u_i^{n+1/2}(y, t_{n+1}-) - k \right| \phi(y, t_{n+1}) \, dy$$

$$+ \frac{1}{2} \int_{\mathbb{R}} \left| u_i^{n+1/2}(y, t_{n+1/2}+) - k \right| \phi(y, t_{n+1/2}) \, dy \geq 0.$$

Integrating the first inequality over y and the second over x and adding the results as well as adding over n gives, where $T = N\Delta t$,

$$\iint_{\mathbb{R}^2} \int_0^T \left(\frac{1}{2} |u_\eta - k| \phi_t + \sum_n \chi_n q^{f_\delta}(u_\eta, k) \phi_x \right.$$

$$\left. + \sum_n \tilde{\chi}_n q^{g_\delta}(u_\eta, k) \phi_y \right) dx \, dy \, dt$$

$$- \frac{1}{2} \left(\iint_{\mathbb{R}^2} |u_\eta(x, y, T) - k| \phi(x, y, T) \, dx \, dy \right.$$

$$\left. + \int_{\mathbb{R}^2} |u_\eta(x, y, 0) - k| \phi(x, y, 0) \, dx \, dy \right)$$

$$\geq \frac{1}{2} \sum_{n=1}^{2N} \iint_{\mathbb{R}^2} \left(\left| u_\eta(x, y, t_{n/2}) - k \right| - \left| u_\eta(x, y, t_{n/2}-) - k \right| \right)$$

$$\times \phi(x, y, t_{n/2}) \, dx \, dy$$

$$=: I,$$

and χ_n and $\tilde{\chi}_n$ as before denote the characteristic functions on $\{(x, y, t) \mid t \in [t_n, t_{n+1/2}]\}$ and $\{(x, y, t) \mid t \in [t_{n+1/2}, t_{n+1}]\}$, respectively. Observe that we have obtained the right-hand side, denoted by I, by using a projection at each time step. As $n \to \infty$ and $\Delta t \to 0$ while keeping T fixed, we have that $\sum_n \chi_n \overset{*}{\rightharpoonup} \frac{1}{2}$. To estimate the term I we write

$$|I| \leq \sum_n \iint \left| u_\eta(x, y, t_{n/2}+) - u_\eta(x, y, t_{n/2}-) \right| \phi(x, y, t_{n/2}) \, dx \, dy$$

$$\leq \|\phi\|_\infty \sum_n \iint \left| \pi u_\eta(x, y, t_{n/2}-) - u_\eta(x, y, t_{n/2}-) \right| dx \, dy$$

$$\leq \mathcal{O}\left(\max\{\Delta x, \Delta y\}/\Delta t \right),$$

using (4.30). In order to conclude that u is an entropy solution, we need that $I \to 0$; that is, we need to assume that

$$\max\{\Delta x, \Delta y\}/\Delta t \to 0$$

as $\eta \to 0$. Under this assumption

$$\iint_{\mathbb{R}^2} \int_0^T \left(|u-k|\phi_t + q^f(u,k)\phi_x + q^g(u,k)\phi_y \right) dt\, dx\, dy$$
$$- \int_{\mathbb{R}^2} |u(x,y,T) - k|\, \phi(x,y,T)\, dx\, dy$$
$$+ \int_{\mathbb{R}^2} |u(x,y,0) - k|\, \phi(x,y,0)\, dx\, dy \geq 0,$$

which shows that u indeed satisfies the Kružkov entropy condition. We summarize the result.

Theorem 4.4. *Let u_0 be a compactly supported function in $L^\infty(\mathbb{R}^m) \cap BV(\mathbb{R}^m)$, and let f_j be Lipschitz continuous functions for $j = 1, \ldots, m$. Construct an approximate solution u_η using front tracking by defining*

$$u^0 = \pi u_0, \quad u^{n+j/m} = \pi \circ S_{\Delta t}^{f_j, \delta, x_j} u^{n+(j-1)/m}, \quad j = 1, \ldots, m, \quad n \in \mathbb{N},$$

and

$$u_\eta(x,t) = \begin{cases} S_{m(t-t_{n+(j-1)/m})}^{f_j, \delta, x_j} u^{n+(j-1)/m}, & \text{for } t \in [t_{n+(j-1)/m}, t_{n+j/m}), \\ u^{n+j/m} & \text{for } t = t_{n+j/m}, \end{cases}$$

where $x = (x_1, \ldots, x_m)$.
For any sequence $\{\eta\}$, with $\eta = (\Delta x_1, \ldots, \Delta x_m, \Delta t, \delta)$ where $\eta \to 0$ and

$$\max_j \{\Delta x_j\} / \Delta t \to 0,$$

we have that $\{u_\eta\}$ converges to the unique solution $u = u(x,t)$ of the initial value problem

$$u_t + \sum_{j=1}^m f_j(u)_{x_j} = 0, \quad u(x,0) = u_0(x), \tag{4.40}$$

that satisfies the Kružkov entropy condition.

4.3 Convergence Rates

> Now I think I'm wrong on account
> of those damn partial integrations.
> I oscillate between right and wrong.
>
> *Letter from Feynman to Walton (1936)*

In this section we show how fast front tracking plus dimensional splitting converges to the exact solution. The analysis is based on Kuznetsov's lemma.

4.3. Convergence Rates

We start by generalizing Kuznetsov's lemma, Theorem 3.11, to the present multidimensional setting. Although the argument carries over, we will present the relevant definitions in arbitrary dimension.

Let the class $\mathcal{K}$ consist of maps $u \colon [0, \infty) \to L^1(\mathbb{R}^m) \cap BV(\mathbb{R}^m) \cap L^\infty(\mathbb{R}^m)$ such that:

(i) The limits $u(t\pm)$ exist.

(ii) The function u is right continuous, i.e., $u(t+) = u(t)$.

(iii) $\|u(t)\|_\infty \leq \|u(0)\|_\infty$.

(iv) T.V. $(u(t)) \leq$ T.V. $(u(0))$.

Recall the following definition of moduli of continuity in time (cf. (3.2)):

$$\nu_t(u, \sigma) = \sup_{|\tau| \leq \sigma} \|u(t + \tau) - u(t)\|_1, \qquad \sigma > 0,$$

$$\nu(u, \sigma) = \sup_{0 \leq t \leq T} \nu_t(u, \sigma).$$

The estimate (3.39) is replaced by

$$\nu(u, \sigma) \leq |\sigma| \, \text{T.V.} \, (u_0) \max_j \{\, \|f_j\|_{\text{Lip}} \,\},$$

for a solution u of (4.21).

In several space dimensions, the Kružkov form reads

$$\Lambda_T(u, \phi, k) = \iint_{\mathbb{R}^m \times [0,T]} \left(|u - k| \phi_t + \sum_j q^{f_j}(u, k) \phi_{x_j} \right) dx_1 \cdots dx_m \, dt$$

$$- \int_{\mathbb{R}^m} |u(x, T) - k| \, \phi(x, T) \, dx_1 \ldots dx_m \, dt$$

$$+ \int_{\mathbb{R}^m} |u_0(x) - k| \, \phi(x, 0) \, dx_1 \cdots dx_m \, dt. \tag{4.41}$$

In this case we use the test function

$$\Omega(x, x', s, s') = \omega_{\varepsilon_0}(s - s') \omega_\varepsilon(x_1 - x_1') \cdots \omega_\varepsilon(x_m - x_m'), \tag{4.42}$$
$$x = (x_1, \ldots, x_m), \quad x' = (x_1', \ldots, x_m').$$

Here ω_ε is the standard mollifier defined by

$$\omega_\varepsilon(x_j) = \frac{1}{\varepsilon} \omega\left(\frac{x_j}{\varepsilon}\right)$$

with

$$0 \leq \omega \leq 1, \quad \text{supp}\,\omega \subseteq [-1, 1], \quad \omega(-x_j) = \omega(x_j), \quad \int_{-1}^1 \omega(z) \, dz = 1.$$

When v is the unique solution of the conservation law (4.23), we introduce

$$\Lambda_{\varepsilon, \varepsilon_0}(u, v) = \int_0^T \int_{\mathbb{R}^m} \Lambda_T(u, \Omega(\,\cdot\,, x', \,\cdot\,, s'), v(x', s')) \, dx' ds'.$$

Kuznetsov's lemma can be formulated as follows.

Theorem 4.5. *Let u be a function in $\mathcal{K}$, and v be an entropy solution of (4.23). If $0 < \varepsilon_0 < T$ and $\varepsilon > 0$, then*
$$\|u(\,\cdot\,,T-) - v(\,\cdot\,,T)\|_1 \leq \|u_0 - v_0\|_1$$
$$+ \operatorname{T.V.}(v_0)\left(2\varepsilon + \varepsilon_0 \max_j\{\|f_j\|_{\operatorname{Lip}}\}\right)$$
$$+ \nu(u,\varepsilon_0) - \Lambda_{\varepsilon,\varepsilon_0}(u,v), \qquad (4.43)$$
where $u_0 = u(\,\cdot\,,0)$ and $v_0 = v(\,\cdot\,,0)$.

The proof of Theorem 3.11 carries over to this setting verbatim. We want to estimate
$$\|S(T)u_0 - u_\eta\|_1 \leq \|S(T)u_0 - S_\delta(T)u_0\|_1 + \|S_\delta(T)u_0 - u_\eta\|_1, \qquad (4.44)$$
where $u = S(T)u_0$ and $S_\delta(T)u_0$ denote the exact solutions of the multidimensional conservation law with flux functions f replaced by their piecewise linear and continuous approximations f_δ. The first term can be estimated by
$$\|S(T)u_0 - S_\delta(T)u_0\|_1 \leq T \max_j\{\|f_j - f_{j,\delta}\|_{\operatorname{Lip}}\}\operatorname{T.V.}(u_0), \qquad (4.45)$$
while we apply Kuznetsov's lemma, Theorem 4.5, for the second term. For the function u we choose u_η, the approximate solution by using front tracking along each dimension and dimensional splitting, while for v we use the exact solution with piecewise linear continuous flux functions f_δ and g_δ and u_0 as initial data, that is, $v = v_\delta = S_\delta(T)u_0$. Thus we find, using (4.39), that
$$\nu(u_\eta,\varepsilon_0) \leq \varepsilon_0\left(C + \mathcal{O}\left(\frac{1}{\Delta t}\max_j\{\Delta x_j\}\right)\right)\operatorname{T.V.}(u_0).$$

Kuznetsov's lemma then reads
$$\|S_\delta(T)u_0 - u_\eta\|_1 \leq \|u_0 - u^0\|_1 + \Big[2\varepsilon + \max_j\{\|f_{j,\delta}\|_{\operatorname{Lip}}\}\varepsilon_0$$
$$+ \varepsilon_0\left(C + \mathcal{O}\left(\frac{\max\{\Delta x_j\}}{\Delta t}\right)\right)\Big]\operatorname{T.V.}(u_0)$$
$$- \Lambda_{\varepsilon,\varepsilon_0}(u_\eta,v_\delta), \qquad (4.46)$$
and the name of the game is to estimate $\Lambda_{\varepsilon,\varepsilon_0}$.

To make the estimates more transparent, we start by rewriting $\Lambda_T(u_\eta,\phi,k)$. Since all the complications of several space dimensions are present in two dimensions, we present the argument in two dimensions only, that is, with $m = 2$, and denote the spatial variables by (x,y). All arguments carry over to arbitrary dimensions without any change. By definition we have (in obvious notation, $q^{f_\delta}(u) = \operatorname{sign}(u-k)(f_\delta(u) - f_\delta(k))$

and similarly for $q^{g\delta}$)

$$\Lambda_T(u_\eta,\phi,k) = \iint \int_0^T \Big(|u_\eta - k|\phi_t + q^{f\delta}(u_\eta,k)\phi_x + q^{g\delta}(u_\eta,k)\phi_y\Big) \, dt\, dx\, dy$$
$$+ \iint |u_\eta - k|\phi|_{t=0+}\, dx\, dy - \iint |u_\eta - k|\phi|_{t=T-}\, dx\, dy$$

$$= \sum_{n=0}^{N-1} \iint \left(\int_{t_n}^{t_{n+1/2}} + \int_{t_{n+1/2}}^{t_{n+1}}\right)\Big(|u_\eta - k|\phi_t$$
$$+ q^{f\delta}(u_\eta,k)\phi_x + q^{g\delta}(u_\eta,k)\phi_y\Big)\, dt\, dx\, dy$$
$$+ \iint |u_\eta - k|\phi|_{t=0+}\, dx\, dy - \iint |u_\eta - k|\phi|_{t=T-}\, dx\, dy$$

$$= \sum_{n=0}^{N-1} \iint \int_{t_n}^{t_{n+1/2}} \Big(|u_\eta - k|\phi_t + 2q^{f\delta}(u_\eta,k)\phi_x\Big)\, dt\, dx\, dy$$
$$+ \sum_n \iint \int_{t_{n+1/2}}^{t_{n+1}} \Big(|u_\eta - k|\phi_t + 2q^{g\delta}(u_\eta,k)\phi_y\Big)\, dt\, dx\, dy$$
$$+ \sum_{n=0}^{N-1} \iint \left(\int_{t_{n+1/2}}^{t_{n+1}} - \int_{t_n}^{t_{n+1/2}}\right) q^{f\delta}(u_\eta,k)\phi_x\, dt\, dx\, dy$$
$$+ \sum_{n=0}^{N-1} \iint \left(\int_{t_n}^{t_{n+1/2}} - \int_{t_{n+1/2}}^{t_{n+1}}\right) q^{g\delta}(u_\eta,k)\phi_y\, dt\, dx\, dy$$
$$+ \iint |u_\eta - k|\phi|_{t=0+}\, dx\, dy - \iint |u_\eta - k|\phi|_{t=T-}\, dx\, dy.$$

We now use that u_η is an exact solution in the x-direction and the y-direction on each strip $[t_n, t_{n+1/2}]$ and $[t_{n+1/2}, t_{n+1}]$, respectively. Thus we can invoke inequalities (4.15) and (4.16), and we conclude that[4]

$$\Lambda_T(u_\eta,\phi,k) \geq \sum_{n=0}^{N-1} \iint \Big(|u_\eta - k||_{t=t_{n+1/2}-}\phi(t_{n+1/2})$$
$$- |u_\eta - k||_{t=t_n+}\phi(t_n)\Big)\, dx\, dy$$
$$+ \sum_{n=0}^{N-1} \iint \Big(|u_\eta - k||_{t=t_{n+1}-}\phi(t_{n+1})$$
$$- |u_\eta - k||_{t=t_{n+1/2}+}\phi(t_{n+1/2})\Big)\, dx\, dy$$

---

[4] Observe that because we employ the projection operator π between each pair of consecutive times we solve a conservation law in one dimension; $u^{n+1/2}$ and u^n are in general discontinuous across $t_{n+1/2}$ and t_n, respectively.

$$+ \sum_{n=0}^{N-1} \iint \left(\int_{t_{n+1/2}}^{t_{n+1}} - \int_{t_n}^{t_{n+1/2}} \right) q^{f_\delta}(u_\eta, k) \phi_x \, dt \, dx \, dy$$

$$+ \sum_{n=0}^{N-1} \iint \left(\int_{t_n}^{t_{n+1/2}} - \int_{t_{n+1/2}}^{t_{n+1}} \right) q^{g_\delta}(u_\eta, k) \phi_y \, dt \, dx \, dy$$

$$+ \iint |u_\eta - k| \phi|_{t=0+} \, dx \, dy - \iint |u_\eta - k| \phi|_{t=T-} \, dx \, dy$$

$$= -2 \sum_{n=0}^{N-1} \iint \int_{t_n}^{t_{n+1/2}} q^{f_\delta}(u_\eta, k) \phi_x \, dt \, dx \, dy$$

$$+ \iint \int_0^T q^{f_\delta}(u_\eta, k) \phi_x \, dt \, dx \, dy$$

$$- 2 \sum_{n=0}^{N-1} \iint \int_{t_{n+1/2}}^{t_{n+1}} q^{g_\delta}(u_\eta, k) \phi_y \, dt \, dx \, dy$$

$$+ \iint \int_0^T q^{g_\delta}(u_\eta, k) \phi_y \, dt \, dx \, dy$$

$$+ \sum_{n=0}^{N-1} \iint \left(|u_\eta - k| \Big|_{t=t_{n+1/2}-} - |u_\eta - k| \Big|_{t=t_{n+1/2}+} \right) \phi(t_{n+1/2}) \, dx \, dy$$

$$+ \sum_{n=1}^{N-1} \iint \left(|u_\eta - k| \Big|_{t=t_n-} - |u_\eta - k| \Big|_{t=t_n+} \right) \phi(t_n) \, dx \, dy$$

$$:= -I_1(u_\eta, k) - I_2(u_\eta, k) - I_3(u_\eta, k) - I_4(u_\eta, k). \quad (4.47)$$

The terms I_1 and I_2 are due to dimensional splitting, while I_3 and I_4 come from the projections.

Choose now for the constant k the function $v_\delta(x', y', s')$, and for ϕ we use Ω given by (4.42). Integrating over the new variables we obtain

$$\Lambda_{\varepsilon,\varepsilon_0}(u_\eta, v_\delta) = \iint \int_0^T \Lambda_T(u_\eta, \Omega(\cdot, x', \cdot, y', \cdot, s'), v_\delta(x', y', s')) \, ds' \, dx' \, dy'$$

$$\geq -I_1^{\varepsilon,\varepsilon_0}(u_\eta, v_\delta) - I_2^{\varepsilon,\varepsilon_0}(u_\eta, v_\delta) - I_3^{\varepsilon,\varepsilon_0}(u_\eta, v_\delta) - I_4^{\varepsilon,\varepsilon_0}(u_\eta, v_\delta),$$

where $I_j^{\varepsilon,\varepsilon_0}$ are given by

$$I_1^{\varepsilon,\varepsilon_0}(u_\eta, v_\delta) = \iint \int_0^T \iint \left(2 \sum_{n=0}^{N-1} \int_{t_n}^{t_{n+1/2}} q^{f_\delta}(u_\eta, v_\delta) \Omega_x \, ds \right.$$

$$\left. - \int_0^T q^{f_\delta}(u_\eta, v_\delta) \Omega_x \, ds \right) dx \, dy \, ds' \, dx' \, dy',$$

4.3. Convergence Rates

$$I_2^{\varepsilon,\varepsilon_0}(u_\eta, v_\delta) = \iint \int_0^T \iint \left(2 \sum_{n=0}^{N-1} \int_{t_{n+1/2}}^{t_{n+1}} q^{g_\delta}(u_\eta, v_\delta)\Omega_y \, ds \right.$$

$$\left. - \int_0^T q^{g_\delta}(u_\eta, v_\delta)\Omega_y \, ds \right) dx \, dy \, ds' \, dx' \, dy',$$

$$I_3^{\varepsilon,\varepsilon_0}(u_\eta, v_\delta) = \sum_{n=1}^{N-1} \iint \int_0^T \iint \Big(|u_\eta - v_\delta| \, |_{s=t_n+}$$

$$- |u_\eta - v_\delta| \, |_{s=t_n-} \Big) \Omega \, dx \, dy \, ds' \, dx' \, dy',$$

$$I_4^{\varepsilon,\varepsilon_0}(u_\eta, v_\delta) = \sum_{n=0}^{N-1} \iint \int_0^T \iint \Big(|u_\eta - v_\delta| \, |_{s=t_{n+1/2}+}$$

$$- |u_\eta - v_\delta| \, |_{s=t_{n+1/2}-} \Big) \Omega \, dx \, dy \, ds' \, dx' \, dy'.$$

We will start by estimating $I_1^{\varepsilon,\varepsilon_0}$ and $I_2^{\varepsilon,\varepsilon_0}$.

Lemma 4.6. *We have the following estimate:*

$$|I_1^{\varepsilon,\varepsilon_0}| + |I_2^{\varepsilon,\varepsilon_0}| \leq T \max\{\|f\|_{\mathrm{Lip}}, \|g\|_{\mathrm{Lip}}\} \, \mathrm{T.V.}(u_0)$$

$$\times \left(\frac{\Delta t}{\varepsilon_0} + \frac{1}{\varepsilon} \Big(\{\|f\|_{\mathrm{Lip}} + \|g\|_{\mathrm{Lip}}\}\Delta t + \Delta x + \Delta y \Big) \right). \quad (4.48)$$

Proof. We will detail the estimate for $|I_1^{\varepsilon,\varepsilon_0}|$. Writing

$$q^{f_\delta}(u_\eta(s), v_\delta(s')) = q^{f_\delta}(u_\eta(t_{n+1/2}), v_\delta(s'))$$

$$+ \big(q^{f_\delta}(u_\eta(s), v_\delta(s')) - q^{f_\delta}(u_\eta(t_{n+1/2}), v_\delta(s')) \big),$$

we rewrite $I_1^{\varepsilon,\varepsilon_0}$ as

$$I_1^{\varepsilon,\varepsilon_0}(u_\eta, v_\delta) = \sum_{n=0}^{N-1} \bigg[\big(J_1(t_n, t_{n+1/2}) - J_1(t_{n+1/2}, t_{n+1}) \big) \qquad (4.49)$$

$$+ \big(J_2(t_n, t_{n+1/2}) - J_2(t_{n+1/2}, t_{n+1}) \big) \bigg],$$

with

$$J_1(\tau_1, \tau_2) = \iint \int_0^T \iint \int_{\tau_1}^{\tau_2} q^{f_\delta}(u_\eta(x, y, t_{n+1/2}), v_\delta(x', y', s'))$$

$$\times \Omega_x(x, x', y, y', s, s') \, ds \, dx \, dy \, ds' \, dx' \, dy',$$

$$J_2(\tau_1, \tau_2) = \iint \int_0^T \iint \int_{\tau_1}^{\tau_2} \Big(q^{f_\delta}(u_\eta(x, y, s), v_\delta(x', y', s'))$$

$$- q^{f_\delta}(u_\eta(x, y, t_{n+1/2}), v_\delta(x', y', s')) \Big)$$

$$\times \Omega_x(x, x', y, y', s, s') \, ds \, dx \, dy \, ds' \, dx' \, dy'.$$

140   4. Multidimensional Scalar Conservation Laws

Here we have written out all the variables explicitly; however, in the following we will display only the relevant variables. All spatial integrals are over the real line unless specified otherwise. Rewriting

$$w_{\varepsilon_0}(s-s') = w_{\varepsilon_0}(t_{n+1/2} - s') + \int_{t_{n+1/2}}^{s} w'_{\varepsilon_0}(\bar{s} - s')\, d\bar{s},$$

we obtain

$$J_1(t_n, t_{n+1/2})$$
$$= \iint \int_0^T \iint q^{f_\delta}(u_\eta(t_{n+1/2}), v_\delta(s'))\Omega_x^\varepsilon \left(\int_{t_n}^{t_{n+1/2}} w_{\varepsilon_0}(t_{n+1/2} - s')\, ds \right.$$
$$\left. + \int_{t_n}^{t_{n+1/2}} \int_{t_{n+1/2}}^{s} w'_{\varepsilon_0}(\bar{s} - s')\, d\bar{s}\, ds \right) dx\, dy\, ds'\, dx'\, dy'$$
$$= \iint \int_0^T \iint q^{f_\delta}(u_\eta(t_{n+1/2}), v_\delta(s'))\Omega_x^\varepsilon \left(\frac{\Delta t}{2} w_{\varepsilon_0}(t_{n+1/2} - s') \right.$$
$$\left. + \int_{t_n}^{t_{n+1/2}} \int_{t_{n+1/2}}^{s} w'_{\varepsilon_0}(\bar{s} - s')\, d\bar{s}\, ds \right) dx\, dy\, ds'\, dx'\, dy',$$

where $\Omega^\varepsilon = w_\varepsilon(x - x')w_\varepsilon(y - y')$ denotes the spatial part of Ω.

If we rewrite $J_1(t_{n+1/2}, t_{n+1})$ in the same way, we obtain

$$J_1(t_{n+1/2}, t_{n+1})$$
$$= \iint \int_0^T \iint q^{f_\delta}(u_\eta(t_{n+1/2}), v_\delta(s'))\Omega_x^\varepsilon \left(\frac{\Delta t}{2} w_{\varepsilon_0}(t_{n+1/2} - s') \right.$$
$$\left. + \int_{t_{n+1/2}}^{t_{n+1}} \int_{t_{n+1/2}}^{s} w'_{\varepsilon_0}(\bar{s} - s')\, d\bar{s}\, ds \right) dx'\, dy'\, ds'\, dx\, dy,$$

and hence

$$J_1(t_n, t_{n+1/2}) - J_1(t_{n+1/2}, t_{n+1})$$
$$= \iint \int_0^T \iint q^{f_\delta}(u_\eta(t_{n+1/2}), v_\delta(s'))\Omega_x^\varepsilon \left(\int_{t_n}^{t_{n+1/2}} \int_{t_{n+1/2}}^{s} w'_{\varepsilon_0}(\bar{s} - s')\, d\bar{s}\, ds \right.$$
$$\left. - \int_{t_{n+1/2}}^{t_{n+1}} \int_{t_{n+1/2}}^{s} w'_{\varepsilon_0}(\bar{s} - s')\, d\bar{s}\, ds \right) dx\, dy\, ds'\, dx'\, dy'. \qquad (4.50)$$

Now using the Lipschitz continuity of q^{f_δ} we can replace variation in q^{f_δ} by variation in u, and obtain, using $\iint w'_{\varepsilon_0}(x - x')\, dx\, dx' = 0$, that

$$\left| \iint q^{f_\delta}(u_\eta(x, y, t_{n+1/2}), v_\delta(s'))w'_{\varepsilon_0}(x - x')\, dx\, dx' \right|$$
$$= \left| \iint w'_{\varepsilon_0}(x - x')\, dx\, dx' \right.$$
$$\left. \times \left[q^{f_\delta}(u_\eta(x, y, t_{n+1/2}), v_\delta(s')) - q^{f_\delta}(u_\eta(x', y, t_{n+1/2}), v_\delta(s')) \right] \right|$$

$$\leq \|f_\delta\|_{\text{Lip}} \iint |\omega'_{\varepsilon_0}(x - x')|$$
$$\times |u_\eta(x, y, t_{n+1/2}) - u_\eta(x', y, t_{n+1/2})| \, dx \, dx'$$
$$= \|f_\delta\|_{\text{Lip}} \iint |u_\eta(x' + z, y, t_{n+1/2}) - u_\eta(x', y, t_{n+1/2})| \, |\omega'_{\varepsilon_0}(z)| \, dx' \, dz$$
$$\leq \|f_\delta\|_{\text{Lip}} \int \frac{1}{|z|} \int |u_\eta(x' + z, y, t_{n+1/2}) - u_\eta(x', y, t_{n+1/2})| \, dx'$$
$$\times |z \omega'_{\varepsilon_0}(z)| \, dz$$
$$\leq \|f_\delta\|_{\text{Lip}} \text{T.V.}_{\cdot x} \left(u_\eta(t_{n+1/2}) \right) \int |z \omega'_{\varepsilon_0}(z)| \, dz$$
$$\leq \|f_\delta\|_{\text{Lip}} \text{T.V.}_{\cdot x} \left(u_\eta(t_{n+1/2}) \right),$$

using that $\int |z \omega'_{\varepsilon_0}(z)| \, dz = 1$. We combine this with (4.50) to get

$$\left| J_1(t_n, t_{n+1/2}) - J_1(t_{n+1/2}, t_{n+1}) \right|$$
$$\leq \|f_\delta\|_{\text{Lip}} \iint \text{T.V.}_{\cdot x} \left(u_\eta(t_{n+1/2}) \right) \omega_{\varepsilon_0}(y - y')$$
$$\times \left(\int_0^T \int_{t_n}^{t_{n+1/2}} \left| \int_{t_{n+1/2}}^s |\omega'_{\varepsilon_0}(\bar{s} - s')| \, d\bar{s} \right| ds \, ds' \right.$$
$$\left. + \int_0^T \int_{t_{n+1/2}}^{t_{n+1}} \left| \int_{t_{n+1/2}}^s |\omega'_{\varepsilon_0}(\bar{s} - s')| \, d\bar{s} \right| ds \, ds' \right) dy' \, dy.$$

Inserting the estimate

$$\int_0^T |\omega'_{\varepsilon_0}(\bar{s} - s')| \, ds' \leq \frac{1}{\varepsilon_0} \int |\omega'(z)| \, dz \leq 2/\varepsilon_0,$$

we obtain

$$\left| J_1(t_n, t_{n+1/2}) - J_1(t_{n+1/2}, t_{n+1}) \right| \leq \frac{\|f_\delta\|_{\text{Lip}} (\Delta t)^2}{2\varepsilon_0} \text{T.V.} \left(u_\eta(t_{n+1/2}) \right).$$
(4.51)

Next we consider the term J_2. We first use the Lipschitz continuity of q^{f_δ}, which yields

$$\left| J_2(t_n, t_{n+1/2}) \right|$$
$$\leq \|f_\delta\|_{\text{Lip}} \iint \int_0^T \iint \int_{t_n}^{t_{n+1/2}} |u_\eta(x, y, s) - u_\eta(x, y, t_{n+1/2})|$$
$$\times |\Omega_x| \, ds \, dx' \, dy' \, ds' \, dx \, dy$$
$$\leq \frac{\|f_\delta\|_{\text{Lip}}}{\varepsilon} \int_{t_n}^{t_{n+1/2}} \iint |u_\eta(x, y, s) - u_\eta(x, y, t_{n+1/2})| \, ds \, dx \, dy$$
$$\leq \frac{\|f_\delta\|_{\text{Lip}}}{\varepsilon} \int_{t_n}^{t_{n+1/2}} \iint |u_\eta(x, y, s) - u_\eta(x, y, t_{n+1/2}-)| \, ds \, dx \, dy$$

$$+ \frac{\|f_\delta\|_{\text{Lip}}\Delta t}{2\varepsilon} \iint \left|u_\eta(x,y,t_{n+1/2}-) - u_\eta(x,y,t_{n+1/2})\right| dx\, dy$$

$$\leq \frac{\|f_\delta\|_{\text{Lip}}\Delta t}{\varepsilon} \left(\|f_\delta\|_{\text{Lip}}\Delta t + \Delta x\right) \text{T.V.}\left(u_\eta\left(t_{n+1/2}\right)\right).$$

Here we integrated to unity in the variables s' and y', and estimated $\int |\omega'_\varepsilon(x-x')|\, dx'$ by $2/\varepsilon$. Finally, we used the continuity in time of the L^1-norm in the x-direction and estimated the error due to the projection. A similar bound can be obtained for $J_2(t_{n+1/2}, t_{n+1})$, and hence

$$\left|J_2(t_n, t_{n+1/2}) - J_2(t_{n+1/2}, t_{n+1})\right|$$
$$\leq \left|J_2(t_n, t_{n+1/2})\right| + \left|J_2(t_{n+1/2}, t_{n+1})\right|$$
$$\leq \frac{\|f\|_{\text{Lip}}\Delta t}{\varepsilon} \left(2\|f\|_{\text{Lip}}\Delta t + \Delta x + \Delta y\right) \text{T.V.}(u_\eta(t_n)), \quad (4.52)$$

where we used that $\text{T.V.}\left(u_\eta(t_{n+1/2})\right) \leq \text{T.V.}(u_\eta(t_n))$. Inserting estimates (4.51) and (4.52) into (4.49) yields

$$|I_1^{\varepsilon,\varepsilon_0}(u_\eta, v_\delta)| \leq \|f_\delta\|_{\text{Lip}} \text{T.V.}(u_\eta(0))$$
$$\times \sum_{n=0}^{N-1} \left(\frac{(\Delta t)^2}{2\varepsilon_0} + \frac{\Delta t}{2\varepsilon}(2\|f_\delta\|_{\text{Lip}}\Delta t + \Delta x + \Delta y)\right)$$
$$\leq T \|f_\delta\|_{\text{Lip}} \text{T.V.}(u_\eta(0))$$
$$\times \left(\frac{\Delta t}{2\varepsilon_0} + \frac{1}{2\varepsilon}(2\|f_\delta\|_{\text{Lip}}\Delta t + \Delta x + \Delta y)\right),$$

where we again used that $\text{T.V.}(u_\eta)$ is nonincreasing. An analogous argument gives the same estimate for $I_2^{\varepsilon,\varepsilon_0}$. Adding the two inequalities, we conclude that (4.48) holds. $\square$

It remains to estimate $I_3^{\varepsilon,\varepsilon_0}$ and $I_4^{\varepsilon,\varepsilon_0}$. We aim at the following result.

Lemma 4.7. *The following estimate holds:*

$$|I_3^{\varepsilon,\varepsilon_0}| + |I_4^{\varepsilon,\varepsilon_0}| \leq \frac{T(\Delta x + \Delta y)^2}{\Delta t\, \varepsilon} \text{T.V.}(u_0).$$

Proof. We discuss the term $I_3^{\varepsilon,\varepsilon_0}$ only. Recall that

$$I_3^{\varepsilon,\varepsilon_0}(u_\eta, v_\delta)$$
$$= \sum_{n=1}^{N-1} \iint \int_0^T \iint \Big(|u_\eta(x,y,t_n) - v_\delta(x',y',s')|$$
$$- |u_\eta(x,y,t_n-) - v_\delta(x',y',s')| \Big)$$
$$\times \Omega(x,x',y,y',t_n,s')\, dx'\, dy'\, ds'\, dx\, dy.$$

The function $u_\eta(x, y, t_n+)$ is the projection of $u_\eta(x, y, t_n-)$, that is,

$$u_\eta(x, y, t_n+) = \frac{1}{\Delta x \Delta y} \iint_{I_{ij}} u_\eta(\bar{x}, \bar{y}, t_n-) \, d\bar{x} \, d\bar{y}. \tag{4.53}$$

If we replace $\iint_{\mathbb{R}^2}$ by $\sum_{i,j} \iint_{I_{ij}}$ and use (4.53), we obtain

$$I_3^{\varepsilon,\varepsilon_0}(u_\eta, v_\delta)$$

$$= \sum_{n=1}^{N-1} \iint \int_0^T \sum_{i,j} \iint_{I_{ij}} \left[\left| \frac{1}{\Delta x \Delta y} \iint_{I_{ij}} u_\eta(\bar{x}, \bar{y}, t_n-) \, d\bar{x} \, d\bar{y} - v_\delta(x', y', s') \right| \right.$$

$$\left. - |u_\eta(x, y, t_n-) - v_\delta(x', y', s')| \right] \Omega(x, x', y, y', t_n, s') \, dx \, dy \, ds' \, dx' \, dy'$$

$$= \frac{1}{\Delta x \Delta y} \sum_{n=1}^{N-1} \iint \int_0^T \Omega(x, x', y, y', t_n, s')$$

$$\times \sum_{i,j} \iint_{I_{ij}} \iint_{I_{ij}} \Big(|u_\eta(\bar{x}, \bar{y}, t_n-) - v_\delta(x', y', s')|$$

$$- |u_\eta(x, y, t_n-) - v_\delta(x', y', s')| \Big) \, d\bar{x} \, d\bar{y} \, dx \, dy \, ds' \, dx' \, dy'$$

$$= \frac{1}{2\Delta x \Delta y} \sum_{n=1}^{N-1} \iint \int_0^T \Omega(x, x', y, y', t_n, s')$$

$$\times \sum_{i,j} \iint_{I_{ij}} \iint_{I_{ij}} \Big(|u_\eta(\bar{x}, \bar{y}, t_n-) - v_\delta(x', y', s')|$$

$$- |u_\eta(x, y, t_n-) - v_\delta(x', y', s')| \Big) \, d\bar{x} \, d\bar{y} \, dx \, dy \, ds' \, dx' \, dy'$$

$$+ \frac{1}{2\Delta x \Delta y} \sum_{n=1}^{N-1} \iint \int_0^T \Omega(\bar{x}, x', \bar{y}, y', t_n, s')$$

$$\times \sum_{i,j} \iint_{I_{ij}} \iint_{I_{ij}} \Big(|u_\eta(x, y, t_n-) - v_\delta(x', y', s')|$$

$$- |u_\eta(\bar{x}, \bar{y}, t_n-) - v_\delta(x', y', s')| \Big) \, dx \, dy \, d\bar{x} \, d\bar{y} \, ds' \, dx' \, dy'$$

$$= \frac{1}{2\Delta x \Delta y} \sum_{n=1}^{N-1} \iint \int_0^T \Big(\Omega(x, x', y, y', t_n, s') - \Omega(\bar{x}, x', \bar{y}, y', t_n, s') \Big)$$

$$\times \sum_{i,j} \iint_{I_{ij}} \iint_{I_{ij}} \Big(|u_\eta(\bar{x}, \bar{y}, t_n-) - v_\delta(x', y', s')|$$

$$- |u_\eta(x, y, t_n-) - v_\delta(x', y', s')| \Big) \, d\bar{x} \, d\bar{y} \, dx \, dy \, ds' \, dx' \, dy'.$$

Estimating $I_3^{\varepsilon,\varepsilon_0}(u_\eta, v_\delta)$ using the inverse triangle inequality we obtain

$$\left|I_3^{\varepsilon,\varepsilon_0}(u_\eta, v_\delta)\right|$$

$$\leq \frac{1}{2\Delta x \Delta y} \sum_{n=1}^{N-1} \iint \int_0^T \sum_{i,j} \iint_{I_{ij}} \iint_{I_{ij}} |u_\eta(\bar{x}, \bar{y}, t_n-) - u_\eta(x, y, t_n-)|$$
$$\times |\Omega(x, x', y, y', t_n, s') - \Omega(\bar{x}, x', \bar{y}, y', t_n, s')| \, d\bar{x} \, d\bar{y} \, dx \, dy \, ds' \, dx' \, dy'. \tag{4.54}$$

The next step is to bound the test functions in (4.54) from above. To this end we first consider for $x, \bar{x} \in \langle i\Delta x, (i+1)\Delta x \rangle$,

$$\int |\omega_\varepsilon(x - x') - \omega_\varepsilon(\bar{x} - x')| \, dx' = \int |\omega(z) - \omega(z + (\bar{x} - x)/\varepsilon)| \, dz$$
$$= \int \left| \int_z^{z+(\bar{x}-x)/\varepsilon} \omega'(\xi) \, d\xi \right| dz$$
$$\leq \int \int_z^{z+(\bar{x}-x)/\varepsilon} |\omega'(\xi)| \, d\xi \, dz$$
$$\leq \int \int_0^{\Delta x/\varepsilon} |\omega'(\alpha + \beta)| \, d\alpha \, d\beta = \frac{2\Delta x}{\varepsilon}.$$

Integrating the time variable to unity we easily see (really, this is easy!) that

$$\iint \int_0^T |\Omega(x, x', y, y', t_n, s') - \Omega(\bar{x}, x', \bar{y}, y', t_n, s')| \, ds' \, dx' dy'$$
$$= \int_0^T \omega_{\varepsilon_0}(s - s') \, ds'$$
$$\times \iint |\omega_\varepsilon(x - x')\omega_\varepsilon(y - y') - \omega_\varepsilon(\bar{x} - x')\omega_\varepsilon(\bar{y} - y')| \, dx' dy'$$
$$\leq \iint |\omega_\varepsilon(x - x') - \omega_\varepsilon(\bar{x} - x')| \omega_\varepsilon(y - y') \, dx' dy'$$
$$+ \iint |\omega_\varepsilon(y - y') - \omega_\varepsilon(\bar{y} - y')| \omega_\varepsilon(\bar{x} - x') \, dx' dy'$$
$$\leq \int |\omega_\varepsilon(x - x') - \omega_\varepsilon(\bar{x} - x')| \, dx' + \int |\omega_\varepsilon(y - y') - \omega_\varepsilon(\bar{y} - y')| \, dy'$$
$$\leq (\Delta x + \Delta y)\frac{2}{\varepsilon}. \tag{4.55}$$

Furthermore,

$$|u_\eta(\bar{x}, \bar{y}, t_n-) - u_\eta(x, y, t_n-)| = |u_\eta(x, \bar{y}, t_n-) - u_\eta(x, y, t_n-)|$$
$$\leq \text{T.V.}_{\langle j\Delta y, (j+1)\Delta y \rangle}(u_\eta(x, \cdot, t_n-)). \tag{4.56}$$

Inserting (4.55) and (4.56) into (4.54) yields

$$|I_3^{\varepsilon,\varepsilon_0}(u_\eta,v_\delta)|$$
$$\leq \frac{1}{2\Delta x\Delta y}\frac{2(\Delta x+\Delta y)}{\varepsilon}$$
$$\times \sum_{n=1}^{N-1}\sum_{i,j}\iint_{I_{ij}}\iint_{I_{ij}}\text{T.V.}_{\langle j\Delta y,(j+1)\Delta y\rangle}(u_\eta(x,\cdot,t_n-))\,d\bar{x}\,d\bar{y}\,dx\,dy$$
$$\leq \frac{\Delta x+\Delta y}{\varepsilon\Delta x\Delta y}\sum_{n=1}^{N-1}\Delta x(\Delta y)^2\sum_{i,j}\int_{i\Delta x}^{(i+1)\Delta x}\text{T.V.}_{\langle j\Delta y,(j+1)\Delta y\rangle}(u_\eta(x,\cdot,t_n-))$$
$$\leq \frac{(\Delta x+\Delta y)}{\varepsilon}\Delta y\sum_{n=1}^{N-1}\text{T.V.}(u_\eta(t_n-))$$
$$\leq \frac{(\Delta x+\Delta y)}{\varepsilon}\Delta y\frac{T}{\Delta t}\text{T.V.}(u_\eta(0))\,, \tag{4.57}$$

where in the final step we used that $\text{T.V.}(u_\eta(t_n-))\leq \text{T.V.}(u_\eta(0))$.
The same analysis provides the following estimate for $I_4^{\varepsilon,\varepsilon_0}(v_\delta,u_\eta)$:

$$|I_4^{\varepsilon,\varepsilon_0}(u_\eta,v_\delta)|\leq \frac{(\Delta x+\Delta y)}{\varepsilon}\Delta x\frac{T}{\Delta t}\text{T.V.}(u_\eta(0))\,. \tag{4.58}$$

Adding (4.57) and (4.58) proves the lemma. □

We now return to the proof of the estimate of $\Lambda_{\varepsilon,\varepsilon_0}(u_\eta,v_\delta)$. Combining Lemma 4.6 and Lemma 4.7 we obtain

$$-\Lambda_{\varepsilon,\varepsilon_0}(u_\eta,v_\delta)$$
$$\leq |I_1^{\varepsilon,\varepsilon_0}(u_\eta,v_\delta)|+|I_2^{\varepsilon,\varepsilon_0}(u_\eta,v_\delta)|+|I_3^{\varepsilon,\varepsilon_0}(u_\eta,v_\delta)|+|I_4^{\varepsilon,\varepsilon_0}(u_\eta,v_\delta)|$$
$$\leq T\bigg[\bigg(\frac{\Delta t}{\varepsilon_0}+\frac{1}{\varepsilon}(\{\|f_\delta\|_{\text{Lip}}+\|g_\delta\|_{\text{Lip}}\}\Delta t+\Delta x+\Delta y)\bigg)$$
$$\times \max\{\|f_\delta\|_{\text{Lip}},\|g_\delta\|_{\text{Lip}}\}+\frac{(\Delta x+\Delta y)^2}{\Delta t\,\varepsilon}\bigg]\text{T.V.}(u_0)$$
$$=:T\,\text{T.V.}(u_0)\,\Lambda(\varepsilon,\varepsilon_0,\eta). \tag{4.59}$$

Returning to (4.44), we combine (4.45), (4.46), as well as (4.59), to obtain

$$\|S(T)u_0-u_\eta(T)\|_1$$
$$\leq \|S(T)u_0-S_\delta(T)u_0\|_1+\|S_\delta(T)u_0-u_\eta(T)\|_1$$
$$\leq T\max\{\|f-f_\delta\|_{\text{Lip}},\|g-g_\delta\|_{\text{Lip}}\}\text{T.V.}(u_0)+\|u_0-u^0\|_1$$
$$+\bigg(2\varepsilon+\max\{\|f_\delta\|_{\text{Lip}},\|g_\delta\|_{\text{Lip}}\}\varepsilon_0+\varepsilon_0\Big(C+\mathcal{O}\Big(\frac{\max\{\Delta x,\Delta y\}}{\Delta t}\Big)\Big)$$
$$+T\Lambda(\varepsilon,\varepsilon_0,\eta)\bigg)\text{T.V.}(u_0)\,. \tag{4.60}$$

Next we take a minimum over ε and ε_0 on the right-hand side of (4.60). This has the form

$$\min_{\varepsilon,\varepsilon_0} \left(a\varepsilon + \frac{b}{\varepsilon} + c\varepsilon_0 + \frac{d}{\varepsilon_0}\right) = 2\sqrt{ab} + 2\sqrt{cd}.$$

The minimum is obtained for $\varepsilon = \sqrt{b/a}$ and $\varepsilon_0 = \sqrt{d/c}$. We obtain

$$\|S(T)u_0 - u_\eta(T)\|_1$$
$$\leq T \max\{\|f - f_\delta\|_{\text{Lip}}, \|g - g_\delta\|_{\text{Lip}}\} \text{T.V.}(u_0) + \|u_0 - u^0\|_1$$
$$+ \mathcal{O}\left(\left((\Delta x + \Delta y) + \Delta t + \frac{(\Delta x + \Delta y)^2}{\Delta t}\right)^{1/2}\right) \text{T.V.}(u_0). \quad (4.61)$$

We may choose the approximation of the initial data such that $\|u_0 - u^0\|_1 = \mathcal{O}(\Delta x + \Delta y) \text{T.V.}(u_0)$. Furthermore, if the flux functions f and g are piecewise C^2 and Lipschitz continuous, then

$$\|f - f_\delta\|_{\text{Lip}} \leq \delta \|f''\|_\infty.$$

We state the final result in the general case.

Theorem 4.8. *Let u_0 be a function in $L^1(\mathbb{R}^m) \cap L^\infty(\mathbb{R}^m)$ with bounded total variation, and let f_j for $j = 1, \ldots, m$ be piecewise C^2 functions that in addition are Lipschitz continuous. Then*

$$\|u(T) - u_\eta(T)\|_1 \leq \mathcal{O}\left(\delta + (\Delta x + \Delta y)^{1/2}\right)$$

as $\eta \to 0$ when

$$\Delta x = K_1 \Delta y = K_2 \Delta t$$

for constants K_1 and K_2.

It is worthwhile to analyze the error terms in the estimate. We are clearly making three approximations with the front-tracking method combined with dimensional splitting. First of all, we are approximating the initial data by step functions. That gives an error of order Δx. Secondly, we are approximating the flux functions by piecewise linear and continuous functions; in this case the error is of order δ. A third source is the intrinsic error in the dimensional splitting, which is of order $(\Delta t)^{1/2}$, and finally, the projection onto the grid gives an error of order $(\Delta x)^{1/2}$.

The advantage of this method over difference methods is the fact that the time step Δt is not bounded by a CFL condition expressed in terms of Δx and Δy. The only relation that must be satisfied is (4.25), which allows for taking large time steps. In practice it is observed that one can choose CFL numbers[5] as high as 10–15 without loss in accuracy. This makes it a very fast method.

---

[5] In several dimensions the CFL number is defined as $\max_i(|f_i'| \Delta t/\Delta x_i)$.

4.4 Operator Splitting: Diffusion

> The answer, my friend, is blowin' in the wind,
> the answer is blowin' in the wind.
>
> Bob Dylan, Blowin' in the Wind (1968)

We show how to use the concept of operator splitting to derive a (weak) solution of the parabolic problem[6]

$$u_t + \sum_{j=1}^{m} f_j(u)_{x_j} = \mu \sum_{j=1}^{m} u_{x_j x_j} \tag{4.62}$$

by solving

$$u_t + f_j(u)_{x_j} = 0, \quad j = 1, \ldots, m, \tag{4.63}$$

and

$$u_t = \mu \Delta u, \tag{4.64}$$

where we employ the notation $\Delta u = \sum_j u_{x_j x_j}$. To this end let $S_j(t)u_0$ and $H(t)u_0$ denotes the solutions of (4.63) and (4.64), respectively, with initial data u_0. Introducing the heat kernel we may write

$$\begin{aligned} u(x,t) &= (H(t)u_0)(x,t) \\ &= \int_{\mathbb{R}^m} K(x-y,t)u_0(y)\,dy \\ &= \frac{1}{(4\pi\mu t)^{m/2}} \int_{\mathbb{R}^m} \exp\left(-\frac{|x-y|^2}{4\mu t}\right) u_0(y)\,dy. \end{aligned}$$

Let Δt be positive and $t_n = n\Delta t$. Define

$$u^0 = u_0, \quad u^{n+1} = (H(\Delta t)S_m(\Delta t) \cdots S_1(\Delta t))\,u^n, \tag{4.65}$$

with the idea that u^n approximates $u(x, t_n)$. We will show that u^n converges to the solution of (4.62) as $\Delta t \to 0$.

Lemma 4.9. *The following estimates hold:*

$$\|u^n\|_\infty \leq \|u^0\|_\infty, \tag{4.66}$$
$$\text{T.V.}(u^n) \leq \text{T.V.}(u^0), \tag{4.67}$$

---

[6] Although we have used the parabolic regularization to motivate the appropriate entropy condition, we have constructed the solution of the multidimensional conservation law per se, and hence it is logically consistent to use the solution of the conservation law in combination with operator splitting to derive the solution of the parabolic problem. A different approach, where we start with a solution of the parabolic equation and subsequently show that in the limit of vanishing viscosity the solution converges to the solution of the conservation law, is discussed in Appendix B.

148   4. Multidimensional Scalar Conservation Laws

$$\|u^{n_1} - u^{n_2}\|_{L^1_{\text{loc}}(\mathbb{R}^m)} \leq C\sqrt{|n_1 - n_2|\,\Delta t}. \tag{4.68}$$

Proof. Equation (4.66) is obvious, since both the heat equation and the conservation law obey the maximum principle.

We know that the solution of the conservation law has the TVD property (4.67); see (4.22). Thus it remains to show that this property is shared by the solution of the heat equation. To this end, we have

$$\int_{\mathbb{R}^m} \left| H(t)u(x+h) - H(t)u(x) \right| dx$$

$$\leq \int_{\mathbb{R}^m}\int_{\mathbb{R}^m} |K(x+h-y,t)u(y) - K(x-y,t)u(y)|\,dy\,dx$$

$$= \int_{\mathbb{R}^m}\int_{\mathbb{R}^m} |K(y,t)u(x+h-y) - K(y,t)u(x-y)|\,dy\,dx$$

$$= \int_{\mathbb{R}^m} K(y,t)\,dy \int_{\mathbb{R}^m} |u(x+h) - u(x)|\,dx$$

$$= \int_{\mathbb{R}^m} |u(x+h) - u(x)|\,dx.$$

Dividing by $|h|$ and letting $h \to 0$ we conclude that

$$\text{T.V.}(H(t)u) \leq \text{T.V.}(u),$$

which proves (4.67).

Finally, we consider (4.68). We will first show that the approximate solution obtained by splitting is weakly Lipschitz continuous in time. More precisely, for each ball $\mathcal{B}_r = \{x \mid |x| \leq r\}$, we will show that

$$\left| \int_{\mathcal{B}_r} (u^{n_1} - u^{n_2})\phi \right| \leq C_r\,|n_1 - n_2|\left(\|\phi\|_\infty + \max_j \|\phi_{x_j}\|_\infty \right), \tag{4.69}$$

for smooth test functions $\phi = \phi(x)$, where C_r is a constant depending on r. It is enough to study the case $m_2 = n_1 + 1$, and we set $n_1 = n$. Furthermore, we can write

$$\left| \int (u^{n+1} - u^n)\phi\,dx \right| \leq \left| \int (H(\Delta t)\tilde{u}^n - \tilde{u}^n)\phi\,dx \right| + \left| \int (\tilde{u}^n - u^n)\phi\,dx \right| \tag{4.70}$$

where $\tilde{u}^n = (S_m(\Delta t)\cdots S_1(\Delta t))\,u^n$. This shows that it suffices to prove this property for the solutions of the conservation law and the heat equation separately. From Theorem 4.3 we know that the solution of the one-dimensional conservation law satisfies the stronger estimate

$$\|S(t)u - u\|_1 \leq C\,|t|.$$

This implies that (for simplicity with $m = 2$)

$$\|S_2(t)S_1(t)u - u\|_1 \leq \|S_2(t)S_1(t)u - S_1(t)u\|_1 + \|S_1(t)u - u\|_1$$
$$\leq C\,|t|,$$

and hence we infer that the last term of (4.70) is of order Δt, that is,

$$\|\tilde{u}^n - u^n\|_1 \leq C\|\phi\|_\infty |\Delta t|.$$

The first term can be estimated as follows (for simplicity of notation we assume $m = 1$). Consider

$$\left|\int (H(t)u_0 - u_0)\phi \, dx\right| = \left|\int \int_0^t u_t \, dt \, \phi \, dx\right| = \left|\int \int_0^t u_{xx} \, dt \, \phi \, dx\right|$$

$$\leq \int \int_0^t |u_x \phi_x| \, dt \, dx$$

$$\leq \|\phi_x\|_\infty \int_0^t \int |u_x| \, dx \, dt$$

$$\leq \|\phi_x\|_\infty \int_0^t \text{T.V.}(u) \, dt \leq \|\phi_x\|_\infty \text{T.V.}(u_0) \, t.$$

Thus we conclude that (4.69) holds.

From the TVD property (4.67), we have that

$$\sup_{|\xi| \leq \rho} \int |u^n(x + \xi, t) - u^n(x, t)| \, dx \leq \rho \, \text{T.V.}(u^n). \tag{4.71}$$

Using Kružkov's interpolation lemma (stated and proved right after this proof) we can infer, using (4.69) and (4.71), that

$$\int_{B_r} |u^{n_1}(x) - u^{n_2}(x)| \, dx \leq C_r \left(\varepsilon + \frac{|n_1 - n_2|\Delta t}{\varepsilon}\right)$$

for all $\varepsilon \leq \rho$. Choosing $\varepsilon = \sqrt{|n - m|\Delta t}$ proves the result. $\square$

We next state and prove Kružkov's interpolation lemma. To do this we need the multi-index notation. A vector of the form $\alpha = (\alpha_1, \ldots, \alpha_m)$, where each component is a nonnegative integer, is called a *multi-index* of order $|\alpha| = \alpha_1 + \cdots + \alpha_m$. Given a multi-index α we define

$$D^\alpha u(x) = \frac{\partial^{|\alpha|} u(x)}{\partial x_1^{\alpha_1} \cdots \partial x_m^{\alpha_m}}.$$

Lemma 4.10 (Kružkov interpolation lemma). *Let $u(x, t)$ be a bounded measurable function defined in the cylinder $\mathcal{B}_{r+\hat{r}} \times [0, T]$, $\hat{r} \geq 0$. For $t \in [0, T]$ and $|\rho| \leq \hat{r}$, assume that u possesses a spatial modulus of continuity*

$$\sup_{|\xi| \leq |\rho|} \int_{\mathcal{B}_r} |u(x + \xi, t) - u(x, t)| \, dx \leq \nu_{r,T,\hat{r}}(|\rho|; u), \tag{4.72}$$

where $\nu_{r,T,\hat{r}}$ does not depend on t. Suppose that for any $\phi \in C_0^\infty(\mathcal{B}_r)$ and any $t_1, t_2 \in [0, T]$,

$$\left| \int_{\mathcal{B}_r} \left(u\left(x, t_2\right) - u\left(x, t_1\right) \right) \phi(x) \, dx \right| \\ \leq \mathrm{Const}_{r,T} \left(\sum_{|\alpha| \leq m} \|D^\alpha \phi\|_{L^\infty(\mathcal{B}_r)} \right) |t_2 - t_1|, \quad (4.73)$$

where α denotes a multi-index.

Then for t and $t + \tau \in [0, T]$ and for all $\varepsilon \in \langle 0, \hat{r}]$,

$$\int_{\mathcal{B}_r} |u(x, t+\tau) - u(x, t)| \, dx \leq \mathrm{Const}_{r,T} \left(\varepsilon + \nu_{r,T,\hat{r}}(\varepsilon; u) + \frac{|\tau|}{\varepsilon^m} \right). \quad (4.74)$$

Proof. Let $\delta \in C_0^\infty$ be a function such that

$$0 \leq \delta(x) \leq 1, \quad \operatorname{supp} \delta \subseteq \mathcal{B}_1, \quad \int \delta(x) \, dx = 1,$$

and define

$$\delta_\varepsilon(x) = \frac{1}{\varepsilon^m} \delta\left(\frac{x}{\varepsilon}\right).$$

Furthermore, write $f(x) = u(x, t + \tau) - u(x, t)$ (suppressing the time dependence in the notation for f),

$$\sigma(x) = \operatorname{sign}(f(x)) \text{ for } |x| \leq r - \varepsilon, \text{ and } 0 \text{ otherwise,}$$

and

$$\sigma_\varepsilon(x) = (\sigma * \delta_\varepsilon)(x) = \int \sigma(x - y) \delta_\varepsilon(y) \, dy.$$

By construction, $\sigma_\varepsilon \in C_0^\infty(\mathbb{R}^m)$ and $\operatorname{supp} \sigma_\varepsilon \subseteq \mathcal{B}_r$. Furthermore, $|\sigma_\varepsilon| \leq 1$ and

$$\left| \frac{\partial}{\partial x_j} \sigma_\varepsilon \right| \leq \frac{1}{\varepsilon^m} \int \left| \frac{\partial}{\partial x_j} \delta\left(\frac{x-y}{\varepsilon}\right) \right| \sigma(y) \, dy$$

$$\leq \frac{1}{\varepsilon^{m+1}} \int \left| \delta_{x_j}\left(\frac{x-y}{\varepsilon}\right) \right| \sigma(y) \, dy \leq \frac{C}{\varepsilon}.$$

This easily generalizes to

$$\|D^\alpha \sigma_\varepsilon\|_\infty \leq \frac{C}{\varepsilon^{|\alpha|}}.$$

Next we have the elementary but important inequality

$$\int_{\mathcal{B}_r} |f(x)| \, dx = \left| \int_{\mathcal{B}_r} |f(x)| \, dx \right|$$

$$= \left| \int_{\mathcal{B}_r} \left(|f(x)| - \sigma_\varepsilon(x) f(x) + \sigma_\varepsilon(x) f(x) \right) dx \right|$$

$$\leq \left| \int_{B_r} (|f(x)| - \sigma_\varepsilon(x) f(x)) \, dx \right| + \left| \int_{B_r} \sigma_\varepsilon(x) f(x) \, dx \right|$$

$$\leq \int_{B_r} ||f(x)| - \sigma_\varepsilon(x) f(x)| \, dx + \left| \int_{B_r} \sigma_\varepsilon(x) f(x) \, dx \right|$$

$$:= I_1 + I_2.$$

We estimate I_1 and I_2 separately. Starting with I_1 we obtain

$$I_1 = \int_{B_r} ||f(x)| - \sigma_\varepsilon(x) f(x)| \, dx$$

$$= \int_{B_r} \left| |f(x)| \frac{1}{\varepsilon^m} \int \delta(\frac{x-y}{\varepsilon}) \, dy - \frac{1}{\varepsilon^m} \int \delta(\frac{x-y}{\varepsilon}) \sigma(y) \, dy \, f(x) \right| dx$$

$$= \frac{1}{\varepsilon^m} \int \int \delta(\frac{x-y}{\varepsilon}) \, ||f(x)| - \sigma(y) f(x)| \, dy \, dx.$$

The integrand is integrated over the domain

$$\{(x, y) \mid |x| \leq r, \, |x - y| \leq \varepsilon\};$$

see Figure 4.2. We further divide this set into two parts: (i) $|y| \geq r - \varepsilon$,

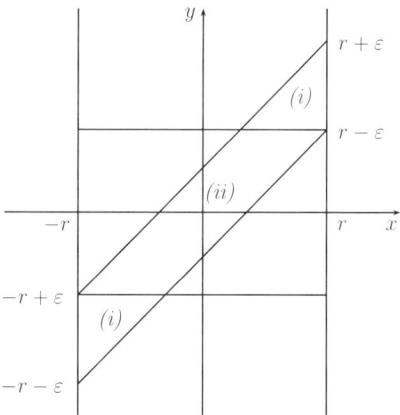

Figure 4.2. The integration domain.

and (ii) $|y| \leq r - \varepsilon$; see Figure 4.2. In case (i) we have

$$||f(x)| - \sigma(y) f(x)| = |f(x)|,$$

since $\sigma(y) = 0$ whenever $|y| \geq r - \varepsilon$. In case (ii)

$$||f(x)| - \sigma(y) f(x)| = ||f(x)| - \text{sign}(f(y)) f(x)| \leq 2 |f(x) - f(y)|$$

152   4. Multidimensional Scalar Conservation Laws

using the elementary inequality
$$||a| - \text{sign}(b)\,a| = ||a| - |b| + \text{sign}(b)(b-a)|$$
$$\leq ||a| - |b|| + |\text{sign}(b)(b-a)|$$
$$\leq 2|a-b|.$$

Thus
$$I_1 \leq \frac{2}{\varepsilon^m} \int_{\mathcal{B}_r}\int_{\mathcal{B}_{r-\varepsilon}} \delta(\frac{x-y}{\varepsilon})|f(x)-f(y)|\,dy\,dx$$
$$+ \frac{1}{\varepsilon^m} \int_{\mathcal{B}_r}\int_{|y|\geq r-\varepsilon} \delta(\frac{x-y}{\varepsilon})|f(x)|\,dy\,dx$$
$$\leq 2\int_{\mathcal{B}_r}\int_{\mathcal{B}_1} \delta(z)|f(x)-f(x-\varepsilon z)|\,dz\,dx$$
$$+ \|f\|_\infty \int_{\mathcal{B}_r}\int_{|y|\geq r-\varepsilon} \delta(\frac{x-y}{\varepsilon})\,dy\,dx$$
$$\leq 2\int_{\mathcal{B}_1} \delta(z)\sup_{|\xi|\leq\varepsilon}\int_{\mathcal{B}_r} |f(x)-f(x+\xi)|\,dx\,dz$$
$$+ \|f\|_\infty \int_{\mathcal{B}_{r+\varepsilon}\setminus\mathcal{B}_{r-\varepsilon}} \frac{1}{\varepsilon^m}\int_{\mathcal{B}_r} \delta(\frac{x-y}{\varepsilon})\,dx\,dy$$
$$\leq 2\nu(\varepsilon;f) + \|f\|_\infty \text{vol}(\mathcal{B}_{r+\varepsilon}\setminus\mathcal{B}_{r-\varepsilon})$$
$$\leq 2\nu(\varepsilon;f) + \|f\|_\infty C_r \varepsilon.$$

Furthermore,
$$\nu(\varepsilon;f) \leq 2\nu(\varepsilon;u).$$

The second term I_2 is estimated by the assumptions of the lemma, namely,
$$I_2 = \left|\int_{\mathcal{B}_r} \sigma_\varepsilon(x)f(x)\,dx\right| \leq \text{Const}_{r,T}\left(\sum_{|\alpha|\leq m} \|D^\alpha \sigma_\varepsilon\|_{L^\infty(\mathcal{B}_r)}\right)|\tau| \leq C\frac{|\tau|}{\varepsilon^m}.$$

Combining the two estimates we conclude that
$$\int_{\mathcal{B}_r} |u(x,t+\tau) - u(x,t)|\,dx \leq C_r\left(\varepsilon + \nu_{r,T,\hat{r}}(\varepsilon;u) + \frac{|\tau|}{\varepsilon^m}\right).$$
□

Next we need to extend the function u^n to all times. First, define
$$u^{n+j/(m+1)} = S_j u^{n+(j-1)/(m+1)}, \quad j=1,\ldots,m.$$

Now let
$$u_{\Delta t}(x,t) = \begin{cases} S_j((m+1)(t-t_{n+(j-1)/(m+1)}))u^{n+(j-1)/(m+1)} \\ \qquad \text{for } t \in [t_{n+(j-1)/(m+1)}, t_{n+j/(m+1)}), \\ H((m+1)(t-t_{n+m/(m+1)}))u^{n+m/(m+1)} \\ \qquad \text{for } t \in [t_{n+m/(m+1)}, t_{n+1}). \end{cases} \quad (4.75)$$

The estimates in Lemma 4.9 carry over to the function $u_{\Delta t}$. Fix $T > 0$. Applying Theorem A.8 we conclude that there exists a sequence of $\Delta t \to 0$ such that for each $t \in [0, T]$ the function $u_{\Delta t}(t)$ converges to a function $u(t)$, and the convergence is in $C([0, T]; L^1_{\text{loc}}(\mathbb{R}^m))$. It remains to show that u is a weak solution of (4.62), or

$$\int_{\mathbb{R}^m} \int_0^t (u\phi_t + f(u) \cdot \nabla\phi - \varepsilon u \Delta\phi) \, dt \, dx = 0 \qquad (4.76)$$

for all smooth and compactly supported test functions ϕ. We have

$$\int_{\mathbb{R}^m} \int_{t_{n+(j-1)/(m+1)}}^{t_{n+j/(m+1)}} \left(\frac{1}{m+1} u_{\Delta t} \phi_t + f(u_{\Delta t}) \cdot \nabla \phi \right) dt \, dx$$

$$= \frac{1}{m+1} \int_{\mathbb{R}^m} \int_0^{\Delta t} \left(u^{n+(j-1)/(m+1)}(x, \tilde{t}) \phi_t\left(x, \frac{\tilde{t} - t_{n+(j-1)/(m+1)}}{m+1}\right) \right.$$
$$\left. + f(u^{n+(j-1)/(m+1)}) \cdot \nabla \phi\left(x, \frac{\tilde{t} - t_{n+(j-1)/(m+1)}}{m+1}\right) \right) d\tilde{t} \, dx$$

$$= \frac{1}{m+1} \int_{\mathbb{R}^m} (u_{\Delta t} \phi) \Big|_{t=t_{n+(j-1)/(m+1)}}^{t=t_{n+j/(m+1)}} dx, \qquad (4.77)$$

for $j = 1, \ldots, m$, where we have used that $u^{n+(j-1)/(m+1)}$ is a solution of the conservation law on the strip $t \in [t_{n+(j-1)/(m+1)}, t_{n+j/(m+1)}\rangle$. Similarly, we find for the solution of the heat equation that

$$\int_{\mathbb{R}^m} \int_{t_{n+m/(m+1)}}^{t_{n+1}} \left(\frac{1}{m+1} u_{\Delta t} \phi_t + \varepsilon u_{\Delta t} \Delta \phi \right) dt \, dx$$
$$= \frac{1}{m+1} \int_{\mathbb{R}^m} \left((u_{\Delta t} \phi)|_{t=t_{n+m/(m+1)}} - (u_{\Delta t} \phi)|_{t=t_{n+1}} \right) dx. \qquad (4.78)$$

Summing (4.77) for $j = 1, \ldots, m$, and adding the result to (4.78), we obtain

$$\int_{\mathbb{R}^m} \int_0^t \left(\frac{1}{m+1} u_{\Delta t} \phi_t + f_{\Delta t}(u_{\Delta t}) \cdot \nabla \phi - \varepsilon \chi_{m+1} u_{\Delta t} \Delta \phi \right) dt \, dx = 0, \qquad (4.79)$$

where

$$f_{\Delta t} = (\chi_1 f_1, \ldots, \chi_m f_m)$$

and

$$\chi_j = \begin{cases} 1 & \text{for } t \in \cup_n [t_{n+(j-1)/(m+1)}, t_{n+j/(m+1)}\rangle, \\ 0 & \text{otherwise.} \end{cases}$$

As $\Delta t \to 0$, we have $\chi_j \overset{*}{\rightharpoonup} 1/(m+1)$, which proves (4.76). We summarize the result as follows.

Theorem 4.11. *Let u_0 be a function in $L^\infty(\mathbb{R}^m) \cap L^1(\mathbb{R}^m) \cap BV(\mathbb{R}^m)$, and assume that f_j are Lipschitz continuous functions for $j = 1, \ldots, m$. Define* the family of functions $\{u_{\Delta t}\}$ by (4.65) and (4.75). Fix $T>0$. Then there exists a sequence of $\Delta t \to 0$ such that $\{u_{\Delta t}(t)\}$ converges to a weak solution u of (4.62). The convergence is in $C([0,T]; L^1_{\text{loc}}(\mathbb{R}^m))$.

One can prove that a weak solution of (4.62) is indeed a classical solution; see [112]. Hence, by uniqueness of classical solutions, the sequence $\{u_{\Delta t}\}$ converges for *any* sequence $\{\Delta t\}$ tending to zero.

4.5 Operator Splitting: Source

> Experience must be our only guide;
> Reason may mislead us.
>
> J. Dickinson, the Constitutional Convention (1787)

We will use operator splitting to study the inhomogeneous conservation law

$$u_t + \sum_{j=1}^m f_j(u)_{x_j} = g(x,t,u), \quad u|_{t=0} = u_0, \tag{4.80}$$

where the source term g is assumed to be continuous in (x,t) and Lipschitz continuous in u. In this case the Kružkov entropy condition reads as follows. The bounded function u is a weak entropy solution on $[0,T]$ if it satisfies

$$\int_0^T \int_{\mathbb{R}^m} \left(|u-k|\varphi_t + \text{sign}\,(u-k) \sum_{j=1}^m (f_j(u)-f_j(k))\,\varphi_{x_j} \right) dx_1 \cdots dx_m\, dt$$

$$+ \int_{\mathbb{R}^m} |u_0 - k|\varphi|_{t=0}\, dx_1 \cdots dx_m - \int_{\mathbb{R}^m} (|u-k|\varphi)|_{t=T}\, dx_1 \cdots dx_m$$

$$\geq \int_0^T \int_{\mathbb{R}^m} \text{sign}\,(u-k)\,\varphi g(x,t,u)\, dx_1 \cdots dx_m\, dt, \tag{4.81}$$

for all constants $k \in \mathbb{R}$ and all nonnegative test functions $\varphi \in C_0^\infty(\mathbb{R}^m \times [0,T])$.

To simplify the presentation we consider only the case with $m=1$, and where $g = g(u)$. Thus

$$u_t + f(u)_x = g(u). \tag{4.82}$$

The case where g also depends on (x,t) is treated in Exercise 4.7. Let $S(t)u_0$ and $R(t)u_0$ denote the solutions of

$$u_t + f(u)_x = 0, \quad u|_{t=0} = u_0, \tag{4.83}$$

and

$$u_t = g(u), \quad u|_{t=0} = u_0, \tag{4.84}$$

4.5. Operator Splitting: Source

respectively. Define the sequence $\{u^n\}$ by (we still use $t_n = n\Delta t$)
$$u^0 = u_0, \quad u^{n+1} = (S(\Delta t) R(\Delta t)) u^n$$
for some positive Δt. Furthermore, we need the extension to all times, defined by[7]

$$u_{\Delta t}(x,t) = \begin{cases} S(2(t-t_n))u^n & \text{for } t \in [t_n, t_{n+1/2}), \\ R\left(2\left(t-t_{n+1/2}\right)\right) u^{n+1/2} & \text{for } t \in [t_{n+1/2}, t_{n+1}), \end{cases} \quad (4.85)$$

with
$$u^{n+1/2} = S(\Delta t) u^n, \quad t_{n+1/2} = \left(n + \frac{1}{2}\right) \Delta t.$$

For this procedure to be well-defined, we must be sure that the ordinary differential equation (4.84) is well-defined. This is the case if g is uniformly Lipschitz continuous in u, i.e.,
$$|g(u) - g(v)| \le \|g\|_{\text{Lip}} |u-v|. \quad (4.86)$$
For convenience, we set $\gamma = \|g\|_{\text{Lip}}$. This assumption also implies that the solution of (4.84) does not "blowup" in finite time, since
$$|g(u)| \le |g(0)| + \gamma |u| \le C_g(1 + |u|), \quad (4.87)$$
for some constant C_g. Under this assumption on g we have the following lemma.

Lemma 4.12. *Assume that u_0 is a function in $L^1_{\text{loc}}(\mathbb{R})$, and that u_0 is of bounded variation. Then for $n\Delta t \le T$, the following estimates hold:*

(i) *There is a constant M_1 independent of n and Δt such that*
$$\|u^n\|_\infty \le M_1. \quad (4.88)$$

(ii) *There is a constant M_2 independent of n and Δt such that*
$$\text{T.V.}(u^n) \le M_2. \quad (4.89)$$

(iii) *There is a constant M_3 independent of n and Δt such that for t_1 and t_2, with $0 \le t_1 \le t_2 \le T$, and for each bounded interval $B \subset \mathbb{R}$,*
$$\int_B |u_{\Delta t}(x,t_1) - u_{\Delta t}(x,t_2)| \, dx \le M_3 |t_1 - t_2|. \quad (4.90)$$

Proof. We start by proving (i). The solution operator S_t obeys a maximum principle, so that $\|u^{n+1/2}\|_\infty \le \|u^n\|_\infty$. Multiplying (4.84) by $\text{sign}(u)$, we find that
$$|u|_t = \text{sign}(u) g(u) \le |g(u)| \le C_g(1+|u|),$$

---
[7] Essentially replacing the operator H used in operator splitting with respect to diffusion by R in the case of a source.

where we have used (4.87). By Gronwall's inequality (see Exercise 4.5), for a solution of (4.84), we have that

$$|u(t)| \le e^{C_g t}(1 + |u_0|) - 1.$$

This means that

$$\|u^{n+1}\|_\infty \le e^{C_g \Delta t}\left(1 + \|u^{n+1/2}\|_\infty\right) - 1 \le e^{C_g \Delta t}\left(1 + \|u^n\|_\infty\right) - 1,$$

which by induction implies

$$\|u^n\|_\infty \le e^{C_g t_n}\left(1 + \|u_0\|_\infty\right) - 1.$$

Setting

$$M_1 = e^{C_g T}\left(1 + \|u_0\|_\infty\right) - 1$$

proves (i).

Next, we prove (ii). The proof is similar to that of the last case, since S_t is TVD, T.V. $(u^{n+1/2}) \le$ T.V. (u^n). As before, let u be a solution of (4.84) and let v be another solution with initial data v_0. Then we have $(u-v)_t = g(u) - g(v)$. Setting $w = u - v$, and multiplying by sign(w), we find that

$$|w|_t = \text{sign}(w)(g(u) - g(v)) \le \gamma |w|.$$

Then by Gronwall's inequality,

$$|w(t)| \le e^{\gamma t}|w(0)|.$$

Hence,

$$|u^{n+1}(x) - u^{n+1}(y)| \le e^{\gamma \Delta t}\left|u^{n+1/2}(x) - u^{n+1/2}(y)\right|.$$

This implies that

$$\text{T.V.}(u^{n+1}) \le e^{\gamma \Delta t}\text{T.V.}(u^{n+1/2}) \le e^{\gamma t}\text{T.V.}(u^n).$$

Inductively, we then have that

$$\text{T.V.}(u^n) \le e^{\gamma t_n}\text{T.V.}(u_0),$$

and setting $M_2 = e^{\gamma T}$ concludes the proof of (ii).

Regarding (iii), we know that

$$\int_B \left|u^{n+1/2}(x) - u^n(x)\right| dx \le C\Delta t.$$

We also have that

$$\int_B \left|u^{n+1}(x) - u^{n+1/2}(x)\right| dx = \int_B \left|\int_0^{\Delta t} g(u_{\Delta t}(x, t - t_n))\, dt\right| dx$$

$$\leq \int_B \int_0^{\Delta t} |g(u_{\Delta t}(x, t - t_n))|\, dt\, dx$$

$$\leq C_g \int_0^{\Delta t} \int_B (1 + M_1)\, dx\, dt$$

$$= |B|\, C_g(1 + M_1)\Delta t,$$

where $|B|$ denotes the length of B. Setting $M_3 = C + |B|\, C_g(1 + M_1)$ shows that

$$\int_B \left|u^{n+1}(x) - u^n(x)\right| \leq M_3 \Delta t,$$

which implies **(iii)**. □

Fix $T > 0$. Theorem A.8 implies the existence of a sequence $\Delta t \to 0$ such that for each $t \in [0, T]$, the function $u_{\Delta t}(t)$ converges in $L^1_{\text{loc}}(\mathbb{R})$ to a bounded function of bounded variation $u(t)$. The convergence is in $C([0, T]; L^1_{\text{loc}}(\mathbb{R}^m))$. It remains to show that u solves (4.82) in the sense of (4.81).

Using that $u_{\Delta t}$ is an entropy solution of the conservation law without source term (4.83) in the interval $[t_n, t_{n+1/2}]$, we obtain[8]

$$2\int_{t_n}^{t_{n+1/2}} \int \left(\frac{1}{2}|u_{\Delta t} - k|\varphi_t + \text{sign}(u_{\Delta t} - k)(f(u_{\Delta t}) - f(k))\varphi_x\right) dx\, dt$$

$$+ \int (|u_{\Delta t} - k|\varphi)\Big|_{t=t_{n+1/2}}^{t=t_n} dx \geq 0. \qquad (4.91)$$

Regarding solutions of (4.84), since $k_t = 0$ for any constant k we find that

$$|u - k|_t = \text{sign}(u - k)(u - k)_t = \text{sign}(u - k)\, g(u).$$

Multiplying this by a test function $\phi(t)$ and integrating over $s \in [0, t]$ we find after a partial integration that

$$\int_0^t \left(|u - k|\phi_s - \text{sign}(u - k)\, g(u)\phi\right) ds - u\phi\Big|_{s=0}^{s=t} = 0.$$

Since $u_{\Delta t}$ is a solution of the ordinary differential equation (4.84) on the interval $[t_{n+1/2}, t_{n+1}]$ (with time running "twice as fast"; see (4.85)), we

---

[8]The constants 2 and $\frac{1}{2}$ come from the fact that time is running "twice as fast" in the solution operators S and R in (4.85) (cf. also (4.14)–(4.15)).

find that

$$2\int_{t_n}^{t_{n+1/2}}\int \left(\frac{1}{2}|u_{\Delta t}-k|\varphi_t - \text{sign}\,(u_{\Delta t}-k)\,g(u_{\Delta t})\varphi\right)dx\,dt$$
$$+\int (|u_{\Delta t}-k|\varphi)\Big|_{t=t_{n+1}}^{t=t_{n+1/2}}dx = 0.$$

Adding this and (4.91), and summing over n, we obtain

$$2\int_0^T\int\left(\frac{1}{2}|u_{\Delta t}-k|\varphi_t + \chi_{\Delta t}\text{sign}\,(u_{\Delta t}-k)\,(f(u_{\Delta t})-f(k))\varphi_x\right.$$
$$\left.-\tilde{\chi}_{\Delta t}\text{sign}\,(u_{\Delta t}-k)\,g(u_{\Delta t})\varphi\right)dx\,dt$$
$$-\int(|u_{\Delta t}-k|\varphi)\,|_{t=0}^{t=T}\,dx \geq 0,$$

where $\chi_{\Delta t}$ and $\tilde{\chi}_{\Delta t}$ denote characteristic functions of the sets $\cup_n[t_n, t_{n+1/2})$ and $\cup_n[t_{n+1/2}, t_{n+1})$, respectively. We have that $\chi_{\Delta t} \stackrel{*}{\rightharpoonup} \frac{1}{2}$ and $\tilde{\chi}_{\Delta t} \stackrel{*}{\rightharpoonup} \frac{1}{2}$, and hence we conclude that (4.81) holds in the limit as $\Delta t \to 0$.

Theorem 4.13. *Let $f(u)$ be Lipschitz continuous, and assume that $g = g(u)$ satisfies the bound (4.86). Let u_0 be a bounded function of bounded variation. Then the initial value problem*

$$u_t + f(u)_x = g(u), \qquad u(x,0) = u_0(x) \qquad (4.92)$$

has a weak entropy solution, which can be constructed as the limit of the sequence $\{u_{\Delta t}\}$ defined by (4.85).

4.6 Notes

Dimensional splitting for hyperbolic equations was first introduced by Bagrinovskiĭ and Godunov [3] in 1957. Crandall and Majda made a comprehensive and systematic study of dimensional splitting (or the fractional steps method) in [38]. In [39] they used dimensional splitting to prove convergence of monotone schemes as well as the Lax–Wendroff scheme and the Glimm scheme, i.e., the random choice method.

There are also methods for multidimensional conservation laws that are intrinsically multidimensional. However, we have here decided to use dimensional splitting as our technique because it is conceptually simple and allows us to take advantage of the one-dimensional analysis.

Another natural approach to the study of multidimensional equations based on the front-tracking concept is first to make the standard front-tracking approximation: Replace the initial data by a piecewise constant function, and replace flux functions by piecewise linear and continuous functions. That gives rise to truly two-dimensional Riemann problems at each grid point $(i\Delta x, j\Delta y)$. However, that approach has turned out to be rather cumbersome even for a single Riemann problem and piecewise linear and continuous flux functions f and g. See Risebro [122].

The one-dimensional front-tracking approach combined with dimensional splitting was first introduced in Holden and Risebro [69]. The theorem on the convergence rate of dimensional splitting was proved independently by Teng [138] and Karlsen [80, 81]. Our presentation here follows Haugse, Lie, and Karlsen [100]. Section 4.4, using operator splitting to solve the parabolic regularization, is taken from Karlsen and Risebro [82]. The Kružkov interpolation lemma, Lemma 4.10, is taken from [87]; see also [82].

The presentation in Section 4.5 can be found in Holden and Risebro [70], where also the case with a stochastic source is treated. The convergence rate in the case of operator splitting applied to a conservation law with a source term is discussed in Langseth, Tveito, and Winther [93].

Exercises

4.1 Consider the initial value problem
$$u_t + f(u)_x + g(u)_y = 0, \quad u|_{t=0} = u_0,$$
where f, g are Lipschitz continuous functions, and u_0 is a bounded, integrable function with finite total variation.

  a. Show that the solution u is Lipschitz continuous in time; that is,
  $$\|u(t) - u(s)\|_1 \leq C\,\text{T.V.}\,(u_0)\,|t - s|.$$

  b. Let v_0 be another function with the same properties as u_0. Show that if $u_0 \leq v_0$, then also $u \leq v$ almost everywhere, where v is the solution with initial data v_0.

4.2 Consider the initial value problem
$$u_t + f(u)_x = 0, \quad u|_{t=0} = u_0, \qquad (4.93)$$
where f is a Lipschitz continuous function and u_0 is a bounded, integrable function with finite total variation. Write
$$f = f_1 + f_2$$
and let $S_j(t)u_0$ denote the solution of
$$u_t + f_j(u)_x = 0, \quad u|_{t=0} = u_0.$$
Prove that operator splitting converges to the solution of (4.93). Determine the convergence rate.

4.3 Consider the heat equation in $\mathbb{R}^m$,

$$u_t = \sum_{i=1}^{m} \frac{\partial^2 u}{\partial x_i^2}, \qquad u(x,0) = u_0(x). \tag{4.94}$$

Let H_t^i denote the solution operator for the heat equation in the ith direction, i.e., we write the solution of

$$u_t = \frac{\partial^2 u}{\partial x_i^2}, \qquad u(x,0) = u_0(x),$$

as $H_t^i u_0$. Define

$$u^n(x) = \left[H_{\Delta t}^m \circ \cdots \circ H_{\Delta t}^1\right]^n u_0(x),$$
$$u^{n+j/m}(x) = H_{\Delta t}^j \circ H_{\Delta t}^{j-1} \circ \cdots \circ H_{\Delta t}^1 u^n(x),$$

for $j = 1, \ldots, m$, and $n \geq 0$.
For t in the interval $[t_n + ((j-1)/m)\Delta t, t_n + (j/m)\Delta t]$ define

$$u_{\Delta t}(x,t) = H^j_{m(t-t_{n+(j-1)/m})} u^{n+(j-1)/m}(x).$$

If the initial function $u_0(x)$ is bounded and of bounded variation, show that $\{u_{\Delta t}\}$ converges in $C([0,T]; L^1_{\text{loc}}(\mathbb{R}^m))$ to a weak solution of (4.94).

4.4 We consider the viscous conservation law in one space dimension

$$u_t + f(u)_x = u_{xx}, \qquad u(x,0) = u_0(x), \tag{4.95}$$

where f satisfies the "usual" assumptions and u_0 is in $L^1 \cap BV$. Consider the following scheme based on operator splitting:

$$U_j^{n+1/2} = \frac{1}{2}\left(U_{j+1}^n + U_{j-1}^n\right) - \lambda\left(f\left(U_{j+1}^n\right) - f\left(U_{j-1}^n\right)\right),$$
$$U_j^{n+1} = U_j^{n+1/2} + \mu\left(U_{j+1}^{n+1/2} - 2U_j^{n+1/2} + U_{j-1}^{n+1/2}\right),$$

for $n \geq 0$, where $\lambda = \Delta t/\Delta x$ and $\mu = \Delta t/\Delta x^2$. Set

$$U_j^0 = \frac{1}{\Delta x}\int_{(j-1/2)\Delta x}^{(j+1/2)\Delta x} u_0(x)\,dx.$$

We see that we use the Lax–Friedrichs scheme for the conservation law and an explicit difference scheme for the heat equation. Let

$$u_{\Delta t}(x,t) = U_j^n$$

for $\left(j-\tfrac{1}{2}\right)\Delta x \leq x < \left(j+\tfrac{1}{2}\right)\Delta x$ and $n\Delta t < t \leq (n+1)\Delta t$.

   a. Show that this gives a monotone and consistent scheme, provided that a CFL condition holds.
   b. Show that there is a sequence of Δt's such that $u_{\Delta t}$ converges to a weak solution of (4.95) as $\Delta t \to 0$.

4.5 We outline a proof of some *Gronwall inequalities*.

    a. Assume that u satisfies
$$u'(t) \leq \gamma u(t).$$
Show that $u(t) \leq e^{\gamma t} u(0)$.

    b. Assume now that u satisfies
$$u'(t) \leq C(1 + u(t)).$$
Show that $u(t) \leq e^{Ct}(1 + u(0)) - 1$.

    c. Assume that u satisfies
$$u'(t) \leq c(t)u(t) + d(t),$$
for $0 \leq t \leq T$, where $c(t)$ and $d(t)$ are in $L^1([0,T])$. Show that
$$u(t) \leq \left(u(0) + \int_0^t d(s) \exp\left(- \int_0^s c(\tilde{s}) \, d\tilde{s} \right) ds \right) \exp\left(\int_0^t c(s) \, ds \right)$$
for $t \leq T$.

    d. Assume that u is in $L^1([0,T])$ and that for $t \in [0,T]$,
$$u(t) \leq C_1 \int_0^t u(s) \, ds + C_2.$$
Show that
$$u(t) \leq C_2 e^{C_1 t}.$$

    e. Assume that u, f, and g are in $L^1([0,T])$, and that g is non-negative, while f is strictly positive and nondecreasing. Assume that
$$u(t) \leq f(t) + \int_0^t g(s) u(s) \, ds, \quad t \in [0,T].$$
Show that
$$u(t) \leq f(t) \exp\left(\int_0^t g(s) \, ds \right), \quad t \in [0,T].$$

4.6 Assume that u and v are entropy solutions of
$$u_t + f(u)_x = g(u), \quad u(x,0) = u_0(x),$$
$$v_t + f(v)_x = g(v), \quad v(x,0) = v_0(x),$$
where u_0 and v_0 are in $L^1(\mathbb{R}) \cap BV(\mathbb{R})$, and f and g satisfy the assumptions of Theorem 4.13.

    a. Use the entropy formulation (4.81) and mimic the arguments used to prove (2.53) to show that for any nonnegative test function ψ,

$$\iint \left(|u(x,t) - v(x,t)|\psi_t + q(u,v)\psi_x \right) dt\, dx$$
$$- \int |u(x,T) - v(x,T)|\psi(x,T)\, dx$$
$$+ \int |u_0(x) - v_0(x)|\psi(x,0)\, dx$$
$$> \iint \text{sign}(u-v)(g(u) - g(v))\psi\, dt\, dx.$$

b. Define $\psi(x,t)$ by (2.54), and set
$$h(t) = \int |u(x,t) - v(x,t)|\psi(x,t)\, dx.$$

Show that
$$h(T) \leq h(0) + \gamma \int_0^T h(t)\, dt,$$

where γ denotes the Lipschitz constant of g. Use the previous exercise to conclude that
$$h(T) \leq h(0)\left(1 + \gamma T e^{\gamma T}\right).$$

c. Show that
$$\|u(\,\cdot\,,t) - v(\,\cdot\,,t)\|_1 \leq \|u_0 - v_0\|_1 \left(1 + \gamma t e^{\gamma t}\right),$$

and hence that entropy solutions of (4.92) are unique. Note that this implies that $\{u_{\Delta t}\}$ defined by (4.85) converges to the entropy solution for *any* sequence $\{\Delta t\}$ such that $\Delta t \to 0$.

4.7 We consider the case where the source depends on (x,t). For $u_0 \in L^1_{\text{loc}} \cap BV$, let u be an entropy solution of
$$u_t + f(u)_x = g(x,t,u), \quad u(x,0) = u_0(x), \tag{4.96}$$

where g bounded for each fixed u and continuous in t, and satisfies
$$|g(x,t,u) - g(x,t,v)| \leq \gamma |u-v|,$$
$$\text{T.V.}\,(g(\,\cdot\,,t,u)) \leq b(t),$$

where the constant γ is independent of x and t, for all u and v and for a bounded function $b(t)$ in $L^1([0,T])$. We let S_t be as before, and let $R(x,t,s)u_0$ denote the solution of
$$u'(t) = g(x,t,u), \quad u(s) = u_0,$$

for $t > s$.

  a. Define an operator splitting approximation $u_{\Delta t}$ using S_t and $R(x,t,s)$.
  b. Show that there is a sequence of Δt's such that $u_{\Delta t}$ converges in $C([0,T]; L^1_{\text{loc}}(\mathbb{R}))$ to a function of bounded variation u.

c. Show that u is an entropy solution of (4.96).

4.8 Show that if the initial data u_0 of the heat equation $u_t = \Delta u$ is smooth, that is, $u_0 \in C_0^\infty$, then
$$\|u(t) - u_0\|_1 \leq C t.$$
Compare this result with (4.4).

5
The Riemann Problem for Systems

> Diese Untersuchung macht nicht darauf Anspruch,
> der experimentellen Forschung nützliche Ergebnisse
> zu liefern; der Verfasser wünscht sie nur als einen
> Beitrag zur Theorie der nicht linearen partiellen
> Differentialgleichungen betrachtet zu sehen.[1]
>
> G. F. B. Riemann [119]

We return to the conservation law (1.2), but now study the case of systems, i.e.,

$$u_t + f(u)_x = 0, \tag{5.1}$$

where $u = u(x,t) = (u_1, \ldots, u_n)$ and $f = f(u) = (f_1, \ldots, f_n) \in C^2$ are vectors in $\mathbb{R}^n$. (We will not distinguish between row and column vectors, and use whatever is more convenient.) Furthermore, in this chapter we will consider only systems on the line; i.e., the dimension of the underlying physical space is still one. In Chapter 1 we proved existence, uniqueness, and stability of the Cauchy problem for the scalar conservation law in one space dimension, i.e., well-posedness in the sense of Hadamard. However, this is a more subtle question in the case of systems of hyperbolic conservation laws. We will here first discuss the basic concepts for systems:

---
[1]The present work does not claim to lead to results in experimental research; the author asks only that it be considered as a contribution to the theory of nonlinear partial differential equations.

fundamental properties of shock waves and rarefaction waves. In particular, we will discuss various entropy conditions to select the right solutions of the Rankine–Hugoniot relations.

Using these results we will eventually be able to prove well-posedness of the Cauchy problem for systems of hyperbolic conservation laws with small variation in the initial data.

5.1 Hyperbolicity and Some Examples

Before we start to define the basic properties of systems of hyperbolic conservation laws we discuss some important and interesting examples. The first example is a model for shallow-water waves and will be used throughout this chapter both as a motivation and an example in which all the basic quantities will be explicitly computed.

◇ **Example 5.1 (Shallow water).**

> Water shapes its course according to the nature
> of the ground over which it flows.
>
> *Sun Tzu, The Art of War (6th–5th century* B.C.*)*

We will now give a brief derivation of the equations governing shallow-water waves in one space dimension, or, if we want, the long-wave approximation. Consider a one-dimensional channel along the x-axis with a perfect, inviscid fluid with constant density ρ, and assume that the bottom of the channel is horizontal. In the long-wave or shallow-

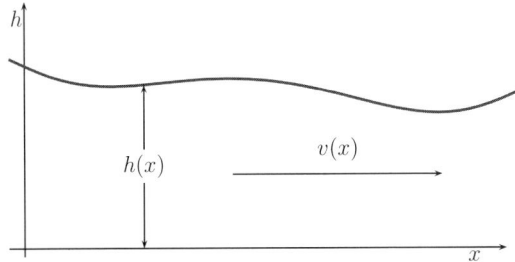

Figure 5.1. A shallow channel.

water approximation we assume that the fluid velocity v is a function only of time and the position along the channel measured along the x-axis. Thus we assume that there is no vertical motion in the fluid. The distance of the surface of the fluid from the bottom is denoted by $h = h(x,t)$. The fluid flow is governed by *conservation of mass* and *conservation of momentum*.

5.1. Hyperbolicity and Some Examples

Consider first the conservation of mass of the system. Let $x_1 < x_2$ be two points along the channel. The change of mass of fluid between these points is given by

$$\frac{d}{dt}\int_{x_1}^{x_2}\int_0^{h(x,t)} \rho\,dy\,dx = -\int_0^{h(x_2,t)} \rho v(x_2,t)\,dy + \int_0^{h(x_1,t)} \rho v(x_1,t)\,dy.$$

Assuming smoothness of the functions and domains involved, we may rewrite the right-hand side as an integral of the derivative of $\rho v h$. We obtain

$$\frac{d}{dt}\int_{x_1}^{x_2}\int_0^{h(x,t)} \rho\,dy\,dx = -\int_{x_1}^{x_2} \frac{\partial}{\partial x}(\rho v(x,t) h(x,t))\,dx,$$

or

$$\int_{x_1}^{x_2}\left[\frac{\partial}{\partial t}(\rho h(x,t)) + \frac{\partial}{\partial x}(\rho v(x,t) h(x,t))\right] dx = 0.$$

Dividing by $(x_2 - x_1)\rho$ and letting $x_2 - x_1 \to 0$, we obtain the familiar

$$h_t + (vh)_x = 0. \qquad (5.2)$$

Observe the similarity in the derivation of (5.2) and (1.19). In fact, in the derivation of (1.19) we started by considering individual cars before we made the continuum assumption corresponding to high traffic densities, thereby obtaining (1.19), while in the derivation of (5.2) we simply assumed a priori that the fluid constituted a continuum, and formulated mass conservation directly in the continuum variables.

For the derivation of the equation describing the conservation of momentum we have to assume that the fluid is in hydrostatic balance. For that we introduce the pressure $P = P(x, y, t)$ and consider a small element of the fluid $[x_1, x_2] \times [y, y + \Delta y]$. Hydrostatic balance means that the pressure exactly balances the effect of gravity, or

$$(P(\tilde{x}, y + \Delta y, t) - P(\tilde{x}, y, t))(x_2 - x_1) = -(x_2 - x_1)\rho g \Delta y$$

for some $\tilde{x} \in [x_1, x_2]$, where g is the acceleration due to gravity. Dividing by $(x_2 - x_1)\Delta y$ and taking $x_1, x_2 \to x$, $\Delta y \to 0$ we find that

$$\frac{\partial P}{\partial y}(x, y, t) = -\rho g.$$

Integrating and normalizing the pressure to be zero at the fluid surface we conclude that

$$P(x, y, t) = \rho g(h(x,t) - y). \qquad (5.3)$$

If we now again study the fluid between two points $x_1 < x_2$ along the channel, and compute the change of momentum for this part of the fluid

168   5. The Riemann Problem for Systems

we obtain

$$\frac{\partial}{\partial t} \int_{x_1}^{x_2} \int_0^{h(x,t)} \rho v(x,t) \, dy \, dx$$

$$= -\int_0^{h(x_2,t)} P(x_2, y, t) \, dy + \int_0^{h(x_1,t)} P(x_1, y, t) \, dy$$

$$- \int_0^{h(x_2,t)} \rho v(x_2, t)^2 \, dy + \int_0^{h(x_1,t)} \rho v(x_1, t)^2 \, dy.$$

In analogy with the derivation of the equation for conservation of mass we may rewrite this, using (5.3), as

$$\frac{\partial}{\partial t} \int_{x_1}^{x_2} \rho v h \, dx = -\rho g \left(h(x_2, t)^2 - \frac{1}{2} h(x_2, t)^2 \right)$$

$$+ \rho g \left(h(x_1, t)^2 - \frac{1}{2} h(x_1, t)^2 \right) - \int_{x_1}^{x_2} \frac{\partial}{\partial x} \left(\rho h v^2 \right) dx$$

$$= -\rho g \int_{x_1}^{x_2} \frac{\partial}{\partial x} \left(\frac{1}{2} h^2 \right) dx - \int_{x_1}^{x_2} \frac{\partial}{\partial x} \left(\rho v^2 h \right) dx.$$

Dividing again by $(x_2 - x_1)\rho$ and letting $x_2 - x_1 \to 0$, scaling g to unity, we obtain

$$(vh)_t + \left(v^2 h + \frac{1}{2} h^2 \right)_x = 0. \tag{5.4}$$

To summarize, we have the following system of conservation laws:

$$h_t + (vh)_x = 0, \quad (vh)_t + \left(v^2 h + \frac{1}{2} h^2 \right)_x = 0, \tag{5.5}$$

where h and v denote the height (depth) and velocity of the fluid, respectively. Introducing the variable q defined by

$$q = vh, \tag{5.6}$$

we may rewrite the shallow-water equations as

$$\begin{pmatrix} h \\ q \end{pmatrix}_t + \begin{pmatrix} q \\ q^2 h + \frac{h^2}{2} \end{pmatrix}_x = 0, \tag{5.7}$$

which is the form we will study in detail later on in this chapter. ◇

◇ **Example 5.2 (The wave equation).**

Let $\phi = \phi(x,t)$ denote the transverse position away from equilibrium of a one-dimensional string. If we assume that the amplitude of the transversal waves is small, we obtain the wave equation

$$\phi_{tt} = (c^2 \phi_x)_x, \tag{5.8}$$

where c denotes the wave speed. Introducing new variables $u = \phi_x$ and $v = \phi_t$, we find that (5.8) may be written as the system

$$\begin{pmatrix} u \\ v \end{pmatrix}_t - \begin{pmatrix} v \\ c^2 u \end{pmatrix}_x = 0. \tag{5.9}$$

If c does not depend on ϕ, we recover the classical linear wave equation $\phi_{tt} = c^2 \phi_{xx}$. ◇

◇ **Example 5.3 (The p-system).**

The p-system is a classical model of an isentropic gas, where one has conservation of mass and momentum, but not of energy. In Lagrangian coordinates it is described by

$$\begin{pmatrix} v \\ u \end{pmatrix}_t + \begin{pmatrix} -u \\ p(v) \end{pmatrix}_x = 0. \tag{5.10}$$

Here v denotes specific volume, that is, the inverse of the density; u is the velocity; and p denotes the pressure. ◇

◇ **Example 5.4 (The Euler equations).**

The Euler equations are commonly used to model gas dynamics. They can be written in several forms, depending on the physical assumptions used and variables selected to describe them. Let it suffice here to describe the case where ρ denotes the density, v velocity, p pressure, and E the energy. Conservation of mass and momentum give $\rho_t + (\rho v)_x = 0$ and $(\rho v)_t + (\rho v^2 + p)_x = 0$, respectively. The total energy can be written as $E = \frac{1}{2}\rho v^2 + \rho e$, where e denotes the specific internal energy. Furthermore, we assume that there is a relation between this quantity and the density and pressure, viz., $e = e(\rho, p)$. Conservation of energy now reads $E_t + (v(E+p))_x = 0$, yielding finally the system

$$\begin{pmatrix} \rho \\ \rho v \\ E \end{pmatrix}_t + \begin{pmatrix} \rho v \\ \rho v^2 + p \\ v(E+p) \end{pmatrix}_x = 0. \tag{5.11}$$

◇

We will have to make assumptions on the (vector-valued) function f so that many of the properties of the scalar case carry over to the case of systems. In order to have finite speed of propagation, which characterizes hyperbolic equations, we have to assume that the Jacobian of f, denoted by df, has n real eigenvalues

$$df(u) r_j(u) = \lambda_j(u) r_j(u), \quad \lambda_j(u) \in \mathbb{R}, \quad j = 1, \ldots, n. \tag{5.12}$$

(We will later normalize the eigenvectors $r_j(u)$.) Furthermore, we order the eigenvalues

$$\lambda_1(u) \leq \lambda_2(u) \leq \cdots \leq \lambda_n(u). \tag{5.13}$$

A system with a full set of eigenvectors with real eigenvalues is called *hyperbolic*, and if all the eigenvalues are distinct, we say that the system is *strictly hyperbolic*.

Let us look at the shallow-water model to see whether that system is hyperbolic.

◇ **Example 5.5 (Shallow water (cont'd.)).**

In case of the shallow-water equations (5.7) we easily find that

$$\lambda_1(u) = \frac{q}{h} - \sqrt{h} < \frac{q}{h} + \sqrt{h} = \lambda_2(u), \tag{5.14}$$

with corresponding eigenvectors

$$r_j(u) = \begin{pmatrix} 1 \\ \lambda_j(u) \end{pmatrix}, \tag{5.15}$$

and thus the shallow-water equations are strictly hyperbolic away from $h = 0$. ◇

5.2 Rarefaction Waves

> Natura non facit saltus.[2]
>
> Carl Linnaeus, Philosophia Botanica (1751)

Let us consider smooth solutions for the initial value problem

$$u_t + f(u)_x = 0, \tag{5.16}$$

with Riemann initial data

$$u(x,0) = \begin{cases} u_l & \text{for } x < 0, \\ u_r & \text{for } x \geq 0. \end{cases} \tag{5.17}$$

First we observe that since both the initial data and the equation are scale-invariant or self-similar, i.e., invariant under the map $x \mapsto kx$ and $t \mapsto kt$, the solution should also have that property. Let us therefore search for solutions of the form

$$u(x,t) = w(x/t) = w(\xi), \quad \xi = x/t. \tag{5.18}$$

---

[2] Nature does not make jumps.

5.2. Rarefaction Waves

Inserting this into the differential equation (5.16) we find that

$$-\frac{x}{t^2}\dot{w} + \frac{1}{t}df(w)\dot{w} = 0, \qquad (5.19)$$

or

$$df(w)\dot{w} = \xi\dot{w}, \qquad (5.20)$$

where $\dot{w}$ denotes the derivative of w with respect to the one variable ξ. Hence we observe that $\dot{w}$ is an eigenvector for the Jacobian $df(w)$ with eigenvalue ξ. From our assumptions on the flux function we know that $df(w)$ has n eigenvectors given by $r_1,\ldots,r_n$, with corresponding eigenvalues $\lambda_1,\ldots,\lambda_n$. This implies

$$\dot{w}(\xi) = r_j(w(\xi)), \quad \lambda_j(w(\xi)) = \xi, \qquad (5.21)$$

for a value of j. Assume in addition that

$$w(\lambda_j(u_l)) = u_l, \quad w(\lambda_j(u_r)) = u_r. \qquad (5.22)$$

Thus for a fixed time t, the function $w(x/t)$ will continuously connect the given left state u_l to the given right state u_r. This means that ξ is increasing, and hence $\lambda_j(w(x/t))$ has to be increasing. If this is the case, we have a solution of the form

$$u(x,t) = \begin{cases} u_l & \text{for } x \leq \lambda_j(u_l)t, \\ w(x/t) & \text{for } t\lambda_j(u_l) \leq x \leq t\lambda_j(u_r), \\ u_r & \text{for } x \geq t\lambda_j(u_r), \end{cases} \qquad (5.23)$$

where $w(\xi)$ satisfies (5.21) and (5.22). We call these solutions *rarefaction waves*, a name that comes from applications to gas dynamics. Furthermore, we observe that the normalization of the eigenvector $r_j(u)$ also is determined from (5.21), namely,

$$\nabla\lambda_j(u) \cdot r_j(u) = 1, \qquad (5.24)$$

which follows by taking the derivative with respect to ξ. But this also imposes an extra condition on the eigenvector fields, since we clearly have to have a nonvanishing scalar product between $r_j(u)$ and $\nabla\lambda_j(u)$ to be able to normalize the eigenvector properly. It so happens that in most applications this can be done. However, the Euler equations of gas dynamics have the property that in one of the eigenvector families the eigenvector and the gradient of the corresponding eigenvalue are orthogonal. We say that the jth family is *genuinely nonlinear* if $\nabla\lambda_j(u) \cdot r_j(u) \neq 0$ and *linearly degenerate* if $\nabla\lambda_j(u) \cdot r_j(u) \equiv 0$ for all u under consideration. We will not discuss mixed cases where a wave family is linearly degenerate only in certain regions in phase space, e.g., along curves or at isolated points.

Before we discuss these two cases separately, we will make a slight but important change in point of view. Instead of considering given left and right states as in (5.17), we will assume only that u_l is given, and consider those states u_r for which we have a rarefaction wave solution. From (5.21) and (5.23) we see that for each point u_l in phase space there are n curves emanating from u_l on which u_r can lie allowing a solution of the form (5.23). Each of these curves is given as integral curves of the vector fields of eigenvectors of the Jacobian $df(u)$. Thus our phase space is now the u_r space.

We may sum up the above discussion in the genuinely nonlinear case by the following theorem.

Theorem 5.6. *Let D be a domain in $\mathbb{R}^n$. Consider the strictly hyperbolic equation $u_t + f(u)_x = 0$ in D and assume that the equation is genuinely nonlinear in the jth wave family in D. Let the jth eigenvector $r_j(u)$ of $df(u)$ with corresponding eigenvalue $\lambda_j(u)$ be normalized so that $\nabla \lambda_j(u) \cdot r_j(u) = 1$ in D.*

Let $u_l \in D$. Then there exists a curve $R_j(u_l)$ in D, emanating from u_l, such that for each $u_r \in R_j(u_l)$ the initial value problem (5.16), (5.17) has weak solution

$$u(x,t) = \begin{cases} u_l & \text{for } x \leq \lambda_j(u_l)t, \\ w(x/t) & \text{for } \lambda_j(u_l)t \leq x \leq \lambda_j(u_r)t, \\ u_r & \text{for } x \geq \lambda_j(u_r)t, \end{cases} \quad (5.25)$$

where w satisfies $\dot{w}(\xi) = r_j(w(\xi))$, $\lambda_j(w(\xi)) = \xi$, $w(\lambda_j(u_l)) = u_l$, and $w(\lambda_j(u_r)) = u_r$.

Proof. The discussion preceding the theorem gives the key computation and the necessary motivation behind the following argument. Assume that we have a strictly hyperbolic, genuinely nonlinear conservation law with appropriately normalized jth eigenvector. Due to the assumptions on f, the system of ordinary differential equations

$$\dot{w}(\xi) = r_j(w(\xi)), \quad w(\lambda_j(u_l)) = u_l \quad (5.26)$$

has a solution for all $\xi \in [\lambda_j(u_l), \lambda_j(u_l)+\eta)$ for some $\eta > 0$. For this solution we have

$$\frac{d}{d\xi}\lambda_j(w(\xi)) = \nabla\lambda_j(w(\xi)) \cdot \dot{w}(\xi) = 1, \quad (5.27)$$

proving the second half of (5.21). We denote the orbit of (5.26) by $R_j(u_l)$. If we define $u(x,t)$ by (5.25), a straightforward calculation shows that u indeed satisfies both the equation and the initial data. $\square$

Observe that we can also solve (5.26) for ξ less than $\lambda_j(u_l)$. However, in that case $\lambda_j(u)$ will be decreasing. We remark that the solution u in (5.25) is continuous, but not necessarily differentiable, and hence is not necessarily a regular, but rather a weak, solution.

We will now introduce a different parameterization of the rarefaction curve $R_j(u_l)$, which will be convenient in Section 5.5 when we construct

5.2. Rarefaction Waves

the wave curves for the solution of the Riemann problem. From (5.27) we see that $\lambda_j(u)$ is increasing along $R_j(u_l)$, and hence we may define the positive parameter ϵ by $\epsilon := \xi - \xi_l = \lambda_j(u) - \lambda_j(u_l) > 0$. We denote the corresponding u by $u_{j,\epsilon}$, that is, $u_{j,\epsilon} = w(\xi) = w(\lambda_j(u)) = w(\epsilon + \lambda_j(u_l))$. Clearly,

$$\frac{du_{j,\epsilon}}{d\epsilon}\bigg|_{\epsilon=0} = r_j(u_l). \tag{5.28}$$

Assume now that the system is linearly degenerate in the family j, i.e., $\nabla \lambda_j(u) \cdot r_j(u) \equiv 0$. Consider the system of ordinary differential equations

$$\frac{du}{d\epsilon} = r_j(u), \quad u|_{\epsilon=0} = u_l, \tag{5.29}$$

with solution $u = u_{j,\epsilon}$ for $\epsilon \in \langle -\eta, \eta \rangle$ for some $\eta > 0$. We denote this orbit by $C_j(u_l)$, along which $\lambda_j(u_{j,\epsilon})$ is constant, since

$$\frac{d}{d\epsilon} \lambda_j(u_{j,\epsilon}) = \nabla \lambda_j(u_{j,\epsilon}) \cdot r_j(u_{j,\epsilon}) = 0.$$

Furthermore, the Rankine–Hugoniot condition is satisfied on $C_j(u_l)$ with speed $\lambda_j(u_l)$, because

$$\frac{d}{d\epsilon}(f(u_{j,\epsilon}) - \lambda_j(u_l)u_{j,\epsilon}) = df(u_{j,\epsilon})\frac{du_{j,\epsilon}}{d\epsilon} - \lambda_j(u_l)\frac{du_{j,\epsilon}}{d\epsilon}$$
$$= (df(u_{j,\epsilon}) - \lambda_j(u_l))r_j(u_{j,\epsilon})$$
$$= (df(u_{j,\epsilon}) - \lambda_j(u_{j,\epsilon}))r_j(u_{j,\epsilon}) = 0,$$

which implies that $f(u_{j,\epsilon}) - \lambda_j(u_l)u_{j,\epsilon} = f(u_l) - \lambda_j(u_l)u_l$.

Let $u_r \in C_j(u_l)$, i.e., $u_r = u_{j,\epsilon_0}$ for some ϵ_0. It follows that

$$u_{j,\epsilon_0}(x,t) = \begin{cases} u_l & \text{for } x < \lambda_j(u_l)t, \\ u_r & \text{for } x \geq \lambda_j(u_l)t, \end{cases}$$

is a weak solution of the Riemann problem (5.16), (5.17). We call this solution a *contact discontinuity*.

We sum up the above discussion concerning linearly degenerate waves in the following theorem.

Theorem 5.7. *Let D be a domain in $\mathbb{R}^n$. Consider the strictly hyperbolic equation $u_t + f(u)_x = 0$ in D. Assume that the equation is linearly degenerate in the jth wave family in D, i.e., $\nabla \lambda_j(u) \cdot r_j(u) \equiv 0$ in D, where $r_j(u)$ is the jth eigenvector of $df(u)$ with corresponding eigenvalue $\lambda_j(u)$.*

Let $u_l \in D$. Then there exists a curve $C_j(u_l)$ in D, passing through u_l, such that for each $u_r \in C_j(u_l)$ the initial value problem (5.16), (5.17) has solution

$$u(x,t) = \begin{cases} u_l & \text{for } x < \lambda_j(u_l)t, \\ u_r & \text{for } x \geq \lambda_j(u_l)t, \end{cases} \tag{5.30}$$

where u_r is determined as follows: Consider the function $\epsilon \mapsto u_\epsilon$ determined by $\frac{du}{d\epsilon} = r_j(u)$, $u|_{\epsilon=0} = u_l$. Then $u_r = u_{\epsilon_0}$ for some ϵ_0.

◊ **Example 5.8 (Shallow water (cont'd.)).**

Let us now consider the actual computation of rarefaction waves in the case of shallow-water waves. Recall that

$$u = \begin{pmatrix} h \\ q \end{pmatrix}, \quad f(u) = \begin{pmatrix} q \\ \frac{q^2}{h} + \frac{h^2}{2} \end{pmatrix},$$

with eigenvalues $\lambda_j = \frac{q}{h} + (-1)^j \sqrt{h}$, and corresponding eigenvectors $r_j(u) = \begin{pmatrix} 1 \\ \lambda_j(u) \end{pmatrix}$. With this normalization of r_j, we obtain

$$\nabla \lambda_j(u) \cdot r_j(u) = \frac{3(-1)^j}{2\sqrt{h}}, \tag{5.31}$$

and hence we see that the shallow-water equations are genuinely nonlinear in both wave families. From now on we will renormalize the eigenvectors to satisfy (5.24), viz.,

$$r_j(u) = \frac{2}{3}(-1)^j \sqrt{h} \begin{pmatrix} 1 \\ \lambda_j(u) \end{pmatrix}. \tag{5.32}$$

For the 1-family we have that

$$\begin{pmatrix} \dot{h} \\ \dot{q} \end{pmatrix} = -\frac{2}{3}\sqrt{h} \begin{pmatrix} 1 \\ \frac{q}{h} - \sqrt{h} \end{pmatrix}, \tag{5.33}$$

implying that

$$\frac{dq}{dh} = \lambda_1 = \frac{q}{h} - \sqrt{h},$$

which can be integrated to yield

$$q = q(h) = q_l \frac{h}{h_l} - 2h(\sqrt{h} - \sqrt{h_l}). \tag{5.34}$$

Since $\lambda_1(u)$ has to increase along the rarefaction wave, we see from (5.14) (inserting the expression (5.34) for q) that we have to use $h \leq h_l$ in (5.34).

For the second family we again obtain

$$\frac{dq}{dh} = \lambda_2 = \frac{q}{h} + \sqrt{h},$$

yielding

$$q = q(h) = q_l \frac{h}{h_l} + 2h(\sqrt{h} - \sqrt{h_l}). \tag{5.35}$$

In this case we see that we have to use $h \geq h_l$. Observe that (5.34) and (5.35) would follow for *any* normalization of the eigenvector $r_j(u)$.

5.2. Rarefaction Waves

Summing up, we obtain the following rarefaction waves expressed in terms of h:

$$R_1: \quad q = R_1(h; u_l) := q_l \frac{h}{h_l} - 2h(\sqrt{h} - \sqrt{h_l}), \quad h \in \langle 0, h_l], \quad (5.36)$$

$$R_2: \quad q = R_2(h; u_l) := q_l \frac{h}{h_l} + 2h(\sqrt{h} - \sqrt{h_l}), \quad h \geq h_l. \quad (5.37)$$

Alternatively, in the (h, v) variables (with $v = q/h$) we have the following:

$$R_1: \quad v = R_1(h; u_l) := v_l - 2(\sqrt{h} - \sqrt{h_l}), \quad h \in \langle 0, h_l], \quad (5.38)$$

$$R_2: \quad v = R_2(h; u_l) := v_l + 2(\sqrt{h} - \sqrt{h_l}), \quad h \geq h_l. \quad (5.39)$$

However, if we want to compute the rarefaction curves in terms of

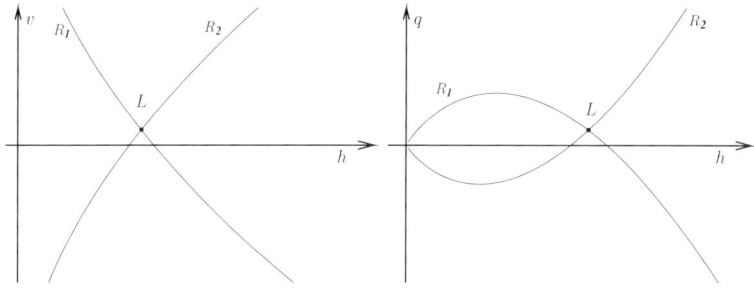

Figure 5.2. Rarefaction curves in the (h, v) and (h, q) planes. We have illustrated the full solution of (5.26) for the shallow-water equations. Only the part given by (5.36) and (5.37) will be actual rarefaction curves.

the parameter ξ or ϵ, we have to use the proper normalization of the eigenvectors given by (5.32). Consider first the 1-family. We obtain

$$\dot{h} = -\frac{2}{3}\sqrt{h}, \quad \dot{q} = \frac{2}{3}\left(-\frac{q}{\sqrt{h}} + h\right). \quad (5.40)$$

Integrating the first equation directly and inserting the result into the second equation, we obtain

$$w_1(\xi) = \begin{pmatrix} h_1 \\ q_1 \end{pmatrix}(\xi) = R_1(\xi; u_l) := \begin{pmatrix} \frac{1}{9}(v_l + 2\sqrt{h_l} - \xi)^2 \\ \frac{1}{27}(v_l + 2\sqrt{h_l} + 2\xi)(v_l + 2\sqrt{h_l} - \xi)^2 \end{pmatrix} \quad (5.41)$$

for $\xi \in [v_l - \sqrt{h_l}, v_l + 2\sqrt{h_l})$. Similarly, for the second family we obtain

$$w_2(\xi) = \begin{pmatrix} h_2 \\ q_2 \end{pmatrix}(\xi) = R_2(\xi; u_l)$$
$$:= \begin{pmatrix} \frac{1}{9}(-v_l + 2\sqrt{h_l} - \xi)^2 \\ \frac{1}{27}(v_l - 2\sqrt{h_l} + 2\xi)(-v_l + 2\sqrt{h_l} - \xi)^2 \end{pmatrix} \quad (5.42)$$

for $\xi \in [\lambda_2(u_l), \infty)$. Hence the actual solution reads

$$u(x,t) = \begin{cases} u_l & \text{for } x \leq \lambda_j(u_l)t, \\ R_j(x/t; u_l) & \text{for } \lambda_j(u_l)t \leq x \leq \lambda_j(u_r)t, \\ u_r & \text{for } x \geq \lambda_j(u_r)t. \end{cases} \quad (5.43)$$

In the (h, v) variables (with $v = q/h$) we obtain

$$v_1(\xi) = \frac{1}{3}(v_l + 2\sqrt{h_l} + 2\xi) \quad (5.44)$$

and

$$v_2(\xi) = \frac{1}{3}(v_l - 2\sqrt{h_l} + 2\xi), \quad (5.45)$$

for the first and the second families, respectively.

In terms of the parameter ϵ we may write (5.41) as

$$u_{1,\epsilon} = \begin{pmatrix} h_{1,\epsilon} \\ q_{1,\epsilon} \end{pmatrix} = R_{1,\epsilon}(u_l) := \begin{pmatrix} (\sqrt{h_l} - \frac{\epsilon}{3})^2 \\ (v_l + \frac{2\epsilon}{3})(\sqrt{h_l} - \frac{\epsilon}{3})^2 \end{pmatrix} \quad (5.46)$$

for $\epsilon \in [0, 3\sqrt{h_l})$, and (5.42) as

$$u_{2,\epsilon} = \begin{pmatrix} h_{2,\epsilon} \\ q_{2,\epsilon} \end{pmatrix} = R_{2,\epsilon}(u_l) := \begin{pmatrix} (\sqrt{h_l} - \frac{\epsilon}{3})^2 \\ (v_l + \frac{2\epsilon}{3})(\sqrt{h_l} - \frac{\epsilon}{3})^2 \end{pmatrix} \quad (5.47)$$

for $\epsilon \in [0, \infty)$. $\diamond$

5.3 The Hugoniot Locus: The Shock Curves

God lives in the details.

Johannes Kepler (1571–1630)

The discussion in Chapter 1 concerning weak solutions, and in particular the Rankine–Hugoniot condition (1.20), carries over to the case of systems without restrictions. However, the concept of entropy is considerably more difficult for systems and is still an area of research. Our main concern in this section is the characterization of solutions of the Rankine–Hugoniot relation. Again, we will take the point of view introduced in the previous

5.3. The Hugoniot Locus: The Shock Curves

section, assuming the left state u_l to be fixed, and consider possible right states u that satisfy the Rankine–Hugoniot condition

$$s(u - u_l) = f(u) - f(u_l), \tag{5.48}$$

for some speed s. We introduce the jump in a quantity ϕ as

$$[\![\phi]\!] = \phi_r - \phi_l,$$

and hence (5.48) takes the familiar form

$$s[\![u]\!] = [\![f(u)]\!].$$

The solutions of (5.48), for a given left state u_l, form a set, which we call the *Hugoniot locus* and write $H(u_l)$, i.e.,

$$H(u_l) := \{u \mid \exists s \in \mathbb{R} \text{ such that } s[\![u]\!] = [\![f(u)]\!]\}. \tag{5.49}$$

We start by computing the Hugoniot locus for the shallow-water equations.

◇ **Example 5.9 (Shallow water (cont'd.)).**

The Rankine–Hugoniot condition reads

$$s(h - h_l) = q - q_l,$$
$$s(q - q_l) = \left(\frac{q^2}{h} + \frac{h^2}{2}\right) - \left(\frac{q_l^2}{h_l} + \frac{h_l^2}{2}\right), \tag{5.50}$$

where s as usual denotes the shock speed between the left state $u_l = \binom{h_l}{q_l}$ and right state $u = \binom{h}{q}$, viz.,

$$\binom{h}{q}(x, t) = \begin{cases} \binom{h_l}{q_l} & \text{for } x < st, \\ \binom{h}{q} & \text{for } x \geq st. \end{cases} \tag{5.51}$$

In the context of the shallow-water equations such solutions are called *bores*. Eliminating s in (5.50), we obtain the equation

$$[\![h]\!]\left(\left[\!\left[\frac{q^2}{h}\right]\!\right] + \frac{1}{2}[\![h^2]\!]\right) = [\![q]\!]^2. \tag{5.52}$$

Introducing the variable v, given by $q = vh$, equation (5.52) becomes

$$[\![h]\!]\left([\![hv^2]\!] + \frac{1}{2}[\![h^2]\!]\right) = [\![vh]\!]^2,$$

with solution

$$v = v_l \pm \frac{1}{\sqrt{2}}(h - h_l)\sqrt{h^{-1} + h_l^{-1}}, \tag{5.53}$$

or, alternatively,

$$q = vh = q_l \frac{h}{h_l} \pm \frac{h}{\sqrt{2}}(h - h_l)\sqrt{h^{-1} + h_l^{-1}}. \tag{5.54}$$

178   5. The Riemann Problem for Systems

For later use, we will also obtain formulas for the corresponding shock speeds. We find that

$$s = \frac{[\![vh]\!]}{[\![h]\!]} = \frac{v(h-h_l)+(v-v_l)h_l}{h-h_l}$$
$$= v + \frac{[\![v]\!]}{[\![h]\!]}h_l = v \pm \frac{h_l}{\sqrt{2}}\sqrt{h^{-1}+h_l^{-1}},$$
(5.55)

or

$$s = v_l + \frac{[\![v]\!]}{[\![h]\!]}h_l = v_l + [\![v]\!] + \frac{[\![v]\!]}{[\![h]\!]}h_l = v_l \pm \frac{h}{\sqrt{2}}\sqrt{h^{-1}+h_l^{-1}}.$$
(5.56)

When we want to indicate the wave family, we write

$$s_j = s_j(h; v_l) = v_l + (-1)^j \frac{h}{\sqrt{2}}\sqrt{h^{-1}+h_l^{-1}}$$
$$= v + (-1)^j \frac{h_l}{\sqrt{2}}\sqrt{h^{-1}+h_l^{-1}}.$$
(5.57)

Thus we see that through a given left state u_l there are two curves on which the Rankine–Hugoniot relation holds, namely,

$$H_1(u_l) := \left\{ \left(q_l \frac{h}{h_l} - \frac{h}{\sqrt{2}}(h-h_l)\sqrt{h^{-1}+h_l^{-1}} \right) \,\middle|\, h > 0 \right\},$$
(5.58)

and

$$H_2(u_l) := \left\{ \left(q_l \frac{h}{h_l} + \frac{h}{\sqrt{2}}(h-h_l)\sqrt{h^{-1}+h_l^{-1}} \right) \,\middle|\, h > 0 \right\}.$$
(5.59)

The Hugoniot locus now reads

$$H(u_l) = \{\, u \mid \exists s \in \mathbb{R} \text{ such that } s[\![u]\!] = [\![f(u)]\!] \,\} = H_1(u_l) \cup H_2(u_l).$$

◇

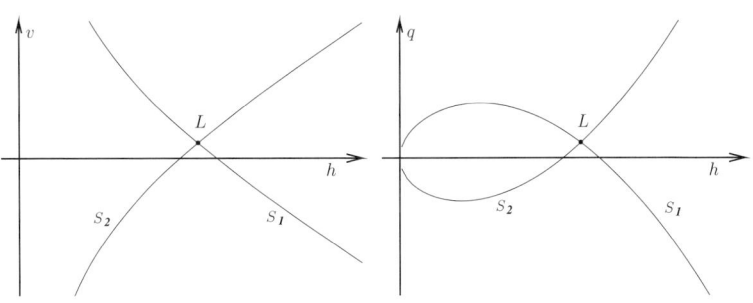

Figure 5.3. Shock curves in the (h, v) and (h, q) planes. Slow (S_1) and fast (S_2) shocks indicated; see Section 5.4.

5.3. The Hugoniot Locus: The Shock Curves

We will soon see that the basic features of the Hugoniot locus of the shallow-water equations carry over to the general case of strictly hyperbolic systems at least for small shocks where u is near u_l. The problem to be considered is to solve implicitly the system of n equations

$$\mathcal{H}(s, u; u_l) := s(u - u_l) - (f(u) - f(u_l)) = 0, \tag{5.60}$$

for the $n+1$ unknowns $u_1, \ldots, u_n$ and s for u close to the given u_l. The major problem comes from the fact that we have one equation fewer than the number of unknowns, and that $\mathcal{H}(s, u_l; u_l) = 0$ for all values of s. Hence the implicit function theorem cannot be used without first removing the singularity at $u = u_l$.

Let us first state the relevant version of the implicit function theorem that we will use.

Theorem 5.10 (Implicit function theorem). *Let the function*

$$\Phi = (\Phi_1, \ldots, \Phi_p) : \mathbb{R}^q \times \mathbb{R}^p \to \mathbb{R}^p \tag{5.61}$$

be C^1 in a neighborhood of a point (x_0, y_0), $x_0 \in \mathbb{R}^q$, $y_0 \in \mathbb{R}^p$ with $\Phi(x_0, y_0) = 0$. Assume that the $p \times p$ matrix

$$\frac{\partial \Phi}{\partial y} = \begin{pmatrix} \frac{\partial \Phi_1}{\partial y_1} & \cdots & \frac{\partial \Phi_1}{\partial y_p} \\ \vdots & \ddots & \vdots \\ \frac{\partial \Phi_p}{\partial y_1} & \cdots & \frac{\partial \Phi_p}{\partial y_p} \end{pmatrix} \tag{5.62}$$

is nonsingular at the point (x_0, y_0).

Then there exist a neighborhood N of x_0 and a unique function $\phi\colon N \to \mathbb{R}^p$ such that

$$\Phi(x, \phi(x)) = 0, \quad \phi(x_0) = y_0. \tag{5.63}$$

We will rewrite equation (5.60) into an eigenvalue problem that we can study locally around each eigenvalue $\lambda_j(u_l)$. This removes the singularity, and hence we can apply the implicit function theorem.

Theorem 5.11. *Let D be a domain in $\mathbb{R}^n$. Consider the strictly hyperbolic equation $u_t + f(u)_x = 0$ in D. Let $u_l \in D$.*

Then there exist n smooth curves $H_1(u_l), \ldots, H_n(u_l)$ locally through u_l on which the Rankine–Hugoniot relation is satisfied.

Proof. Writing

$$\begin{aligned} f(u) - f(u_l) &= \int_0^1 \frac{\partial}{\partial \alpha} f((1-\alpha)u_l + \alpha u) \, d\alpha \\ &= \int_0^1 df((1-\alpha)u_l + \alpha u)(u - u_l) \, d\alpha \\ &= M(u, u_l)(u - u_l), \end{aligned} \tag{5.64}$$

where $M(u, u_l)$ is the averaged Jacobian

$$M(u, u_l) = \int_0^1 df((1-\alpha)u_l + \alpha u)\, d\alpha,$$

we see that (5.60) takes the form

$$\mathcal{H}(s, u, u_l) = (s - M(u, u_l))(u - u_l) = 0. \tag{5.65}$$

Here $u - u_l$ is an eigenvector of the matrix M with eigenvalue s. The matrix $M(u_l, u_l) = df(u_l)$ has n distinct eigenvalues $\lambda_1(u_l), \ldots, \lambda_n(u_l)$, and hence we know that there exists an open set N such that the matrix $M(u, u_l)$ has twice-differentiable eigenvectors and distinct eigenvalues, namely,

$$(\mu_j(u, u_l) - M(u, u_l))\, v_j(u, u_l) = 0, \tag{5.66}$$

for all $u, u_l \in N$.[3] Let $w_j(u, u_l)$ denote the corresponding left eigenvectors normalized so that

$$w_k(u, u_l) \cdot v_j(u, u_l) = \delta_{jk}. \tag{5.67}$$

In this terminology u and u_l satisfy the Rankine–Hugoniot relation with speed s if and only if there exists a j such that

$$w_k(u, u_l) \cdot (u - u_l) = 0 \text{ for all } k \neq j, \quad s = \mu_j(u, u_l), \tag{5.68}$$

and $w_j(u, u_l) \cdot (u - u_l)$ is nonzero. We define functions $F_j : \mathbb{R}^n \times \mathbb{R} \to \mathbb{R}^n$ by

$$F_j(u, \epsilon) = \bigl(w_1(u, u_l) \cdot (u - u_l) - \epsilon \delta_{1j}, \ldots, w_n(u, u_l) \cdot (u - u_l) - \epsilon \delta_{nj}\bigr). \tag{5.69}$$

The Rankine–Hugoniot relation is satisfied if and only if $F_j(u, \epsilon) = 0$ for some ϵ and j. Furthermore, $F_j(u_l, 0) = 0$. A straightforward computation shows that

$$\frac{\partial F_j}{\partial u}(u_l, 0) = \begin{pmatrix} \ell_1(u_l) \\ \vdots \\ \ell_n(u_l) \end{pmatrix},$$

which is nonsingular. Hence the implicit function theorem implies the existence of a unique solution $u_j(\epsilon)$ of

$$F_j(u_j(\epsilon), \epsilon) = 0 \tag{5.70}$$

for ϵ small. □

Occasionally, in particular in Chapter 7, we will use the notation

$$H_j(\epsilon) u_l = u_j(\epsilon)$$

---

[3] The properties of the eigenvalues follow from the implicit function theorem used on the determinant of $\mu I - M(u, u_l)$, and for the eigenvectors by considering the one-dimensional eigenprojections $\int (M(u, u_l) - \mu)^{-1} d\mu$ integrated around a small curve enclosing each eigenvalue $\lambda_j(u_l)$.

5.3. The Hugoniot Locus: The Shock Curves

for a point on the Hugoniot locus.

We will have the opportunity later to study in detail properties of the parameterization of the Hugoniot locus. Let it suffice here to observe that by differentiating each component of $F_j(u_j(\epsilon), \epsilon) = 0$ at $\epsilon = 0$, we find that

$$\ell_k(u_l) \cdot u_j'(0) = \delta_{jk} \tag{5.71}$$

for all $k = 1, \ldots, n$, showing that indeed

$$u_j'(0) = r_j(u_l). \tag{5.72}$$

From the definition of M we see that $M(u, u_l) = M(u_l, u)$, and this symmetry implies that

$$\begin{aligned}
\mu_j(u, u_l) &= \mu_j(u_l, u), & \mu_j(u_l, u_l) &= \lambda_j(u_l), \\
v_j(u, u_l) &= v_j(u_l, u), & v_j(u_l, u_l) &= r_j(u_l), \\
w_j(u, u_l) &= w_j(u_l, u), & w_j(u_l, u_l) &= \ell_j(u_l).
\end{aligned} \tag{5.73}$$

Let $\nabla_k h(u_1, u_2)$ denote the gradient of a function $h\colon \mathbb{R}^n \times \mathbb{R}^n \to \mathbb{R}$ with respect to the kth variable $u_k \in \mathbb{R}^n$, $k = 1, 2$. Then the symmetries (5.73) imply that

$$\nabla_1 \mu_j(u, u_l) = \nabla_2 \mu_j(u, u_l). \tag{5.74}$$

Hence

$$\nabla \lambda_j(u_l) = \nabla_1 \mu_j(u_l, u_l) + \nabla_2 \mu_j(u_l, u_l) = 2\nabla_1 \mu_j(u_l, u_l). \tag{5.75}$$

For a vector-valued function $\phi(u) = (\phi_1(u), \ldots, \phi_n(u))$ we let $\nabla \phi(u)$ denote the Jacobian matrix, viz.,

$$\nabla \phi(u) = \begin{pmatrix} \nabla \phi_1 \\ \vdots \\ \nabla \phi_n \end{pmatrix}. \tag{5.76}$$

Now the symmetries (5.73) imply that

$$\nabla \ell_k(u_l) = 2\nabla_1 w_k(u_l, u_l) \tag{5.77}$$

in obvious notation.

5.4 The Entropy Condition

> ...and now remains
> That we find out the cause of this effect,
> Or rather say, the cause of this defect ...
>
> W. Shakespeare, Hamlet (1603)

Having derived the Hugoniot loci for a general class of conservation laws in the previous section, we will have to select the parts of these curves that give admissible shocks, i.e., satisfy an entropy condition. This will be considerably more complicated in the case of systems than in the scalar case. To guide our intuition we will return to the example of shallow-water waves.

◇ **Example 5.12 (Shallow water (cont'd.)).**

Let us first study the points on $H_1(u_l)$; a similar analysis will apply to $H_2(u_l)$. We will work with the variables h, v rather than h, q. Consider the Riemann problem where we have a high-water bank at rest to the left of the origin, and a lower-water bank to the right of the origin, with a positive velocity; or in other words, the fluid from the lower-water bank moves away from the high-water bank. More precisely, for $h_l > h_r$ we let

$$\begin{pmatrix} h \\ v \end{pmatrix}(x, 0) = \begin{cases} \begin{pmatrix} h_l \\ 0 \end{pmatrix} & \text{for } x < 0, \\ \begin{pmatrix} h_r \\ \frac{h_l - h_r}{\sqrt{2}} \sqrt{h_r^{-1} + h_l^{-1}} \end{pmatrix} & \text{for } x \geq 0, \end{cases}$$

where we have chosen initial data so that the right state is on $H_1(u_l)$, i.e., the Rankine–Hugoniot is already satisfied for a certain speed s. This implies that

$$\begin{pmatrix} h \\ v \end{pmatrix}(x, t) = \begin{cases} \begin{pmatrix} h_l \\ 0 \end{pmatrix} & \text{for } x < st, \\ \begin{pmatrix} h_r \\ \frac{h_l - h_r}{\sqrt{2}} \sqrt{h_r^{-1} + h_l^{-1}} \end{pmatrix} & \text{for } x \geq st, \end{cases}$$

for $h_l > h_r$, where the negative shock speed s given by

$$s = -\frac{h_r \sqrt{h_r^{-1} + h_l^{-1}}}{\sqrt{2}}$$

is a perfectly legitimate weak solution of the initial value problem. However, we see that this is not at all a reasonable solution, since the solution predicts a high-water bank being pushed by a lower one! If we change

5.4. The Entropy Condition   183

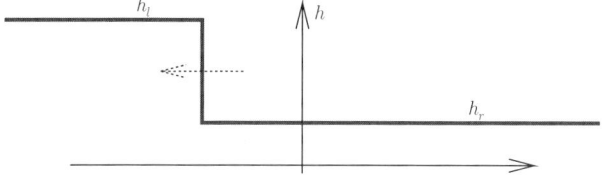

Figure 5.4. Unphysical solution.

the initial conditions so that the right state is on the other branch of $H_1(u_l)$, i.e., we consider a high-water bank moving into a lower-water bank at rest, or

$$\begin{pmatrix} h \\ v \end{pmatrix}(x,0) = \begin{cases} \begin{pmatrix} h_l \\ 0 \end{pmatrix} & \text{for } x < 0, \\ \begin{pmatrix} h_r \\ \frac{h_l-h_r}{\sqrt{2}}\sqrt{h_r^{-1}+h_l^{-1}} \end{pmatrix} & \text{for } x \geq 0, \end{cases}$$

for $h_l < h_r$, we see that the weak solution

$$\begin{pmatrix} h \\ v \end{pmatrix}(x,t) = \begin{cases} \begin{pmatrix} h_l \\ 0 \end{pmatrix} & \text{for } x < st, \\ \begin{pmatrix} h_r \\ \frac{h_l-h_r}{\sqrt{2}}\sqrt{h_r^{-1}+h_l^{-1}} \end{pmatrix} & \text{for } x \geq st, \end{cases}$$

for $h_l < h_r$ and with speed $s = -h_r\sqrt{h_r^{-1}+h_l^{-1}}/\sqrt{2}$, is reasonable physically since the high-water bank now is pushing the lower one. If

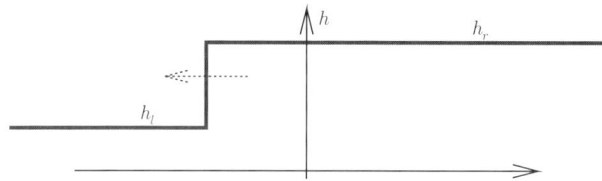

Figure 5.5. Reasonable solution.

you are worried about the fact that the shock is preserved, i.e., that there is no deformation of the shock profile, this is due to the fact that the right state is carefully selected. In general we will have both a shock wave and a rarefaction wave in the solution. This will be clear when we solve the full Riemann problem.

Let us also consider the above examples with energy conservation in mind. In our derivation of the shallow-water equations we used conservation of mass and momentum only. For smooth solutions of these equations, conservation of mechanical energy will follow. Indeed, the kinetic energy of a vertical section of the shallow-water system at a point x is given by $h(x,t)v(x,t)^2/2$ in dimensionless variables, and the potential energy of the same section is given by $h(x,t)^2/2$, and hence the total mechanical energy reads $(h(x,t)v(x,t)^2 + h(x,t)^2)/2$. Consider now a section of the channel between points $x_1 < x_2$ and assume that we have a smooth (classical) solution of the shallow-water equations. The rate of change of mechanical energy is given by the energy flow across x_1 and x_2 plus the work done by the pressure. Energy conservation yields

$$\begin{aligned} 0 &= \frac{d}{dt}\int_{x_1}^{x_2}\left(\frac{1}{2}hv^2 + \frac{1}{2}h^2\right)dx + \int_{x_1}^{x_2}\frac{\partial}{\partial x}\left(\frac{1}{2}hv^3 + \frac{1}{2}h^2 v\right)dx \\ &+ \int_0^{h(x_2,t)} P(x_2,y,t)v(x_2,t)\,dy - \int_0^{h(x_1,t)} P(x_1,y,t)v(x_1,t)\,dy \\ &= \frac{d}{dt}\int_{x_1}^{x_2}\left(\frac{1}{2}hv^2 + \frac{1}{2}h^2\right)dx + \int_{x_1}^{x_2}\frac{\partial}{\partial x}\left(\frac{1}{2}hv^3 + \frac{1}{2}h^2 v\right)dx \\ &+ \int_{x_1}^{x_2}\frac{\partial}{\partial x}\left(\frac{1}{2}h^2 v\right) dx \\ &= \int_{x_1}^{x_2}\frac{\partial}{\partial t}\left(\frac{1}{2}hv^2 + \frac{1}{2}h^2\right)dx + \int_{x_1}^{x_2}\frac{\partial}{\partial x}\left(\frac{1}{2}hv^3 + h^2 v\right)dx, \end{aligned}$$

using that $P(x,y,t) = h(x,t) - y$ in dimensionless variables. Hence we conclude that

$$\left(\frac{1}{2}hv^2 + \frac{1}{2}h^2\right)_t + \left(\frac{1}{2}hv^3 + h^2 v\right)_x = 0.$$

This equation follows easily directly from (5.5) for smooth solutions.

However, for weak solutions, mechanical energy will in general not be conserved. Due to dissipation we expect an energy loss across a bore. Let us compute this change in energy ΔE across the bore in the two examples above, for a time t such that $x_1 < st < x_2$. We obtain

$$\begin{aligned} \Delta E &= \frac{d}{dt}\int_{x_1}^{x_2}\left(\frac{1}{2}hv^2 + \frac{1}{2}h^2\right) dx + \left.\left(\frac{1}{2}hv^3 + h^2 v\right)\right|_{x_1}^{x_2} \\ &= -s\left[\tfrac{1}{2}hv^2 + \tfrac{1}{2}h^2\right] + \left[\tfrac{1}{2}hv^3 + h^2 v\right] \\ &= \frac{1}{2}h_r\delta(\llbracket h\rrbracket^2\delta^2 h_r + h_r^2 - h_l^2) + (-\llbracket h\rrbracket^3\delta^3 h_r - 2\llbracket h\rrbracket\delta h_r^2) \\ &= -\frac{1}{4}\llbracket h\rrbracket^3\delta, \end{aligned} \qquad (5.78)$$

where we have introduced

$$\delta := \frac{\sqrt{h_r^{-1} + h_l^{-1}}}{\sqrt{2}}. \qquad (5.79)$$

(Recall that $v_l = 0$ and $v_r = [\![v]\!] = -[\![h]\!]\delta$ from the Rankine–Hugoniot condition.) Here we have used that we have a smooth solution with energy conservation on each interval $[x_1, st]$ and $[st, x_2]$. In the first case, where we had a low-water bank pushing a high-water bank with $h_r < h_l$, we find indeed that $\Delta E > 0$, while in the other case we obtain the more reasonable $\Delta E < 0$.

From these two simple examples we get a hint that only one branch of $H_1(u_l)$ is physically acceptable. We will now translate this into conditions on existence of viscous profiles, and conditions on the eigenvalues of $df(u)$ at $u = u_l$ and $u = u_r$, conditions we will use in cases where our intuition will be more blurred.

In Chapter 2 we discussed the notion of traveling waves. Recall from (2.7) that a shock between two fixed states u_l and u_r with speed s, viz.,

$$u(x,t) = \begin{cases} u_l & \text{for } x < st, \\ u_r & \text{for } x \geq st, \end{cases} \tag{5.80}$$

admits a viscous profile if $u(x,t)$ is the limit as $\epsilon \to 0$ of $u^\epsilon(x,t) = U((x - st)/\epsilon) = U(\xi)$ with $\xi = (x - st)/\epsilon$, which satisfies

$$u^\epsilon_t + f(u^\epsilon)_x = \epsilon u^\epsilon_{xx}.$$

Integrating this equation, using $\lim_{\epsilon \to 0} U(\xi) = u_l$ if $\xi < 0$, we obtain

$$\dot{U} = A(h, q) := f(U) - f(u_l) - s(U - u_l), \tag{5.81}$$

where the differentiation is with respect to ξ. We will see that it is possible to connect the left state with a viscous profile to a right state only for the branch with $h_r > h_l$ of $H_1(u_l)$, i.e., the physically correct solution.

Computationally it will be simpler to work with viscous profiles in the (h, v) variables rather than (h, q). Using $\dot{q} = \dot{v}h + v\dot{h}$ and (5.81), we find that there is a viscous profile in (h, q) if and only if (h, v) satisfies

$$\begin{pmatrix} \dot{h} \\ \dot{v} \end{pmatrix} = B(h, v) := \begin{pmatrix} vh - v_l h_l - s(h - h_l) \\ (v - v_l)(v_l - s)\frac{h_l}{h} + \frac{h^2 - h_l^2}{2h} \end{pmatrix}$$

$$= \begin{pmatrix} vh - v_l h_l - s(h - h_l) \\ (v - v_l)\frac{h_l h_r}{h}\delta + \frac{h^2 - h_l^2}{2h} \end{pmatrix}. \tag{5.82}$$

We will analyze the vector field $B(h, v)$ carefully. The Jacobian of B reads

$$dB(h, v) = \begin{pmatrix} v - s & h \\ \frac{h^2 + h_l^2}{2h^2} - (v - v_l)\frac{h_l h_r}{h^2}\delta & \frac{h_l h_r}{h}\delta \end{pmatrix}. \tag{5.83}$$

At the left state u_l we obtain

$$dB(h_l, v_l) = \begin{pmatrix} v_l - s & h_l \\ 1 & h_r \delta \end{pmatrix} = \begin{pmatrix} h_r \delta & h_l \\ 1 & h_r \delta \end{pmatrix}, \tag{5.84}$$

using the value of the shock speed s, equation (5.56). The eigenvalues of $dB(h_l, v_l)$ are $h_r \delta \pm \sqrt{h_l}$, which both are easily seen to be positive when $h_r > h_l$; thus (h_l, v_l) is a source. Similarly, we obtain

$$dB(h_r, v_r) = \begin{pmatrix} h_l \delta & h_r \\ 1 & h_l \delta \end{pmatrix}, \tag{5.85}$$

with eigenvalues $h_l \delta \pm \sqrt{h_r}$. In this case, one eigenvalue is positive and one negative, thus (h_r, v_r) is a saddle point. However, we still have to establish an orbit connecting the two states. To this end we construct a region K with (h_l, v_l) and (h_r, v_r) at the boundary of K such that a connecting orbit has to connect the two points within K. The region K will have two curves as boundaries where the first and second component of B vanish, respectively. The first curve, denoted by C_h, is defined by the first component being zero, viz.,

$$vh - v_l h_l - s(h - h_l) = 0, \quad h \in [h_l, h_r],$$

which can be simplified to yield

$$v = v_l - (h - h_l) \frac{h_r}{h} \delta, \quad h \in [h_l, h_r]. \tag{5.86}$$

For the second curve, C_v, we have

$$(v - v_l)(v_l - s) \frac{h_l}{h} + \frac{h^2 - h_l^2}{2h} = 0, \quad h \in [h_l, h_r],$$

which can be rewritten as

$$v = v_l - \frac{h^2 - h_l^2}{2 h_l h_r \delta}, \quad h \in [h_l, h_r]. \tag{5.87}$$

Let us now study the behavior of the second component of B along the curve C_h where the first component vanishes, i.e.,

$$\left[(v - v_l) \frac{h_l h_r}{h} \delta + \frac{h^2 - h_l^2}{2h} \right] \Big|_{C_h}$$
$$= -\frac{h_l}{2h^2}(h_r - h)(h - h_l)\left(1 + \frac{h + h_r}{h_l}\right) < 0. \tag{5.88}$$

Similarly, for the first component of B along C_v, we obtain

$$\left[vh - v_l h_l - s(h - h_l) \right] \Big|_{C_v}$$
$$= \frac{h - h_l}{2 h_r h_l \delta} \left(h_r(h_l + h_r) - h(h + h_l) \right) > 0, \tag{5.89}$$

which is illustrated in Figure 5.6. The flow of the vector field is leaving the region K along the curves C_h and C_v. Locally, around (h_r, v_r)

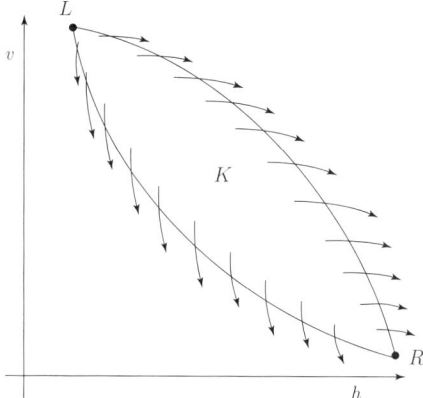

Figure 5.6. The vector field B.

there has to be an orbit entering K as ξ decreases from ∞. This curve cannot escape K and has to connect to a curve coming from (h_l, v_l). Consequently, we have proved existence of a viscous profile.

We saw that the relative values of the shock speed and the eigenvalues of B, and hence of A, at the left and right states were crucial for this analysis to hold. Let us now translate these assumptions into assumptions on the eigenvalues of dA. The Jacobian of A reads

$$dA(h,q) = \begin{pmatrix} -s & 1 \\ h - \frac{q^2}{h^2} & \frac{2q}{h} - s \end{pmatrix}.$$

Hence the eigenvalues are $-s + \frac{q}{h} \pm \sqrt{h} = -s + \lambda(u)$. At the left state both eigenvalues are positive, and thus u_l is a source, while at u_r one is positive and one negative, and hence u_r is a saddle. We may write this as

$$\lambda_1(u_r) < s < \lambda_1(u_l), \quad s < \lambda_2(u_r). \tag{5.90}$$

We call these the *Lax inequalities*, and say that a shock satisfying these inequalities is a *Lax 1-shock* or a *slow Lax shock*. We have proved that for the shallow-water equations with $h_r > h_l$ there exists a viscous profile, and that the Lax shock conditions are satisfied.

Let us now return to the unphysical shock "solution." In this case we had $h_r < h_l$ with the eigenvalues at the left state (h_l, v_l) of different sign. Thus (h_l, v_l) is a saddle, both eigenvalues are positive for the right state, and hence that point is a source. Accordingly, there cannot be any orbit connecting the left state with the right state.

A similar analysis can be performed for $H_2(u_l)$, giving that there exists a viscous profile for a shock satisfying the Rankine–Hugoniot relation if

and only if the following *Lax entropy conditions* are satisfied:
$$\lambda_2(u_r) < s < \lambda_2(u_l), \quad s > \lambda_1(u_l). \tag{5.91}$$
In that case we have a *fast Lax shock* or a *Lax 2-shock*.

We may sum up the above argument as follows. A shock has a viscous profile if and only if the Lax shock conditions are satisfied. We call such shocks *admissible* and denote the part of the Hugoniot locus where the Lax j conditions are satisfied by S_j. In the case of shallow-water equations we obtain

$$S_1(u_l) := \left\{ \left(q_l \frac{h}{h_l} - \frac{h}{\sqrt{2}} (h - h_l) \sqrt{h^{-1} + h_l^{-1}} \right) \middle| h \geq h_l \right\}, \tag{5.92}$$

$$S_2(u_l) := \left\{ \left(q_l \frac{h}{h_l} + \frac{h}{\sqrt{2}} (h - h_l) \sqrt{h^{-1} + h_l^{-1}} \right) \middle| h \leq h_l \right\}. \tag{5.93}$$

(These curves are depicted in Section 5.3.) We may also want to parameterize the admissible shocks differently. For the slow Lax shocks let
$$h_{1,\epsilon} := h_l - \frac{2}{3}\sqrt{h_l}\,\epsilon, \quad \epsilon < 0. \tag{5.94}$$

This gives
$$q_{1,\epsilon} := q_l \left(1 - \frac{2\epsilon}{3\sqrt{h_l}}\right) + \frac{\epsilon}{9}\sqrt{2h_l \left(6\sqrt{h_l} - 2\epsilon\right)\left(3\sqrt{h_l} - \epsilon\right)} \tag{5.95}$$

such that
$$\left.\frac{d}{d\epsilon}\begin{pmatrix} h_{1,\epsilon} \\ q_{1,\epsilon} \end{pmatrix}\right|_{\epsilon=0} = r_1(u_l), \tag{5.96}$$

where $r_1(u_l)$ is given by (5.32). Similarly, for the fast Lax shocks let
$$h_{2,\epsilon} := h_l + \frac{2}{3}\sqrt{h_l}\,\epsilon, \quad \epsilon < 0. \tag{5.97}$$

Then
$$q_{2,\epsilon} := q_l \left(1 + \frac{2\epsilon}{3\sqrt{h_l}}\right) + \frac{\epsilon}{9}\sqrt{2h_l \left(6\sqrt{h_l} + 2\epsilon\right)\left(3\sqrt{h_l} + 2\epsilon\right)}, \tag{5.98}$$

such that
$$\left.\frac{d}{d\epsilon}\begin{pmatrix} h_{2,\epsilon} \\ q_{2,\epsilon} \end{pmatrix}\right|_{\epsilon=0} = r_2(u_l), \tag{5.99}$$

where $r_2(u_l)$ is given by (5.32). $\diamondsuit$

In the above example we have seen the equivalence between the existence of a viscous profile and the Lax entropy conditions for the shallow-water equations. This analysis has yet to be carried out for general systems. We will use the above example as a motivation for the following definition, stated for general systems.

Definition 5.13. *We say that a shock*

$$u(x,t) = \begin{cases} u_l & \text{for } x < st, \\ u_r & \text{for } x \geq st, \end{cases} \quad (5.100)$$

is a Lax j-shock if the shock speed s satisfies

$$\lambda_{j-1}(u_l) < s < \lambda_j(u_l), \quad \lambda_j(u_r) < s < \lambda_{j+1}(u_r). \quad (5.101)$$

(Here $\lambda_0 = -\infty$ and $\lambda_{n+1} = \infty$.)

Observe that for strictly hyperbolic systems, where the eigenvalues are distinct, it suffices to check the inequalities $\lambda_j(u_r) < s < \lambda_j(u_l)$ for small Lax j-shocks if the eigenvalues are continuous in u.

The following result follows from Theorem 5.11.

Theorem 5.14. *Consider the strictly hyperbolic equation $u_t + f(u)_x = 0$ in a domain D in $\mathbb{R}^n$. Let $u_l \in D$. A state $u_{j,\epsilon} \in H_j(u_l)$ is a Lax j-shock near u_l if $|\epsilon|$ is sufficiently small and ϵ negative. If ϵ is positive, the shock is not a Lax j-shock.*

Proof. Using the ϵ parameterization of the Hugoniot locus we see that the shock is a Lax j-shock if and only if

$$\lambda_{j-1}(0) < s(\epsilon) < \lambda_j(0), \quad \lambda_j(\epsilon) < s(\epsilon) < \lambda_{j+1}(\epsilon), \quad (5.102)$$

where for simplicity we write $u(\epsilon) = u_{j,\epsilon}$, $s(\epsilon) = s_{j,\epsilon}$, and $\lambda_k(\epsilon) = \lambda_k(u_{j,\epsilon})$. The observation following the definition of Lax shocks shows that it suffices to check the inequalities

$$\lambda_j(\epsilon) < s(\epsilon) < \lambda_j(0). \quad (5.103)$$

Assume first that $u(\epsilon) \in H_j(u_l)$ and that ϵ is negative. We know from the implicit function theorem that $s(\epsilon)$ tends to $\lambda_j(0)$ as ϵ tends to zero. From the fact that also $\lambda_j(\epsilon) \to \lambda_j(0)$ as $\epsilon \to 0$, and

$$\left. \frac{d\lambda_j(\epsilon)}{d\epsilon} \right|_{\epsilon=0} = \nabla \lambda_j(0) \cdot r_j(u_l) = 1,$$

it suffices to prove that $0 < s'(0) < 1$. We will in fact prove that $s'(0) = \frac{1}{2}$. Recall from (5.68) that s is an eigenvalue of the matrix $M(u, u_l)$, i.e., $s(\epsilon) = \mu_j(u(\epsilon), u_l)$. Then

$$s'(0) = \nabla_1 \mu_j(u_l, u_l) \cdot u'(0) = \frac{1}{2} \nabla \lambda_j(u_l) \cdot r_j(u_l) = \frac{1}{2}, \quad (5.104)$$

using the symmetry (5.75) and the normalization of the right eigenvalue.

If $\epsilon > 0$, we immediately see that $s(\epsilon) > s(0) = \lambda_j(0)$, and in this case we cannot have a Lax j-shock. □

5.5 The Solution of the Riemann Problem

> Wie für die Integration der linearen partiellen Differentialgleichungen die fruchtbarsten Methoden nicht durch Entwicklung des allgemeinen Begriffs dieser Aufgabe gefunden worden, sondern vielmehr aus der Behandlung specieller physikalischer Probleme hervorgegangen sind, so scheint auch die Theorie der nichtlinearen partiellen Differentialgleichungen durch eine eingehende, alle Nebenbedingungen berücksichtigende, Behandlung specieller physikalischer Probleme am meisten gefördert zu werden, und in der That hat die Lösung der ganz speciellen Aufgabe, welche den Gengstand dieser Abhandlung bildet, neue Methoden und Auffassungen erfordert, und zu Ergebnissen geführt, welche wahrscheinlich auch bei allgemeineren Aufgaben eine Rolle spielen werden.[4]
>
> G. F. B. Riemann [119]

In this section we will combine the properties of the rarefaction waves and shock waves from the previous sections to derive the unique solution of the Riemann problem for small initial data. Our approach will be the following. Assume that the left state u_l is given, and consider the space of all right states u_r. For each right state we want to describe the solution of the corresponding Riemann problem. (We could, of course, reverse the picture and consider the right state as fixed and construct the solution for all possible left states.)

To this end we start by defining *wave curves*. If the jth wave family is genuinely nonlinear, we define

$$W_j(u_l) := R_j(u_l) \cup S_j(u_l), \qquad (5.105)$$

and if the jth family is linearly degenerate, we let

$$W_j(u_l) := C_j(u_l). \qquad (5.106)$$

Recall that we have parameterized the shock and rarefaction curves separately with a parameter ϵ such that ϵ positive (negative) corresponds to a rarefaction (shock) wave solution in the case of a genuinely nonlinear wave family. The important fact about the wave curves is that they almost form

---

[4]The theory of nonlinear equations can, it seems, achieve the most success if our attention is directed to special problems of physical content with thoroughness and with a consideration of all auxiliary conditions. In fact, the solution of the very special problem that is the topic of the current paper requires new methods and concepts and leads to results which probably will also play a role in more general problems.

a local coordinate system around u_l, and this will make it possible to prove existence of solutions of the Riemann problem for u_r close to u_l.

We will commence from the left state u_l and connect it to a nearby intermediate state $u_{m_1} = u_{1,\epsilon_1} \in W_1(u_l)$ using either a rarefaction wave solution ($\epsilon_1 > 0$) or a shock wave solution ($\epsilon_1 < 0$) if the first family is genuinely nonlinear. If the first family is linearly degenerate, we use a contact discontinuity for all ϵ_1. From this state we find another intermediate state $u_{m_2} = u_{2,\epsilon_2} \in W_2(u_{m_1})$. We continue in this way until we have reached an intermediate state $u_{m_{n-1}}$ such that $u_r = u_{n,\epsilon_n} \in W_n(u_{m_{n-1}})$. The problem is to show existence of a unique n-tuple of $(\epsilon_1, \ldots, \epsilon_n)$ such that we "hit" u_r starting from u_l using this construction.

As usual, we will start by illustrating the above discussion for the shallow-water equations. This example will contain the fundamental description of the solution that in principle will carry over to the general case.

◇ **Example 5.15 (Shallow water (cont'd.)).**

Fix u_l. For each right state u_r we have to determine one middle state u_m on the first-wave curve through u_l such that u_r is on the second-wave curve with left state u_m, i.e., $u_m \in W_1(u_l)$ and $u_r \in W_2(u_m)$. (In the special case that $u_r \in W_1(u_l) \cup W_2(u_l)$ no middle state u_m is required.) For 2×2 systems of conservation laws it is easier to consider the "backward" second-wave curve $W_2^-(u_r)$ consisting of states u_m that can be connected to u_r on the right with a *fast* wave. The Riemann problem with left state u_l and right state u_r has a unique solution if and only if $W_1(u_l)$ and $W_2^-(u_r)$ have a unique intersection. In that case, clearly the intersection will be the middle state u_m. The curve $W_1(u_l)$ is given by

$$v = v(h) = \begin{cases} v_l - 2(\sqrt{h} - \sqrt{h_l}) & \text{for } h \in [0, h_l], \\ v_l - \frac{h - h_l}{\sqrt{2}}\sqrt{h^{-1} + h_l^{-1}} & \text{for } h \geq h_l, \end{cases} \quad (5.107)$$

and we easily see that $W_1(u_l)$ is strictly decreasing, unbounded, and starting at $v_l + 2\sqrt{h_l}$. Using (5.37) and (5.93) we find that $W_2^-(u_r)$ reads

$$v = v(h) = \begin{cases} v_r + 2(\sqrt{h} - \sqrt{h_r}) & \text{for } h \in [0, h_r], \\ v_r + \frac{h - h_r}{\sqrt{2}}\sqrt{h^{-1} + h_r^{-1}} & \text{for } h \geq h_r, \end{cases} \quad (5.108)$$

which is strictly increasing, unbounded, with minimum $v_r - 2\sqrt{h_r}$. Thus we conclude that the Riemann problem for shallow water has a unique solution in the region where

$$v_l + 2\sqrt{h_l} \geq v_r - 2\sqrt{h_r}. \quad (5.109)$$

To obtain explicit equations for the middle state u_m we have to make case distinctions, depending on the type of wave curves that intersect, i.e., rarefaction waves or shock curves. This gives rise to four regions, denoted by I,...,IV. For completeness we give the equations for the middle state u_m in all cases.

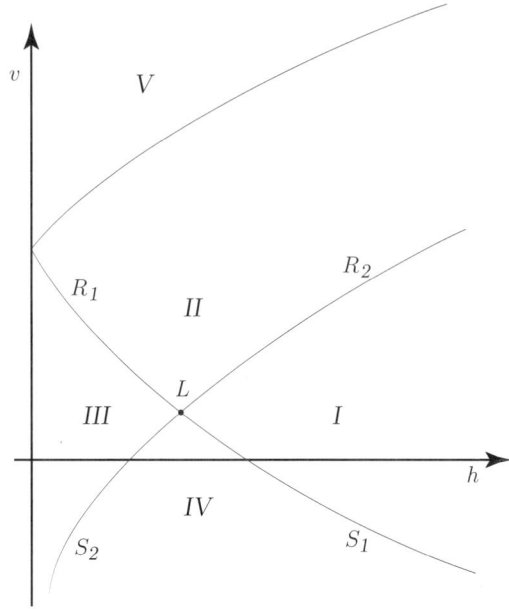

Figure 5.7. The partition of the (h, v) plane; see (5.114) and (5.133).

Assume first that $u_r \in$ I. We will determine a unique intermediate state $u_m \in S_1(u_l)$ such that $u_r \in R_2(u_m)$. These requirements give the following equations to be solved for h_m, v_m such that $u_m = (h_m, q_m) = (h_m, h_m v_m)$:

$$v_m = v_l - \frac{1}{\sqrt{2}}(h_m - h_l)\sqrt{\frac{1}{h_m} + \frac{1}{h_l}}, \quad v_r = v_m + 2(\sqrt{h_r} - \sqrt{h_m}).$$

Summing these equations we obtain the equation

$$\sqrt{2}[\![v]\!] = 2\sqrt{2}(\sqrt{h_r} - \sqrt{h_m}) - (h_m - h_l)\sqrt{\frac{1}{h_m} + \frac{1}{h_l}} \quad \text{(I)} \quad (5.110)$$

to determine h_m. Consider next the case with $u_r \in$ III. Here $u_m \in R_1(u_l)$ and $u_r \in S_2(u_m)$, and in this case we obtain

$$\sqrt{2}[\![v]\!] = (h_r - h_m)\sqrt{\frac{1}{h_r} + \frac{1}{h_m}} - 2\sqrt{2}(\sqrt{h_m} - \sqrt{h_l}), \quad \text{(III)} \quad (5.111)$$

5.5. The Solution of the Riemann Problem

while in the case $u_r \in$ IV, we obtain (here $u_m \in S_1(u_l)$ and $u_r \in S_2(u_m)$)

$$\sqrt{2}[\![v]\!] = (h_r - h_m)\sqrt{\frac{1}{h_r} + \frac{1}{h_m}} - (h_m - h_l)\sqrt{\frac{1}{h_m} + \frac{1}{h_l}}. \quad \text{(IV)} \quad (5.112)$$

The case $u_r \in$ II is special. Here $u_m \in R_1(u_l)$ and $u_r \in R_2(u_m)$. The intermediate state u_m is given by

$$v_m = v_l - 2(\sqrt{h_m} - \sqrt{h_l}), \quad v_r = v_m + 2(\sqrt{h_r} - \sqrt{h_m}),$$

which can easily be solved for h_m to yield

$$\sqrt{h_m} = \frac{2(\sqrt{h_r} + \sqrt{h_l}) - [\![v]\!]}{4}. \quad \text{(II)} \quad (5.113)$$

This equation is solvable only for right states such that the right-hand side of (5.113) is nonnegative. Observe that this is consistent with what we found above in (5.109). Thus we find that for

$$u_r \in \left\{ u \in \langle 0, \infty \rangle \times \mathbb{R} \mid 2(\sqrt{h} + \sqrt{h_l}) \geq [\![v]\!] \right\} \quad (5.114)$$

the Riemann problem has a unique solution consisting of a slow wave followed by a fast wave. Let us summarize the solution of the Riemann problem for the shallow-water equations. First of all, we were not able to solve the problem globally, but only locally around the left state. Secondly, the general solution consists of a composition of elementary waves. More precisely, let $u_r \in \left\{ u \in \langle 0, \infty \rangle \times \mathbb{R} \mid 2(\sqrt{h} + \sqrt{h_l}) \geq [\![v]\!] \right\}$. Let $w_j(x/t; h_m, h_l)$ denote the solution of the Riemann problem for $u_m \in W_j(u_l)$, here as in most of our calculations on the shallow-water equations we use h rather than ϵ as the parameter. We will introduce the notation $\sigma_j^{\pm}$ for the slowest and fastest wave speed in each family to simplify the description of the full solution. Thus we have that for $j = 1$ $(j = 2)$ and $h_r < h_l$ $(h_r > h_l)$, w_j is a rarefaction-wave solution

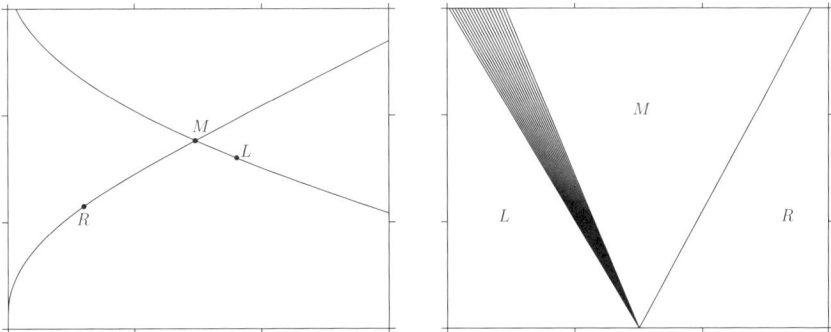

Figure 5.8. The solution of the Riemann problem in phase space (left) and in (x, t) space (right).

with slowest speed $\sigma_j^- = \lambda_j(u_l)$ and fastest speed $\sigma_j^+ = \lambda_j(u_r)$. If $j = 1$ ($j = 2$) and $h_r > h_l$ ($h_r < h_l$), then w_j is a shock-wave solution with speed $\sigma_j^- = \sigma_j^+ = s_j(h_r, h_l)$. The solution of the Riemann problem reads

$$u(x,t) = \begin{cases} u_l & \text{for } x \leq \sigma_1^- t, \\ w_1(x/t; u_m, u_l) & \text{for } \sigma_1^- t \leq x \leq \sigma_1^+ t, \\ u_m & \text{for } \sigma_1^+ t \leq x \leq \sigma_2^- t, \\ w_2(x/t; u_r, u_m) & \text{for } \sigma_2^- t \leq x \leq \sigma_2^+ t, \\ u_r & \text{for } x \geq \sigma_2^+ t. \end{cases} \quad (5.115)$$

We will show later in this chapter how to solve the Riemann problem globally for the shallow-water equations. ◇

Before we turn to the existence and uniqueness theorem for solutions of the Riemann problem we will need a certain property of the wave curves that we can explicitly verify for the shallow-water equations.

Recall from (5.72) and (5.28) that $\frac{du_\epsilon}{d\epsilon}\big|_{\epsilon=0} = r_j(u_l)$; thus $W_j(u_l)$ is at least differentiable at u_l. In fact, one can prove that $W_j(u_l)$ has a continuous second derivative across u_l.

We introduce the following notation for the *directional derivative* of a quantity $h(u)$ in the direction r (not necessarily normalized) at the point u, which is defined as

$$D_r h(u) = \lim_{\epsilon \to 0} \frac{1}{\epsilon} (h(u + \epsilon r) - h(u)) = (\nabla h \cdot r)(u). \quad (5.116)$$

(When h is a vector, ∇h denotes the Jacobian.)

Theorem 5.16. *The wave curve $W_j(u_l)$ has a continuous second derivative across u_l. In particular,*

$$u_{j,\epsilon} = u_l + \epsilon r_j(u_l) + \frac{1}{2}\epsilon^2 D_{r_j} r_j(u_l) + \mathcal{O}(\epsilon^3).$$

Proof. In our proof of the admissibility of parts of the Hugoniot loci, Theorem 5.14, we derived most of the ingredients required for the proof of this theorem. The rarefaction curve $R_j(u_l)$ is the integral curve of the right eigenvector $r_j(u)$ passing through u_l, and thus we have (when for simplicity we have suppressed the j dependence in the notation for u, and write $u(\epsilon) = u_{j,\epsilon}$, etc.)

$$u(0+) = u_l, \quad u'(0+) = r_j(u_l), \quad u''(0+) = \nabla r_j(u_l) r_j(u_l). \quad (5.117)$$

(Here $\nabla r_j(u_l) r_j(u_l)$ denotes the product of the $n \times n$ matrix $\nabla r_j(u_l)$, cf. (5.76), with the (column) vector $r_j(u_l)$.) Recall that the Hugoniot locus is determined by the relation (5.70), i.e.,

$$w_k(u(\epsilon), u_l) \cdot (u(\epsilon) - u_l) = \epsilon \delta_{jk}, \quad k = 1, \ldots, n. \quad (5.118)$$

We know already from (5.72) that $u'(0-) = r_j(u_l)$. To find the second derivative of $u(\epsilon)$ at $\epsilon = 0$, we have to compute the second derivative of

(5.118). Here we find that[5]
$$2r_j(u_l)\nabla_1 w_k(u_l, u_l)r_j(u_l) + w_k(u_l, u_l) \cdot u''(0-) = 0, \quad k = 1, \ldots, n. \quad (5.119)$$
(A careful differentiation of each component may be helpful here; at least we thought so.) In the first term, the matrix $\nabla_1 w_k(u_l, u_l)$ is multiplied from the right by the (column) vector $r_j(u_l)$ and by the (row) vector $r_j(u_l)$ from the left.) Using (5.77), i.e., $\nabla_1 w_k(u_l, u_l) = \frac{1}{2}\nabla \ell_k(u_l)$, we find that
$$r_j(u_l)\nabla \ell_k(u_l)r_j(u_l) + \ell_k(u_l) \cdot u''(0-) = 0. \quad (5.120)$$
The orthogonality of the left and the right eigenvectors, $\ell_k(u_l) \cdot r_j(u_l) = \delta_{jk}$, shows that
$$r_j(u_l)\nabla \ell_k(u_l) = -\ell_k(u_l)\nabla r_j(u_l). \quad (5.121)$$
Inserting this into (5.120) we obtain
$$\ell_k(u_l) \cdot u''(0-) = \ell_k(u_l)\nabla r_j(u_l)r_j(u_l) \text{ for all } k = 1, \ldots, n.$$
From this we conclude that also $u''(0-) = \nabla r_j(u_l)r_j(u_l)$, thereby proving the theorem. □

We will now turn to the proof of the classical Lax theorem about existence of a unique entropy solution of the Riemann problem for small initial data. The assumption of strict hyperbolicity of the system implies the existence of a full set of linearly independent eigenvectors. Furthermore, we have proved that the wave curves are C^2, and hence intersect transversally at the left state. This shows, in a heuristic way, that it is possible to solve the Riemann problem locally. Indeed, we saw that we could write the solution of the corresponding problem for the shallow-water equations as a composition of individual elementary waves that do not interact, in the sense that the fastest wave of one family is slower than the slowest wave of the next family. This will enable us to write the solution in the same form in the general case. In order to do this, we introduce some notation. Let $u_{j,\epsilon_j} = u_{j,\epsilon_j}(x/t; u_r, u_l)$ denote the unique solution of the Riemann problem with left state u_l and right state u_r that consists of a single elementary wave (i.e., shock wave, rarefaction wave, or contact discontinuity) of family j with strength ϵ_j. Furthermore, we need to define notation for speeds corresponding to the fastest and slowest waves of a fixed family. Let

$$\begin{aligned}\sigma_j^+ = \sigma_j^- &= s_{j,\epsilon_j} & \text{if } \epsilon_j < 0, \\ \sigma_j^- &= \lambda_j(u_{j-1,\epsilon_{j-1}}) = \lambda_j(u_{m_{j-1}}), \\ \sigma_j^+ &= \lambda_j(u_{j,\epsilon_j}) = \lambda_j(u_{m_j}) \end{aligned} \right\} \quad \text{if } \epsilon_j > 0, \quad (5.122)$$

---

[5]Lo and behold; the second derivative of $w_k(u(\epsilon), u_l)$ is immaterial, since it is multiplied by $u(\epsilon) - u_l$ at $\epsilon = 0$.

if the jth wave family is genuinely nonlinear, and

$$\sigma_j^+ = \sigma_j^- = \lambda_j(u_{j,\epsilon_j}) = \lambda_j(u_{m_j}) \tag{5.123}$$

if the jth wave family is linearly degenerate. With these definitions we are ready to write the solution of the Riemann problem as

$$u(x,t) = \begin{cases} u_l & \text{for } x \leq \sigma_1^- t, \\ u_{1,\epsilon_1}(x/t; u_{m_1}, u_l) & \text{for } \sigma_1^- t \leq x \leq \sigma_1^+ t, \\ u_{m_1} & \text{for } \sigma_1^+ t \leq x \leq \sigma_2^- t, \\ u_{2,\epsilon_2}(x/t; u_{m_2}, u_{m_1}) & \text{for } \sigma_2^- t \leq x \leq \sigma_2^+ t, \\ u_{m_2} & \text{for } \sigma_2^+ t \leq x \leq \sigma_3^- t, \\ \vdots \\ u_{n,\epsilon_n}(x/t; u_r, u_{m_{n-1}}) & \text{for } \sigma_n^- t \leq x \leq \sigma_n^+ t, \\ u_r & \text{for } x \geq \sigma_n^+ t. \end{cases} \tag{5.124}$$

Theorem 5.17 (Lax's theorem). *Assume that $f_j \in C^2(\mathbb{R}^n)$, $j = 1, \ldots, n$. Let D be a domain in $\mathbb{R}^n$ and consider the strictly hyperbolic equation $u_t + f(u)_x = 0$ in D. Assume that each wave family is either genuinely nonlinear or linearly degenerate.*

Then for $u_l \in D$ there exists a neighborhood $\tilde{D} \subset D$ of u_l such that for all $u_r \in \tilde{D}$ the Riemann problem

$$u(x,0) = \begin{cases} u_l & \text{for } x < 0, \\ u_r & \text{for } x \geq 0, \end{cases} \tag{5.125}$$

has a unique solution in $\tilde{D}$ consisting of up to n elementary waves, i.e., rarefaction waves, shock solutions satisfying the Lax entropy condition, or contact discontinuities. The solution is given by (5.124).

Proof. Consider the map $W_{j,\epsilon} \colon u \mapsto u_{j,\epsilon} \in W_j(u)$. We may then write the solution of the Riemann problem using the composition

$$W_{(\epsilon_1,\ldots,\epsilon_n)} = W_{n,\epsilon_n} \circ \cdots \circ W_{1,\epsilon_1} \tag{5.126}$$

as

$$W_{(\epsilon_1,\ldots,\epsilon_n)} u_l = u_r, \tag{5.127}$$

and we want to prove the existence of a unique $(\epsilon_1, \ldots, \epsilon_n)$ (near the origin) such that (5.127) is satisfied for $|u_l - u_r|$ small. In our proof we will need the two leading terms in the Taylor expansion for W. We summarize this in the following lemma.

5.5. The Solution of the Riemann Problem

Lemma 5.18. *We have*

$$W_{(\epsilon_1,\ldots,\epsilon_n)}(u_l) = u_l + \sum_{i=1}^{n} \epsilon_i r_i(u_l) + \frac{1}{2}\sum_{i=1}^{n} \epsilon_i^2 D_{r_i} r_i(u_l)$$
$$+ \sum_{\substack{i,j=1\\i<j}}^{n} \epsilon_i \epsilon_j D_{r_i} r_j(u_l) + \mathcal{O}\left(|\epsilon|^3\right). \tag{5.128}$$

Proof (of Lemma 5.18). We shall show that for $k = 0, \ldots, n$,

$$W_{(\epsilon_1,\ldots,\epsilon_k,0,\ldots,0)}(u_l) = u_l + \sum_{i=1}^{k} \epsilon_i r_i(u_l) + \frac{1}{2}\sum_{i=1}^{k} \epsilon_i^2 D_{r_i} r_i(u_l)$$
$$+ \sum_{\substack{i,j=1\\i<j}}^{k} \epsilon_i \epsilon_j D_{r_i} r_j(u_l) + \mathcal{O}\left(|\epsilon|^3\right) \tag{5.129}$$

by induction on k. It is clearly true for $k = 0$. Assume (5.129). Now,

$$W_{(\epsilon_1,\ldots,\epsilon_{k+1},0,\ldots,0)}(u_l)$$
$$= u_{k+1,\epsilon_{k+1}}\left(W_{(\epsilon_1,\ldots,\epsilon_n)}(u_l)\right)$$
$$= u_l + \sum_{i=1}^{k} \epsilon_i r_i(u_l) + \frac{1}{2}\sum_{i=1}^{k} \epsilon_i^2 D_{r_i} r_i(u_l)$$
$$+ \sum_{\substack{i,j=1\\i<j}}^{k} \epsilon_i \epsilon_j D_{r_i} r_j(u_l) + \epsilon_{k+1} r_{k+1}\left(W_{(\epsilon_1,\ldots,\epsilon_n)}(u_l)\right)$$
$$+ \frac{1}{2}\epsilon_{k+1}^2 D_{r_{k+1}} r_{k+1}\left(W_{(\epsilon_1,\ldots,\epsilon_n)}(u_l)\right) + \mathcal{O}\left(|\epsilon|^3\right)$$
$$= u_l + \sum_{i=1}^{k+1} \epsilon_i r_i(u_l) + \frac{1}{2}\sum_{i=1}^{k+1} \epsilon_i^2 D_{r_i} r_i(u_l)$$
$$+ \sum_{\substack{i,j=1\\i<j}}^{k+1} \epsilon_i \epsilon_j D_{r_i} r_j(u_l) + \mathcal{O}\left(|\epsilon|^3\right)$$

by Theorem 5.16. □

Let $u_l \in D$ and define the map

$$\mathcal{L}(\epsilon_1,\ldots,\epsilon_n, u) = W_{(\epsilon_1,\ldots,\epsilon_n)} u_l - u. \tag{5.130}$$

This map $\mathcal{L}$ satisfies

$$\mathcal{L}(0,\ldots,0, u_l) = 0, \quad \nabla_\epsilon \mathcal{L}(0,\ldots,0, u_l) = (r_1(u_l),\ldots,r_n(u_l)),$$

where the matrix $\nabla \mathcal{L}$ has the right eigenvectors r_j evaluated at u_l as columns. This matrix is nonsingular by the strict hyperbolicity assumption.

The implicit function theorem then implies the existence of a neighborhood N around u_l and a unique function $(\epsilon_1,\ldots,\epsilon_n) = (\epsilon_1(u),\ldots,\epsilon_n(u))$ such that $\mathcal{L}(\epsilon_1,\ldots,\epsilon_n,u) = 0$. If $u_r \in N$, then there exists unique $(\epsilon_1,\ldots,\epsilon_n)$ with $W_{(\epsilon_1,\ldots,\epsilon_n)}u_l = u_r$, which proves the theorem. □

Observe that we could rephrase the Lax theorem as saying that we may use $(\epsilon_1,\ldots,\epsilon_n)$ to measure distances in phase space, and that we indeed have

$$A\,|u_r - u_l| \leq \sum_{j=1}^{n} |\epsilon_j| \leq B\,|u_r - u_l| \tag{5.131}$$

for constants A and B.

Let us now return to the shallow-water equations and prove existence of a global solution of the Riemann problem.

◇ **Example 5.19 (Shallow water (cont'd.)).**

We will construct a global solution of the Riemann problem for the shallow-water equations for all left and right states in $D = \{(h,v) \mid h \in [0,\infty), v \in \mathbb{R}\}$. Of course, we will maintain the same solution in the region where we already have constructed a solution, so it remains to construct a solution in the region

$$u_r \in V := \left\{ u_r \in D \mid 2\left(\sqrt{h_r} + \sqrt{h_l}\right) < [\![v]\!] \right\} \cup \{h = 0\}. \tag{5.132}$$

We will work in the h,v variables rather than (h,q). Assume first that $u_r = (h_r, v_r)$ in V with h_r positive. We first connect u_l, using a slow rarefaction wave, with a state u_m on the "vacuum line" $h = 0$. This state is given by

$$v_m = v_l + 2\sqrt{h_l}, \tag{5.133}$$

using (5.36). From this state we jump to the unique point v^* on $h = 0$ such that the fast rarefaction starting at $h^* = 0$ and v^* hits u_r. Thus we see from (5.37) that $v^* = v_r - 2\sqrt{h_r}$, which gives the following solution:

$$u(x,t) = \begin{cases} \binom{h_l}{v_l} & \text{for } x < \lambda_1(u_l)t, \\ R_1(x/t; u_l) & \text{for } \lambda_1(u_l)t < x < (\sqrt{h_l} + v_l)t, \\ \binom{0}{\tilde{v}(x,t)} & \text{for } (\sqrt{h_l} + v_l)t < x < v^*t, \\ R_2(x/t; (0,v^*)) & \text{for } v^*t < x < \lambda_2(u_r)t, \\ \binom{h_r}{v_r} & \text{for } x > \lambda_2(u_r)t. \end{cases} \tag{5.134}$$

Physically, it does not make sense to give a value of the speed v of the water when there is no water, i.e., $h = 0$, and mathematically we see that any v will satisfy the equations when $h = 0$. Thus we do not have to associate any value with $\tilde{v}(x,t)$.

If u_r is on the vacuum line $h = 0$, we still connect to a state u_m on $h = 0$ using a slow rarefaction, and subsequently we connect to u_r along the vacuum line. By considering a nearby state $\tilde{u}_r$ with $\tilde{h} > 0$ we see that with this construction we have continuity in data.

Finally, we have to solve the Riemann problem with the left state on the vacuum line $h = 0$. Now let $u_l = (0, v_l)$, and let $u_r = (h_r, v_r)$ with $h_r > 0$. We now connect u_l to an intermediate state u_m on the vacuum line given by $v_m = v_r - 2\sqrt{h_r}$ and continue with a fast rarefaction to the right state u_r. ◇

We will apply the above theory to one old and two ancient problems:

◇ **Example 5.20 (Dam breaking).**

For this problem we consider Riemann initial data of the form (in (h, v) variables)

$$u(x,0) = \begin{pmatrix} h(x,0) \\ v(x,0) \end{pmatrix} = \begin{cases} \begin{pmatrix} h_l \\ 0 \end{pmatrix} & \text{for } x < 0, \\ \begin{pmatrix} 0 \\ 0 \end{pmatrix} & \text{for } x \geq 0. \end{cases}$$

From the above discussion we know that the solution consists of a slow rarefaction; thus

$$u(x,t) = \begin{pmatrix} h(x,t) \\ v(x,t) \end{pmatrix}$$

$$= \begin{cases} \begin{pmatrix} h_l \\ 0 \end{pmatrix} & \text{for } x < -\sqrt{h_l}t, \\ \begin{pmatrix} \frac{1}{9}(2\sqrt{h_l} - \frac{x}{t})^2 \\ \frac{2}{3}(\sqrt{h_l} + \frac{x}{t}) \end{pmatrix} & \text{for } -\sqrt{h_l}t < x < 2\sqrt{h_l}t, \\ \begin{pmatrix} 0 \\ 0 \end{pmatrix} & \text{for } x > 2\sqrt{h_l}t. \end{cases}$$

◇

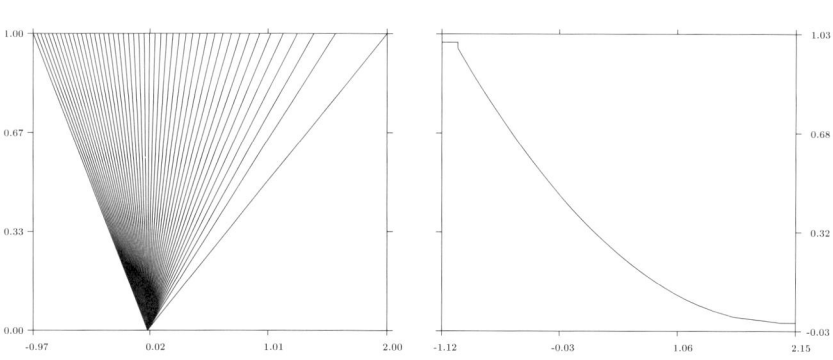

Figure 5.9. The solution of the dam breaking problem in (x, t) space (left), and the h component (right).

200   5. The Riemann Problem for Systems

We shall call the two ancient problems Moses' first and second problems.

> And Moses stretched out his hand over the sea; and the Lord caused the sea to go back by a strong east wind all that night, and made the sea dry land, and the waters were divided. And the children of Israel went into the midst of the sea upon the dry ground: and the waters were a wall unto them on their right hand, and on their left.
>
> Exodus (14:21–22)

◇ **Example 5.21 (Moses' first problem).**

For the first problem we consider initial data of the form (in (h,v) variables)

$$u(x,0) = \begin{cases} \binom{h_0}{-v_0} & \text{for } x < 0, \\ \binom{h_0}{v_0} & \text{for } x \geq 0, \end{cases}$$

for a positive speed v_0. By applying the above analysis we find that in this case we connect to an intermediate state u_1 on the vacuum line using a slow rarefaction. This state is connected to another state u_2 also on the vacuum line, which subsequently is connected to the right state using a fast rarefaction wave. More precisely, the state u_1 is determined by $v_1 = v(x_1, t_1)$, where $h(x,t) = \frac{1}{9}\left(-v_0 + 2\sqrt{h_0} - \frac{x}{t}\right)^2$ along the slow rarefaction wave (cf. (5.41)) and $h(x_1, t_1) = 0$. We find that $x_1 = (2\sqrt{h_0} - v_0)t_1$ and thus $v_1 = 2\sqrt{h_0} - v_0$. The second intermediate state u_2 is such that a fast rarefaction wave with left state u_2 hits u_r. This implies that $v_0 = v_2 + 2\sqrt{h_0}$ from (5.39), or $v_2 = v_0 - 2\sqrt{h_0}$. In order for this construction to be feasible, we will have to assume that $v_2 > v_1$ or $v_0 \geq 2\sqrt{h_0}$. If this condition does not hold, we will not get a region without water, and thus the original problem of Moses will not be solved. Combining the above waves in one solution we obtain

$$h(x,t) = \begin{cases} h_0 & \text{for } x < -(v_0 + \sqrt{h_0})t, \\ \frac{1}{9}\left(-v_0 + 2\sqrt{h_0} - \frac{x}{t}\right)^2 & \text{for } -(v_0 + \sqrt{h_0})t < x < (2\sqrt{h_0} - v_0)t, \\ 0 & \text{for } (2\sqrt{h_0} - v_0)t < x < (v_0 - 2\sqrt{h_0})t, \\ \frac{1}{9}\left(v_0 - 2\sqrt{h_0} - \frac{x}{t}\right)^2 & \text{for } (v_0 - 2\sqrt{h_0})t < x < (v_0 + \sqrt{h_0})t, \\ h_0 & \text{for } x > (v_0 + \sqrt{h_0})t, \end{cases}$$

$$v(x,t) = \begin{cases} -v_0 & \text{for } x < -(v_0 + \sqrt{h_0})t, \\ \frac{1}{3}(-v_0 + 2\sqrt{h_0} + 2\frac{x}{t}) & \text{for } -(v_0 + \sqrt{h_0})t < x < (2\sqrt{h_0} - v_0)t, \\ 0 & \text{for } (2\sqrt{h_0} - v_0)t < x < (v_0 - 2\sqrt{h_0})t, \\ \frac{1}{3}(v_0 - 2\sqrt{h_0} + 2\frac{x}{t}) & \text{for } (v_0 - 2\sqrt{h_0})t < x < (v_0 + \sqrt{h_0})t, \\ v_0 & \text{for } x > (v_0 + \sqrt{h_0})t. \end{cases}$$

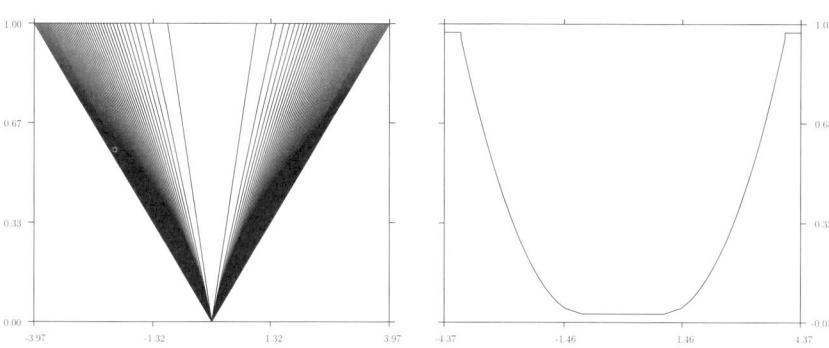

Figure 5.10. The solution of Moses' first problem in (x,t) space (left), and the h component (right).

◇ **Example 5.22 (Moses' second problem).**

> And Moses stretched forth his hand over the sea, and the sea returned to his strength when the morning appeared; and the Egyptians fled against it; and the Lord overthrew the Egyptians in the midst of the sea.
>
> Exodus (14:27)

Here we study the multiple Riemann problem given by (in (h,v) variables)

$$u(x,0) = \begin{cases} \binom{h_0}{0} & \text{for } x < 0, \\ \binom{0}{0} & \text{for } 0 < x < L, \\ \binom{h_0}{0} & \text{for } x > L. \end{cases}$$

For small times t, the solution of this problem is found by patching together the solution of two dam breaking problems. The left problem is solved by a fast rarefaction wave, and the right by a slow rarefaction. At some positive time, these rarefactions will interact, and thereafter explicit computations become harder.

202   5. The Riemann Problem for Systems

In place of explicit computation we therefore present the numerical solution constructed by front tracking. This method is a generalization of the front-tracking method presented in Chapter 2, and will be the subject of the next chapter.

In the left part of Figure 5.11 we see the fronts in (x,t) space. These fronts are similar to the fronts for the scalar front tracking, and the approximate solution is discontinuous across the lines shown in the figure. Looking at the figure, it is not hard to see why explicit computations become difficult as the two rarefaction waves interact. The right part of the figure shows the water level as it engulfs the Egyptians. The lower figure shows the water level before the two rarefaction waves interact, and the two upper ones show that two shock waves result from the interaction of the two rarefaction waves.   ◇

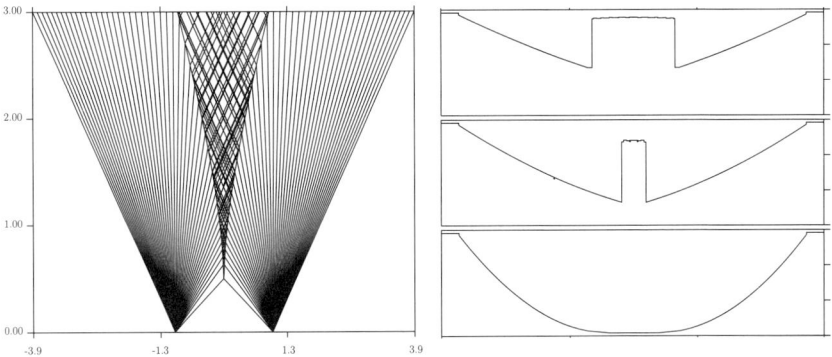

Figure 5.11. The solution of Moses' second problem in (x,t) space (left), and the h component (right).

5.6 Notes

The fundamentals of the Riemann problem for systems of conservation laws were presented in the seminal paper by Lax [95], where also the Lax entropy condition was introduced. We refer to Smoller [130] as a general reference for this chapter. Our proof of Theorem 5.11 follows Schatzman [127]. This also simplifies the proof of the classical result that $s'(0) = \frac{1}{2}$ in Theorem 5.14. The parameterization of the Hugoniot locus introduced in Theorem 5.11 makes the proof of the smoothness of the wave curves, Theorem 5.16, quite simple.

We have used shallow-water equations as our prime example in this chapter. This model can be found in many sources; a good presentation is in Kevorkian [83]. See also the paper by Gerbeau and Perthame [51] for a more rigourous derivation of the model. Our treatment of the vacuum for these equations can be found in Liu and Smoller [105].

There is extensive literature on the Euler equations; see, e.g., [37], [130], and [29].

Our version of the implicit function theorem, Theorem 5.10, was taken from Renardy and Rogers [117]. See Exercise 5.11 for a proof.

Exercises

5.1 What assumption on p is necessary for the p-system to be hyperbolic?

5.2 Solve the Riemann problem for the p-system in the case where $p(v) = 1/v$. For what left and right states does this Riemann problem have a solution?

5.3 Repeat Exercise 5.2 in the general case where $p = p(v)$ is such that p' is negative and p'' is positive.

5.4 Solve the following Riemann problem for the shallow-water equations:

$$u(x,0) = \begin{pmatrix} h(x,0) \\ v(x,0) \end{pmatrix} = \begin{cases} \binom{h_l}{0} & \text{for } x < 0, \\ \binom{h_r}{0} & \text{for } x \geq 0, \end{cases}$$

with $h_l > h_r > 0$.

5.5 Let $w = (u, v)$ and let $\varphi(w)$ be a smooth scalar function. Consider the system of conservation laws

$$w_t + (\varphi(w)w)_x = 0. \tag{5.135}$$

a. Find the characteristic speeds λ_1 and λ_2 and the associated eigenvectors r_1 and r_2 for the system (5.135).
b. Let $\varphi(w) = |w|^2/2$. Then find the solution of the Riemann problem for (5.135).
c. Now let

$$\varphi(w) = \frac{1}{1+u+v},$$

and assume that u and v are positive. Find the solution of the Riemann problem of (5.135) in this case.

5.6 Let us consider the Lax–Friedrichs scheme for systems of conservation laws. As in Chapter 3 we write this as

$$U_j^{n+1} = \frac{1}{2}\left(U_{j-1}^n + U_{j+1}^n\right) - \frac{\lambda}{2}\left(f\left(U_{j+1}^n\right) - f\left(U_{j-1}^n\right)\right),$$

where $\lambda = \Delta t/\Delta x$, and we assume that the CFL condition

$$\lambda \leq \max_k |\lambda_k|$$

holds. Let $u_j^n(x,t)$ denote the solution of the Riemann problem with initial data

$$\begin{cases} U_{j-1}^n & \text{for } x < j\Delta x, \\ U_{j+1}^n & \text{for } x \geq j\Delta x. \end{cases}$$

Show that

$$U_j^{n+1} = \frac{1}{2\Delta x} \int_{(j-1)\Delta x}^{(j+1)\Delta x} u_j^n(x, \Delta t)\, dx.$$

5.7 A smooth function $w : \mathbb{R}^n \to \mathbb{R}$ is called a k-Riemann invariant if

$$\nabla w(u) \cdot r_k(u) = 0,$$

where r_k is the kth right eigenvector of the Jacobian matrix df, which is assumed to be strictly hyperbolic.

   a. Show that there locally exist precisely $(n-1)$ k-Riemann invariants whose gradients are linearly independent.

   b. Let $R_k(u_l)$ denote the kth rarefaction curve through a point u_l. Then show that all $(n-1)$ k-Riemann invariants are constant on $R_k(u_l)$. This gives an alternative definition of the rarefaction curves.

   c. We say that we have a coordinate system of Riemann invariants if there exist n scalar-valued functions $w_1, \ldots, w_n$ such that w_j is a k-Riemann invariant for $j, k = 1, \ldots, n$, $j \neq k$, and

$$\nabla w_j(u) \cdot r_k(u) = \gamma_j(u)\delta_{j,k}, \tag{5.136}$$

   for some nonzero function g_j. Why cannot we expect to find such a coordinate system if $n > 2$?

   d. Find the Riemann invariants for the shallow-water system, and verify parts **b** and **c** in this case.

5.8 We study the p-system with $p(v) = 1/v$ as in Exercise 5.2.

   a. Find the two Riemann invariants w_1 and w_2 in this case.

   b. Introduce coordinates

$$\mu = w_1(v,u) \quad \text{and} \quad \tau = w_2(v,u),$$

   and find the wave curves in (μ, τ) coordinates.

   c. Find the solution of the Riemann problem in (μ, τ) coordinates.

   d. Show that the wave curves W_1 and W_2 are *stiff* in the sense that if a point (μ, τ) is on a wave curve through (μ_l, τ_l), then the point $(\mu + \Delta\mu, \tau + \Delta\tau)$ is on a wave curve through $(\mu_l + \Delta\mu, \tau_l + \Delta\tau)$. Hence the solution of the Riemann problem can be said to be translation-invariant in (μ, τ) coordinates.

   e. Show that the 2-shock curve through a point (μ_l, τ_l) is the reflection about the line $\mu - \mu_l = \tau - \tau_l$ of the 1-shock curve through (μ_l, τ_l).

5.9 As for scalar equations, we define an entropy/entropy flux pair (η, q) as scalar functions of u such that for smooth solutions,
$$u_t + f(u)_x = 0 \quad \Rightarrow \quad \eta_t + q_x = 0,$$
and η is supposed to be a convex function.

   a. Show that η and q are related by
$$\nabla_u q = \nabla_u \eta \, df. \tag{5.137}$$

   b. Why cannot we expect to find entropy/entropy flux pairs if $n > 2$?

   c. Find an entropy/entropy flux pair for the p-system if $p(v) = 1/v$.

   d. Find an entropy/entropy flux pair for the shallow-water equations.

5.10 Let A be a constant $n \times n$ matrix with real and distinct eigenvalues $\lambda_1 < \lambda_2 < \cdots < \lambda_n$. Consider the Riemann problem for the linear system of equations
$$u_t + A u_x = 0, \quad u(x, 0) = \begin{cases} u_l & \text{for } x < 0, \\ u_r & \text{for } x \geq 0. \end{cases}$$

   a. Find the solution of this Riemann problem.

   b. Extend this solution to the general Cauchy problem.

   c. What do you obtain for the linear wave equation $\phi_{tt} = c^2 \phi_{xx}$ with initial data $\phi(x, 0) = f(x)$ and $\phi_t(x, 0) = g(x)$ (cf. Example 5.2)?

5.11 This exercise outlines a proof of the implicit function theorem, Theorem 5.10.

   a. Define T to be a mapping $\mathbb{R}^p \to \mathbb{R}^p$ such that for y_1 and y_2,
$$|T(y_1) - T(y_2)| \leq c \, |y_1 - y_2|, \quad \text{for some constant } c < 1.$$
   Such mappings are called contractions. Show that there exists a unique y such that $T(y) = y$.

   b. Let $u : \mathbb{R}^p \to \mathbb{R}^p$, and assume that u is C^1 in some neighborhood of a point y_0, and that $du(y_0)$ is nonsingular. We are interested in solving the equation
$$u(y) = u(y_0) + v \tag{5.138}$$
   for some v where $|v|$ is sufficiently small. Define
$$T(y) = y - du(y_0)^{-1} \left(u(y) - u(y_0) - v \right).$$
   Show that T is a contraction in a neighborhood of y_0, and consequently that (5.138) has a unique solution $x = \varphi(v)$ for small v, and that $\varphi(0) = y_0$.

c. Now let $\Phi(x, y)$ be as in Theorem 5.10. Show that for x close to x_0 we can find $\varphi(x, v)$ such that
$$\Phi(x, \varphi(x, v)) = \Phi(x, y_0) + v$$
for small v.
d. Choose a suitable $v = v(x)$ to conclude the proof of the theorem.

6
Existence of Solutions of the Cauchy Problem for Systems

> Faith is an island in the setting sun. But proof, yes.
> Proof is the bottom line for everyone.
>
> Paul Simon, Proof (1990)

In this chapter we study the generalization of the front-tracking algorithm to systems of conservation laws, and how this generalization generates a convergent sequence of approximate weak solutions. We shall then proceed to show that the limit is a weak solution. Thus we shall study the initial value problem

$$u_t + f(u)_x = 0, \quad u|_{t=0} = u_0, \tag{6.1}$$

where $f\colon \mathbb{R}^n \to \mathbb{R}^n$ and u_0 is a function in $L^1(\mathbb{R})$.

In doing this, we are in the setting of Lax's theorem (Theorem 5.17); we have a system of strictly hyperbolic conservation laws where each characteristic field is either genuinely nonlinear or linearly degenerate, and the initial data are close to a constant. This restriction is necessary, since the Riemann problem may fail to have a solution for initial states far apart, which is analogous to the appearance of a "vacuum" in the solution of the shallow-water equations.

The convergence part of the argument follows the traditional method of proving compactness in the context of conservation laws, namely, via Kolmogorov's compactness theorem or Helly's theorem.

Again, the basic ingredient in front tracking is the solution of Riemann problems, or in this case, the approximate solution of Riemann problems. Therefore, we start by defining these approximations.

6.1 Front Tracking for Systems

> Nisi credideritis, non intelligetis.[1]
>
> Saint Augustine, De Libero Arbitrio (387/9)

In order for us to define front tracking in the scalar case, the solution of the Riemann problem had to be a piecewise constant function. For systems, this is possible only if all waves are shock waves or contact discontinuities. Consequently, we need to approximate the continuous parts of the solution, the rarefaction waves, by piecewise constants.

There are, of course, several ways to make this approximation. We use the following: Let δ be a small parameter. For the rest of this chapter, δ will always denote a parameter that controls the accuracy of the approximation. We start with the system of conservation laws (6.1), and the Riemann problem

$$u(x,0) = \begin{cases} u_l & \text{for } x < 0, \\ u_r & \text{for } x \geq 0. \end{cases} \tag{6.2}$$

We have seen (Theorem 5.17) that the solution of this Riemann problem consists of at most $n+1$ constant states, separated by either shock waves, contact discontinuities, or rarefaction waves. We wish to approximate this solution by a piecewise constant function in (x/t).

When the solution has shocks or contact discontinuities, it is already a step function for some range of (x/t), and we set the approximation equal to the exact solution u for such x and t.

Thus, if the jth wave is a shock or a contact discontinuity, we let

$$u^\delta_{j,\epsilon_j}(x,t) = u_{j,\epsilon_j}(x,t), \qquad t\sigma_j^+ < x < t\sigma_{j+1}^-,$$

where the right-hand side is given by (5.124).

A rarefaction wave is a smooth transition between two constant states, and we will replace this by a step function whose "steps" are no further apart than δ and lie on the correct rarefaction curve R_j. The discontinuity between two steps is defined to move with a speed equal to the characteristic speed of the left state.

More precisely, let the solution to (6.2) be given by (5.124). Assume that the jth wave is a rarefaction wave; that is, the solutions u and u_{m_j} lie on the jth rarefaction curve $R_j\left(u_{m_{j-1}}\right)$ through $u_{m_{j-1}}$, or

$$u(x,t) = u_{j,\epsilon_j}\left(x,t;\, u_{m_j}, u_{m_{j-1}}\right), \qquad \text{for } t\sigma_j^- \leq x \leq t\sigma_j^+.$$

Let $k = \text{rnd}\left(\epsilon_j/\delta\right)$ for the moment, where $\text{rnd}\,(z)$ denotes the integer closest[2] to z, and let $\hat{\delta} = \epsilon_j/k$. The step values of the approximation are

---

[1] Soft on Latin? It says, "If you don't believe it, you won't understand it."
[2] Such that $z \leq \text{rnd}\,(z) < z + \frac{1}{2}$.

now defined as

$$u_{j,l} = R_j\left(l\hat{\delta};\, u_{m_{j-1}}\right), \qquad \text{for } l = 0, \ldots, k. \tag{6.3}$$

We have that $u_{j,0} = u_{m_{j-1}}$ and $u_{j,k} = u_{m_j}$. We set the speed of the steps equal to the characteristic speed to the left, and hence the piecewise constant approximation we make is the following:

$$u^{\delta}_{j,\epsilon_j}(x,t) := u_{j,0} + \sum_{l=1}^{k} (u_{j,l} - u_{j,l-1})\, H\left(x - \lambda_j\left(u_{j,l-1}\right)t\right), \tag{6.4}$$

where H now denotes the Heaviside function. Equation (6.4) is to hold for $t\sigma_j^+ < x < \sigma_{j+1}^- t$. Loosely speaking, we step along the rarefaction curve with steps of size at most δ. Observe that the discontinuities that occur as a result of the approximation of the rarefaction wave will not satisfy the Rankine–Hugoniot condition, and hence the function will not be a weak solution. However, we will prove that u^{δ} converges to a weak solution as $\delta \to 0$. Figure 6.1 illustrates this in phase space and in (x,t) space. The

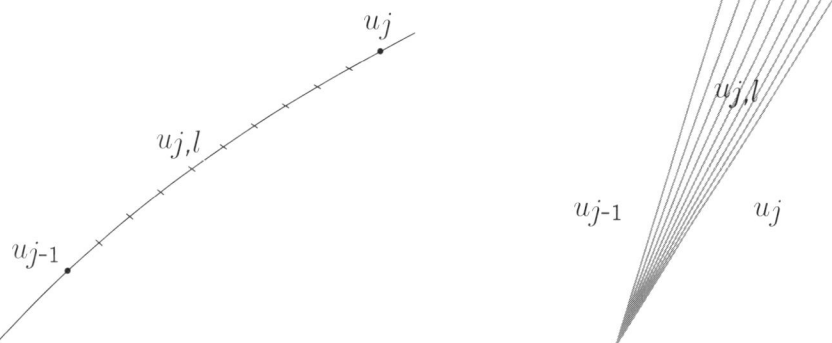

Figure 6.1. An approximated rarefaction wave in phase space and in (x,t) space.

approximate solution to the Riemann problem is then found by inserting a superscript δ at the appropriate places in equation (5.124), resulting in

$$u^{\delta}(x,t) = \begin{cases} u_l & \text{for } x \leq \sigma_1^- t, \\ u^{\delta}_{1,\epsilon_1}(x/t;\, u_{m_1}, u_l) & \text{for } \sigma_1^- t \leq x \leq \sigma_1^+ t, \\ u_{m_1} & \text{for } \sigma_1^+ t \leq x \leq \sigma_2^- t, \\ u^{\delta}_{2,\epsilon_2}(x/t;\, u_{m_2}, u_{m_1}) & \text{for } \sigma_2^- t \leq x \leq \sigma_2^+ t, \\ u_{m_2} & \text{for } \sigma_2^+ t \leq x \leq \sigma_3^- t, \\ \vdots & \\ u^{\delta}_{n,\epsilon_n}(x/t;\, u_r, u_{m_{n-1}}) & \text{for } \sigma_n^- t \leq x \leq \sigma_n^+ t, \\ u_r & \text{for } x \geq \sigma_n^+ t. \end{cases} \tag{6.5}$$

It is clear that u^δ converges to the exact solution given by (5.124) pointwise. Indeed,
$$\left|u^\delta(x,t) - u(x,t)\right| = \mathcal{O}(\delta).$$
Therefore, we also have that $\left\|u^\delta - u\right\|_1 = \mathcal{O}(\delta)$.

Now we are ready to define the front-tracking procedure to (approximately) solve the initial value problem (6.1).

Our first step is to approximate the initial function u_0 by a piecewise constant function u_0^δ (we let δ denote this approximation parameter as well) such that
$$\lim_{\delta \to 0} \left\|u_0^\delta - u_0\right\|_1 = 0. \tag{6.6}$$

We then generate approximations, given by (6.5), to the solutions of the Riemann problems defined by the discontinuities of u_0^δ. Already here we see one reason why we must assume T.V. (u_0) to be small: The initial Riemann problems must be solvable. Therefore, we assume our initial data u_0, as well as the approximation u_0^δ, to be in some small neighborhood D of a constant $\bar{u}$. Without loss of generality, $\bar{u}$ can be chosen to be zero.

As the initial discontinuities interact at some later time, we can solve the Riemann problems defined by the states immediately to the left and right of the collisions. These solutions are then replaced by approximations, and we may continue to propagate the front-tracking construction until the next interaction. However, as in the scalar case, it is not obvious that this procedure will take us up to any predetermined time. A priori, it is not even clear whether the number of discontinuities will blow up at some finite time, that is, that the collision times will converge to some finite time. This problem is much more severe in the case of a system of conservation laws than in the scalar case, since a collision of two discontinuities generically will result in at least $n - 2$ new discontinuities. So for $n > 2$, the number of discontinuities seems to be growing without bound as t increases. As in the scalar case, the key to the solution of these problems lies in the study of interactions of discontinuities. To keep the number of waves finite we shall eliminate small waves emanating from Riemann problems. However, there is a trade-off: The more waves we eliminate, the easier it is to prove convergence, but the less likely it is that the limit is a solution of the differential equation.

From now on, we shall call all discontinuities in the front-tracking construction *fronts*. We distinguish between *waves* and *fronts*. In general, a solution of a Riemann problem consists of exactly n waves coming from distinct wave families. Each approximation of a rarefaction wave is one wave, but will consist of several fronts. Hence a front is an object with a left and a right state, labeled L and R respectively, and an associated *family*. The family of a front separating states L and R is the (unique) number

j such that

$$R \in W_j(L),$$

where, as in Chapter 5, $W_j(u)$ denotes the jth wave curve through the point u. These are parameterized as in Theorem 5.16. (Observe that we still have this relation for fronts approximating a rarefaction wave.) The *strength* of a front is defined to be $|\epsilon|$, where $R = u_{j,\epsilon}(L)$. The strength of a front coming from an approximate rarefaction wave of the jth family is $\hat{\delta} = \epsilon_j/k$ in the terminology of equation (6.3), (6.4).

We wish to estimate the strengths of the fronts resulting from a collision in terms of the strengths of the colliding fronts. With some abuse of notation we shall refer to both the front itself and its strength by ϵ_i.

Consider therefore N fronts $\gamma_N, \ldots, \gamma_1$ interacting at a single point as in Figure 6.2. We will have to keep track of the associated family of each front. We denote by $\hat{\imath}$ the family of wave γ_i. Thus if $\gamma_1, \ldots, \gamma_4$ all come from the first family, we have $\hat{1} = \cdots = \hat{4} = 1$. Since the speed of γ_i is greater than the speed of γ_j for $j > i$, we have $\hat{\jmath} \geq \hat{\imath}$. We label the waves resulting from the collision $\beta_1, \ldots, \beta_n$. Let β denote the vector of waves in the front-

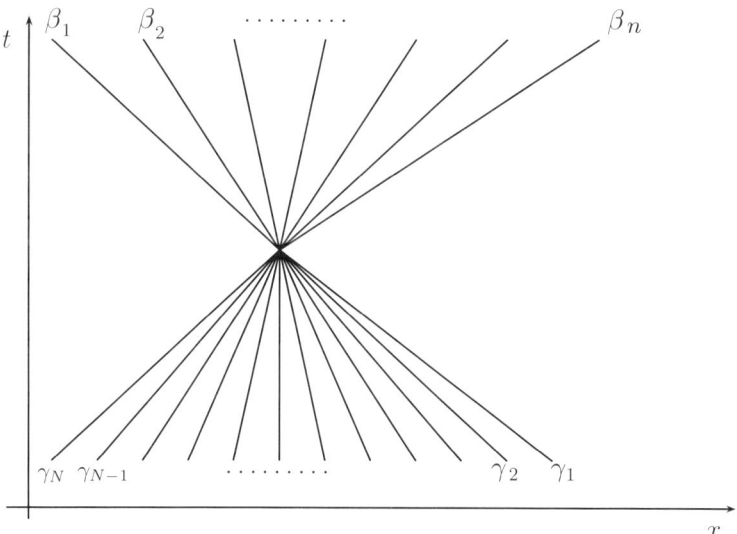

Figure 6.2. A collision of N fronts.

tracking approximation resulting from the solution of the Riemann problem defined by the collision of $\gamma_1, \ldots, \gamma_N$, i.e., $\beta = (\beta_1, \ldots, \beta_n)$, and let

$$\alpha = \Big(\sum_{\hat{\imath}=1} \gamma_i, \sum_{\hat{\imath}=2} \gamma_i, \ldots, \sum_{\hat{\imath}=n} \gamma_i \Big).$$

For simplicity, also set $\gamma = (\gamma_1, \ldots, \gamma_N)$. Note that β is a function of γ, that is, $\beta = \beta(\gamma)$. For $i < j$ we define
$$\beta_{i,j}(\sigma, \tau) := \frac{\partial^2 \beta}{\partial \gamma_i \partial \gamma_j}(\gamma_1, \ldots, \gamma_{i-1}, \sigma \gamma_i, 0, \ldots, 0, \tau \gamma_j, 0, \ldots, 0).$$

Then
$$\gamma_i \gamma_j \int_0^1 \int_0^1 \beta_{i,j}(\sigma, \tau) \, d\sigma \, d\tau \qquad (6.7)$$
$$= \beta(\gamma_1, \ldots, \gamma_i, 0, \ldots, 0, \gamma_j, 0, \ldots, 0) + \beta(\gamma_1, \ldots, \gamma_{i-1}, 0, \ldots, 0)$$
$$- \beta(\gamma_1, \ldots, \gamma_i, 0, \ldots, 0) - \beta(\gamma_1, \ldots, \gamma_{i-1}, 0, \ldots, 0, \gamma_j, 0, \ldots, 0).$$

Furthermore,
$$\beta(0, \ldots, 0, \gamma_k, 0, \ldots, 0) = (0, \ldots, 0, \gamma_k, 0, \ldots, 0), \qquad (6.8)$$

where γ_k on the right is at the $\hat{k}$th place, since in this case we have no collision. Summing (6.7) for all $i < j$, we obtain

$$\sum_{i<j}^N \gamma_i \gamma_j \int_0^1 \int_0^1 \beta_{i,j}(\sigma, \tau) \, d\sigma \, d\tau$$
$$= \beta(\gamma_1, \ldots, \gamma_N) - \sum_{i=1}^N \beta(0, \ldots, 0, \gamma_i, 0, \ldots, 0) = \beta - \alpha. \qquad (6.9)$$

By the solution of the general Riemann problem, we have that $\beta_{i,j}$ is bounded; hence
$$|\beta - \alpha| \le \mathcal{O}(1) \sum_{i,j; i<j}^N |\gamma_i \gamma_j|, \qquad (6.10)$$

or
$$\beta = \alpha + \mathcal{O}(1) \sum_{\substack{i,j \\ i<j}}^N |\gamma_i \gamma_j|. \qquad (6.11)$$

Note that if the incoming fronts γ_k are small, then the fronts resulting from the collision will be very small for those families that are not among the incoming fronts.

◇ **Example 6.1 (Higher-order estimates).**

The estimate (6.10) is enough for our purposes, but we can extract some more information from (6.9) by considering higher-order terms. Firstly, note that
$$\beta = \alpha + \sum_{i<j} \gamma_i \gamma_j \beta_{i,j}(0,0) + \mathcal{O}(1) \sum_{i<j} |\gamma_i \gamma_j| \, |\gamma|. \qquad (6.12)$$

Therefore, we evaluate $\beta_{i,j}(0,0)$. To do this, observe that
$$u_r = \mathcal{R}_{\beta(\gamma)} u_l = \mathcal{R}_{\gamma_1} \circ \mathcal{R}_{\gamma_2} \circ \cdots \circ \mathcal{R}_{\gamma_N} u_l, \tag{6.13}$$
where $\mathcal{R}_\beta$ is defined as (5.128), and u_l and u_r are the states to the left and right of the collision, respectively. Introducing
$$\beta_{\gamma_j} := \frac{\partial \beta}{\partial \gamma_j},$$
equation (6.8) implies
$$\beta_{\gamma_j}(0,\ldots,0) = e_{\hat{j}},$$
where e_k denotes the kth standard basis vector in $\mathbb{R}^n$. Also note that
$$\frac{\partial}{\partial \gamma_i} \mathcal{R}_{\beta(\gamma)} = \nabla_\beta \mathcal{R}_\beta \cdot \beta_{\gamma_i}.$$
Furthermore, from Lemma 5.18, (5.128), we have that
$$\nabla_\beta \mathcal{R}_\beta = \left(\ldots, r_k + \sum_{j=1}^n \beta_j D_{r_{\min(j,k)}} r_{\max(j,k)}, \ldots \right) + \mathcal{O}\left(|\beta|^2\right).$$
Here the first term on the right-hand side is the $n \times n$ matrix with the kth column equal to $r_k + \sum_{j=1}^n \beta_j D_{r_{\min(j,k)}} r_{\max(j,k)}$. Consequently,[3]
$$\frac{\partial}{\partial \gamma_j} \nabla_\beta \mathcal{R}_{(0,\ldots,0)} = \left(D_{r_1} r_{\hat{j}}, D_{r_2} r_{\hat{j}}, \ldots, D_{r_{\hat{j}}} r_{\hat{j}}, D_{r_{\hat{j}}} r_{\hat{j}+1}, \ldots, D_{r_{\hat{j}}} r_n \right)$$
evaluated at u_l. Differentiating (6.13) with respect to γ_i we find
$$(\nabla_\beta \mathcal{R}_\beta \cdot \beta_{\gamma_i})|_{\gamma=(0,\ldots,0,\gamma_j,0\ldots,0)}(u_l) = r_{\hat{i}}\left(\mathcal{R}_{\gamma_j} u_l\right)$$
for $j > i$. Differentiating this with respect to γ_j we obtain
$$\left(\frac{\partial}{\partial \gamma_j} \nabla_\beta \mathcal{R}_\beta\right)|_{\gamma=(0,\ldots,0)} e_{\hat{j}} + \nabla_\beta \mathcal{R}_{(0,\ldots,0)} \beta_{i,j}(0,0) = D_{r_{\hat{j}}} r_{\hat{i}}(u_l).$$
Inserting this in (6.12), we finally obtain
$$\beta = \alpha + \sum_{i<j}^N \gamma_i \gamma_j \left(\nabla_\beta \mathcal{R}_\beta\right)^{-1} \left(D_{r_{\hat{j}}} r_{\hat{i}} - D_{r_{\hat{i}}} r_{\hat{j}}\right) + \mathcal{O}(1) \sum_{i<j} |\gamma_i \gamma_j| |\gamma|, \tag{6.14}$$
which we call the *interaction estimate*. One can also use (6.9) to obtain estimates of higher order.

In passing, we note that if the integral curves of the eigenvectors form a coordinate system near M, then
$$\left(D_{r_{\hat{j}}} r_{\hat{i}} - D_{r_{\hat{i}}} r_{\hat{j}}\right) = 0$$

---

[3]The right-hand side denotes the $n \times n$ matrix where the first $\hat{j}$ columns equal $D_{r_k} r_{\hat{j}}$, $k = 1,\ldots,\hat{j}$, and the remaining $(n-\hat{j})$ columns equal $D_{r_{\hat{j}}} r_k$, $k = \hat{j}+1,\ldots,n$.

214   6. Existence of Solutions of the Cauchy Problem

for all i and j, and we obtain a third-order estimate. The estimate (6.10) will prove to be the key ingredient in our analysis of front tracking.

For the reader with knowledge of differential geometry, the estimate (6.14) is no surprise. Assume that only two fronts collide, ϵ_l and ϵ_r, separating states L, M, and R. Let the families of the two fronts be l and r, respectively. The states L, M, and R are almost connected by the integral curves of r_l and r_r, respectively. If we follow the integral curve of r_l a (parameter) distance $-\epsilon_l$ from R, and then follow the integral curve of r_r a distance $-\epsilon_r$, we end up with, up to third order in ϵ_l and ϵ_r, half the Lie bracket of $\epsilon_l r_l$ and $\epsilon_r r_r$ away from L. This Lie bracket is given by

$$[\epsilon_l r_l, \epsilon_r r_r] := \epsilon_l \epsilon_r \left(D_{r_l} r_r - D_{r_r} r_l \right).$$

This means that if we start from L and follow r_r a distance ϵ_r, and then r_l a distance ϵ_l, we finish a distance $\mathcal{O}\left([\epsilon_l r_l, \epsilon_r r_r]\right)$ away from R. Consequently, up to $\mathcal{O}\left([\epsilon_l r_l, \epsilon_r r_r]\right)$, the solution of the Riemann problem with right state R and left state L is given by a wave of family r of strength ϵ_r, followed by a wave of family l of strength ϵ_l. While not a formal proof, these remarks illuminate the mechanism behind the calculation leading up to (6.14). See Figure 6.3.   ◇

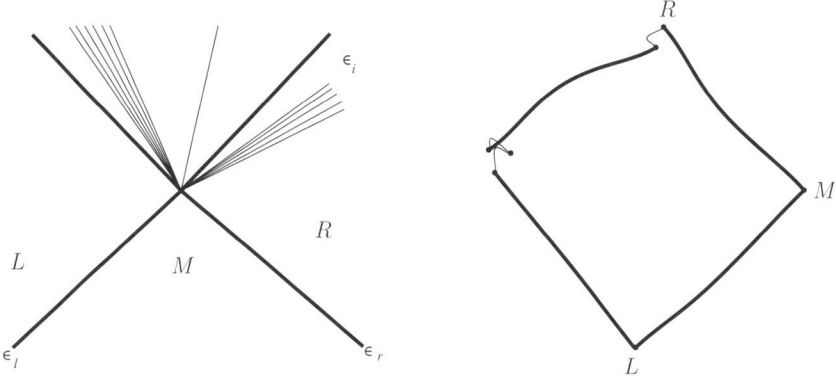

Figure 6.3. An interaction in (x, t) space and in phase space.

Before we proceed further, we introduce some notation. Front tracking will produce a piecewise constant function labeled $u^\delta(x, t)$ that has, at least initially, some finite number N of fronts. These fronts have strengths ϵ_i, $i = 1, \ldots, N$. We will refer to the ith front by its strength ϵ_i, and label the left and right states L_i and R_i, respectively. The position of ϵ_i is denoted by $x_i(t)$, and with a slight abuse of notation we have that

$$x_i(t) = x_i + s_i \left(t - t_i \right), \qquad (6.15)$$

where s_i is the speed of the front, and (x_i, t_i) is the position and time it originated. In this notation, u^δ can be written

$$u^\delta(x,t) = L_1 + \sum_{i=1}^{N}(R_i - L_i)H(x - x_i(t)). \tag{6.16}$$

The interaction estimate (6.10) shows that the "amount of change" produced by a collision is proportional to the product of the strengths of the colliding fronts. Therefore, in order to obtain some estimate of what will happen as fronts collide, we define the *interaction potential* Q. The idea is that Q should (over)estimate the amount of change in u^δ caused by all future collisions. Hence, by (6.10) Q should involve terms of type $|\epsilon_l \epsilon_r|$. We say that two fronts are approaching if the front to the left has a larger family than the front to the right, or if both fronts are of the same family and at least one of the fronts is a shock wave. We collect all pairs of approaching fronts in the *approaching set* $\mathcal{A}$, that is,

$$\mathcal{A} := \{(\epsilon_i, \epsilon_j) \text{ such that } \epsilon_i \text{ and } \epsilon_j \text{ are approaching}\}. \tag{6.17}$$

The set $\mathcal{A}$ will, of course, depend on time. All future collisions will now involve two fronts from $\mathcal{A}$ due to the hyperbolicity of the equation. Observe that two approximate rarefaction waves of the same family never collide unless there is another front between, all colliding at the same point (x,t). Therefore, we define Q as

$$Q := \sum_{\mathcal{A}} |\epsilon_i \epsilon_j|. \tag{6.18}$$

For scalar equations we saw that the total variation of the solution of the conservation law was not greater than the total variation of the initial data. From the solution of the Riemann problem, we know that this is not true for systems. Nevertheless, we shall see that if the initial total variation is small enough, the total variation of the solution is bounded. To measure the total variation we use another time-dependent functional T defined by

$$T := \sum_{i=1}^{N} |\epsilon_i|, \tag{6.19}$$

where N is the number of fronts. Lax's theorem (Theorem 5.17) implies that T is equivalent to the total variation as long as the total variation is small.

Let t_1 denote the first time two fronts collide. At this time we will have another Riemann problem, which can be solved up to the next collision time t_2, etc. In this way we obtain an increasing sequence of collision times t_i, $i \in \mathbb{N}$. To show that front tracking is well-defined, we need to show that the sequence $\{t_i\}$ is finite, or if infinite, not convergent. In the scalar case we saw that indeed this sequence is finite.

216   6. Existence of Solutions of the Cauchy Problem

We will analyze more closely the changes in Q and T when fronts collide. Clearly, they change only at collisions. Let t_c be some fixed collision time.

Assume then that the situation is as in Figure 6.4: N fronts $\epsilon_1, \ldots, \epsilon_N$ colliding at some point (x_c, t_c), giving n waves $\epsilon'_1, \ldots, \epsilon'_n$ in the exact solution. Let I be a small interval containing x_c, and let J be the complement

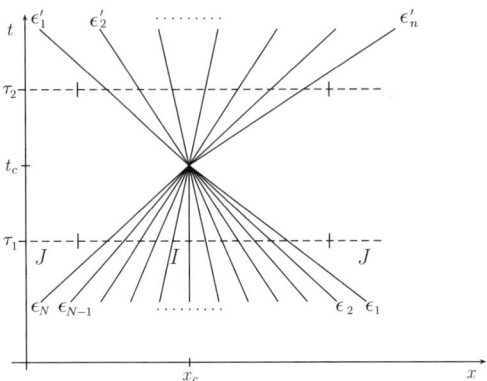

Figure 6.4. A collision of N fronts.

of I. Then we may write $Q = Q(I) + Q(J) + Q(I, J)$, where $Q(I)$ and $Q(J)$ indicate that the summation is restricted to those pairs of fronts that both lie in I and J, respectively. Similarly, $Q(I, J)$ means that the summation is over those pairs where one front is in I and the other in J. Let $\tau_1 < t_c < \tau_2$ be two times, chosen such that no other collisions occur in the interval $[\tau_1, \tau_2]$, and such that no fronts other than $\epsilon_1, \ldots, \epsilon_N$ are crossing the interval I at time τ_1, and only waves emanating from the collision at t_c, i.e., waves denoted by $\epsilon'_1, \ldots, \epsilon'_n$, cross I at time τ_2. Let Q_i and T_i denote the values of Q and T at time τ_i. By construction $Q_2(I) = 0$ and $Q_2(J) = Q_1(J)$, and hence

$$Q_2 - Q_1 = Q_2(I, J) - Q_1(I, J) - Q_1(I). \tag{6.20}$$

We now want to bound the increase in $Q(I, J)$ from time τ_1 to τ_2. More precisely, we want to prove that

$$Q_2(I, J) \leq Q_1(I, J) + \mathcal{O}\left(Q_1(I) T_1(J)\right). \tag{6.21}$$

Let $|\epsilon \epsilon'_i|$ be a term in $Q_2(I, J)$, i.e., $(\epsilon, \epsilon'_i) \in \mathcal{A}$ at time τ_2. Note that by the interaction estimate (6.10) we have

$$\epsilon'_i = \sum_{\hat{j} = i} \epsilon_j + \mathcal{O}\left(Q_1(I)\right).$$

Assume now that $\hat{j} = i$. If the family of ϵ equals i, then $(\epsilon, \epsilon_j) \in \mathcal{A}$ if and only if $\operatorname{sign}(\epsilon_j) = \operatorname{sign}(\epsilon'_i)$. If the family of ϵ is different from i, then

6.1. Front Tracking for Systems

$(\epsilon, \epsilon_j) \in \mathcal{A}$ independently of the sign of ϵ_j. Writing

$$|\epsilon'_i| = \sum_{\substack{\hat{j}=i \\ \operatorname{sign}(\epsilon_j)=\operatorname{sign}(\epsilon'_i)}} |\epsilon_j| - \sum_{\substack{\hat{j}=i \\ \operatorname{sign}(\epsilon_j)\neq\operatorname{sign}(\epsilon'_i)}} |\epsilon_j| + \mathcal{O}\left(Q_1(I)\right)$$

if the family of ϵ is i, and

$$|\epsilon'_i| \leq \sum_{\hat{j}=i} |\epsilon_j| + \mathcal{O}\left(Q_1(I)\right)$$

if the family of ϵ is different from i, we see that this implies (6.21).

Inserting (6.21) into (6.20), using the constant K to replace the order symbol, we obtain

$$Q_2 - Q_1 \leq K Q_1(I) T_1 - Q_1(I) = Q_1(I)(K T_1 - 1) \leq -\frac{1}{2} Q_1(I) \quad (6.22)$$

if T_1 is smaller than $1/(2K)$. We summarize the above discussion in the following lemma.

Lemma 6.2. *Assume that $T_1 \leq \frac{1}{2} K$. Then*

$$Q_2 - Q_1 \leq -\frac{1}{2} Q_1(I).$$

We will use this lemma to deduce that the total variation remains bounded if it initially is sufficiently small, or, in other words, if the initial data are sufficiently close to a constant state.

Lemma 6.3. *If T is sufficiently small at $t = 0$, then there is some constant c independent of δ such that*

$$G = T + c Q$$

is nonincreasing. We call G the Glimm functional. Consequently, T and T.V.(u^δ) are bounded independently of δ.

Proof. Let T_n and Q_n denote the value of T and Q, respectively, before the nth collision of fronts at t_n, with $0 < t_1 < t_2 < \cdots$. Using the interaction estimate (6.10) we first infer that

$$T_{n+1} = \sum_j |\epsilon'_j| \leq T_n + K Q_n(I). \quad (6.23)$$

Let $c \geq 2K$ and assume that $T_1 + c T_1^2 \leq 1/(2K)$. Assume furthermore that $T + cQ$ is nonincreasing for all t less than t_n, and that $T_n \leq 1/(2K)$. Lemma 6.2 and (6.23) imply that

$$T_{n+1} + c Q_{n+1} \leq T_n + K Q_c(I) + c Q_n - \frac{c}{2} Q_n(I)$$
$$= T_n + c Q_n + \left(K - \frac{c}{2}\right) Q_n(I)$$
$$\leq T_n + c Q_n,$$

since $K - \frac{c}{2} \leq 0$. Consequently,
$$T_{n+1} \leq T_{n+1} + cQ_{n+1} \leq \cdots \leq T_1 + cQ_1 \leq T_1 + cT_1^2 \leq 1/(2K),$$
which by induction proves the result. □

We still have not shown that the front-tracking approximation can be continued up to any desired time. We saw that one problem was that the number of fronts may not be bounded, and one could envisage that the number of fronts approaches infinity at some finite time. In order to circumvent this, we will change the approximate Riemann solution by removing certain weak fronts. The advantage will be that we obtain a hyperfast method that converges as $\delta \to 0$. However, there is a trade-off; if we remove too much, the limit will not be a weak solution of the original conservation law.

First we introduce the concept of *generation* of a front. We say that each initial front starting at $t = 0$ belongs to the first generation. Consider two first-generation fronts of families l and r, respectively, that collide. The resulting fronts of families l and r will still belong to the first generation, while all the remaining fronts resulting from this collision will be called second-generation fronts. More generally, if a front of family l and generation m interacts with a front of family r and generation n, the resulting fronts of families l and r are still assigned generations m and n, respectively, while the remaining fronts resulting from this collision are given generation $n + m$. The idea behind this concept is that fronts of high generation will be weak.

Given the initial approximation parameter δ, we will remove all fronts with generation higher than N, where
$$N = \text{rnd}\left(\ln_{4KT}(\delta)\right). \tag{6.24}$$

More precisely, if two fronts of generation n and m collide, at most two waves will retain their generation. If $n + m$ is greater than N, then the remaining waves will be removed, but if $n + m \leq N$, we use the full original solution. When we remove fronts, we let the function u^δ be equal to the value it has to the left of the removed fronts, provided that the removed fronts are not the rightmost fronts in the solution of the Riemann problem. If the rightmost fronts are removed, then u^δ is set equal to the value immediately to the right of the removed fronts.

We will now show that there is only a finite number of fronts of generation less than or equal to N, and that for a fixed δ there is only a finite number of collisions, hence that the method will be what we called *hyperfast*. To see this, we first assume that $T \leq 1/(4K)$, such that the strength of each individual front is also bounded by $1/(4K)$. For later reference we also note that by (6.24)
$$\delta \geq (4KT)^{N+1}. \tag{6.25}$$

First we consider the number of fronts of first generation. This number can increase when first-generation rarefaction fronts split up into several rarefaction fronts. By the term *rarefaction front* we mean a front approximating a rarefaction wave. Note that by the construction of the approximate Riemann problem, the strength of each split rarefaction front is at least $\frac{3}{4}\delta$. Remembering that T is uniformly bounded, we have that

$$\#(\text{first-generation fronts}) \leq \#(\text{initial fronts}) + \frac{4T}{3\delta}. \tag{6.26}$$

Thus the number of first-generation fronts is finite. This also means that there will be only a finite number of collisions between first-generation fronts. To see this, note that due to the strict hyperbolicity of the conservation law, each wave family will have speeds that are distinct; cf. (5.122)–(5.124). Hence each first-generation front will remain in a wedge in the (x, t) plane determined by the slowest and fastest speeds of that family. Eventually, all first-generation fronts will have interacted, and we can conclude that there can be only a finite number of collisions between first-generation fronts globally.

Assuming now that for some m there will be only a finite number of fronts of generation i, for all $i < m$, and that there will only be a finite number of interactions between fronts of generation less than m, then, in analogy to (6.26), we find that

$$\#(m\text{th generation fronts}) \tag{6.27}$$
$$\leq \#(\text{collisions of } j\text{th- and } i\text{th-generation fronts}; i+j=m) + \frac{4T}{3\delta}.$$

Consequently, the number of fronts of generation less than or equal to m is finite. We can now repeat the argument above showing that there is only a finite number of collisions between first-generation fronts, replacing "first generation" by "of generation less than or equal to m," and show that there is only a finite number of collisions producing fronts of generation $m+1$. Thus we can conclude that there is only a finite number of fronts of generation less than $N+1$, and that these interact only a finite number of times.

Summing up our results so far we have proved the following result.

Theorem 6.4. *Let $f_j \in C^2(\mathbb{R}^n)$, $j = 1, \ldots, n$. Let D be a domain in $\mathbb{R}^n$ and consider the strictly hyperbolic equation $u_t + f(u)_x = 0$ in D. Assume that f is such that each wave family is either genuinely nonlinear or linearly degenerate. Assume also that the function $u_0(x)$ has sufficiently small total variation.*

Then the front-tracking approximation $u^\delta(x, t)$, defined by (6.5), (6.6) and constructed by the front-tracking procedure described above, is well-defined. Furthermore, the method is hyperfast, i.e., requires only a finite number of computations to define $u^\delta(x,t)$ for all t. The total variation of

u^δ is uniformly bounded, and there is a finite constant C such that
$$\text{T.V.}\left(u^\delta(\,\cdot\,,t)\right) \leq C$$
for all $t > 0$ and all $\delta > 0$.

(i) Given a one-dimensional strictly hyperbolic system of conservation laws,
$$u_t + f(u)_x = 0, \quad u|_{t=0} = u_0, \tag{6.28}$$
where u_0 has small total variation.

(ii) Approximate initial data u_0 by a piecewise constant function u_0^δ.

(iii) Approximate the solution of each Riemann problem by a piecewise constant function by sampling points with distance δ on the rarefaction curve and using the exact shocks and contact discontinuities.

(iv) Track discontinuities ("fronts").

(v) Solve new Riemann problems as in (iii).

(vi) Eliminate weak waves ("high generation").

(vii) Continue to solve Riemann problems approximately as in (iii), eliminating weak waves. Denote approximate solution by u^δ.

(viii) As $\delta \to 0$, the approximate solution u^δ will converge to u, the solution of (6.28).

Front tracking in a box (systems).

6.2 Convergence

The Devil is in the details.

English proverb

At this point we could proceed as in the scalar case, by showing that front tracking is stable with respect to L^1 perturbations of the initial data. This would then imply that the sequence of approximations $\{u^\delta\}$ has a unique limit as $\delta \to 0$. For systems, however, this analysis is rather complicated. In this section we shall instead prove that the sequence $\{u^\delta\}$ is compact and that any (there is really only one) limit is a weak solution. The reader willing to accept this, or primarily interested in front tracking, may skip ahead to the next chapter.

6.2. Convergence

To show that a subsequence of the sequence $\{u^\delta\}_{\delta>0}$ converges in $L^1_{\text{loc}}(\mathbb{R} \times [0,T])$ we use Theorem A.8 from Appendix A. We have already shown that that $u^\delta(x,t)$ is bounded, and we have that

$$\int_\mathbb{R} |u^\delta(x+\rho,t) - u^\delta(x,t)|\, dt \leq \rho\, \text{T.V.}\,(u^\delta(\,\cdot\,,t)) \leq C\rho,$$

for some C independent of δ. Hence, by Theorem A.8, to conclude that a subsequence of $\{u^\delta\}$ converges, we must show that

$$\int_{-R}^{R} |u^\delta(x,t) - u^\delta(x,s)|\, dx \leq C(t-s),$$

where $t \geq s \geq 0$, for any $R > 0$, and for some C independent of δ. Since u^δ is bounded, we have that (recall that $\lambda_1 < \cdots < \lambda_n$)

$$L := \max_{|u|\leq \sup|u^\delta|} \{|\lambda_n(u)|, |\lambda_1(u)|\}$$

is bounded. Let t_i and t_{i+1} be two consecutive collision times. For $t \in \langle t_i, t_{i+1}]$ we write u^δ in the form

$$u^\delta(x,t) = \sum_{k=1}^{N_i} (u_{k-1}^i - u_k^i)\, H(x - x_k^i(t)) + u_{N_i}, \qquad (6.29)$$

where $x_k^i(t)$ denotes the position of the kth front from the left, and H the Heaviside function. Here $u^\delta(x,t) = u_k^i$ for x between x_k^i and x_{k+1}^i. Assume now that $t \in [t_i, t_{i+1}]$ and $s \in [t_j, t_{j+1}]$, where $j \leq i$ and $s \leq t$. Then

$$\int_\mathbb{R} |u^\delta(x,t) - u^\delta(x,t_i)|\, dx$$

$$= \int_\mathbb{R} \left|\int_{t_i}^{t} \frac{d}{d\tau} u^\delta(x,\tau)\, d\tau\right| dx$$

$$\leq \int_\mathbb{R} \int_{t_i}^{t} \sum_{k=1}^{N_i} |u_{k-1}^i - u_k^i|\, |x_k^{i\,\prime}(\tau)|\, |H'(x - x_k^i(\tau))|\, d\tau\, dx$$

$$\leq L \int_{t_i}^{t} \sum_{k=1}^{N_i} |u_{k-1}^i - u_k^i| \int_\mathbb{R} |H'(x - x_k^i(\tau))|\, dx\, d\tau$$

$$\leq L(t-t_i)\, \text{T.V.}\,(u^\delta(\,\cdot\,,t))$$

$$\leq LC(t-t_i),$$

since $|x_k^{i\,\prime}| \leq L$. Similarly, we show that

$$\int_\mathbb{R} |u^\delta(x,t_i) - u^\delta(x,t_{j+1})|\, dx \leq LC(t_i - t_{j+1}) \quad \text{if } j+1 < i,$$

and

$$\int_\mathbb{R} |u^\delta(x,t_{j+1}) - u^\delta(x,s)|\, dx \leq LC(t_{j+1}-s).$$

222   6. Existence of Solutions of the Cauchy Problem

Therefore,
$$\left\|u^\delta(\,\cdot\,,t) - u^\delta(\,\cdot\,,s)\right\|_1 \le C\,|t-s|,$$
for some constant C independent of t and δ. Hence, we can use Theorem A.8 to conclude that there exists a function $u(x,t)$ and a subsequence $\{\delta_j\} \subset \{\delta\}$ such that $u^{\delta_j} \to u(x,t)$ in L^1_{loc} as $j \to \infty$.

As in the scalar case, it is by no means obvious that the limit function $u(x,t)$ is a weak solution of the original initial value problem (6.1). For a single conservation law, this was not difficult to show, using that the approximations were weak solutions of approximate problems. This is not so in the case of systems, so we must analyze how close the approximations are to being weak solutions.

There are three sources of error in the front-tracking approximation. Firstly, the initial data are approximated by a step function. Secondly, there is the approximation of rarefaction waves by step functions, and finally, weak waves are ignored in the solution of some Riemann problems.

We start the error analysis by estimating how much we "throw away" by ignoring fronts with a generation larger than N. To do this, we estimate the total variation of the fronts belonging to a given generation. Let $\mathcal{G}_m$ denote the set of all fronts of generation m, and let $\mathcal{T}_m$ denote the sum of the strengths of fronts of generation m. Thus
$$\mathcal{T}_m = \sum_{\epsilon_j \in \mathcal{G}_m} |\epsilon_j|.$$

Clearly,
$$T = \sum_{m=1}^{N} \mathcal{T}_m.$$

Lemma 6.5. *We have that*
$$\mathcal{T}_m \le 2\frac{(2m-3)!}{(m-2)!\,m!}K^{m-1}T^n \le C(4KT)^m$$
for some constant C. In particular, for $m = N+1$,
$$\mathcal{T}_{N+1} \le C\delta.$$

Proof. Clearly,
$$\mathcal{T}_{m+1} = \sum_{j=1}^{m} \sum_{\epsilon_l \in \mathcal{G}_{m+1-j}} \sum_{\epsilon_r \in \mathcal{G}_j} \mathcal{O}\left(|\epsilon_l|\,|\epsilon_r|\right)$$
$$\le K \sum_{j=1}^{m} \sum_{\epsilon_l \in \mathcal{G}_{m+1-j}} \sum_{\epsilon_r \in \mathcal{G}_j} |\epsilon_l|\,|\epsilon_r|$$
$$= K \sum_{j=1}^{m} \mathcal{T}_{m+1-j}\mathcal{T}_j.$$

By introducing $\tilde{T}_m = T_m/(T^m K^{m-1})$, we see that $\tilde{T}_m$ satisfies

$$\tilde{T}_{m+1} \leq \sum_{j=1}^{m} \tilde{T}_{m+1-j}\tilde{T}_j.$$

By choosing T sufficiently small, we may assume that $\tilde{T}_1 \leq 1$. Furthermore, we see that $\tilde{T}_m \leq a_m$, where

$$a_1 = 1, \quad a_m = \sum_{j=1}^{m-1} a_{m-j}a_j, \quad m = 2, 3, \ldots. \tag{6.30}$$

If we define the function

$$y(x) = \sum_{m=1}^{\infty} a_m x^m,$$

then, using (6.30),

$$y^2 = \sum_{m=2}^{\infty} \left(\sum_{j=1}^{m-1} a_{m-j}a_j \right) x^m = y - x,$$

and we infer that (recall that $y(0) = 0$)

$$y(x) = \frac{1}{2}\left(1 - \sqrt{1-4x}\right) = \sum_{m=1}^{\infty} (-1)^{m+1} \binom{1/2}{m} 2^{2m-1} x^m,$$

which implies

$$a_m = (-1)^{m+1}\binom{1/2}{m} 2^{2m-1}.$$

We may rewrite this as

$$a_m = (-1)^{m+1} \frac{\frac{1}{2} \cdot \left(\frac{1}{2}-1\right) \cdots \left(\frac{1}{2}-m+1\right)}{m!} 2^{2m-1} = 2\frac{(2m-3)!}{m!(m-2)!}.$$

To estimate a_m as $m \to \infty$, we apply Stirling's formula [144, p. 253]

$$n! = \sqrt{2\pi} \exp\left(\left(n - \frac{1}{2}\right)\ln(n+1) - (n+1) + \frac{\theta}{12(n+1)}\right),$$

for $0 \leq \theta \leq 1$. We obtain

$$a_m = 2\frac{(2m-3)!}{m!(m-2)!} = \mathcal{O}(1)\, 4^m\, m^{-1/2}.$$

Using

$$T_m = \tilde{T}_m T^m K^{m-1} \leq a_m T^m K^{m-1},$$

the lemma follows. □

224   6. Existence of Solutions of the Cauchy Problem

Equipped with this lemma, we can start to estimate how well u^δ approximates a weak solution of (6.1). First we shall determine how far the front-tracking approximation is from being a weak solution between two fixed consecutive collision times t_j and t_{j+1}. For convenience we temporarily call these t_1 and t_2.

Recall that u is a weak solution in a strip $[t, s]$ if

$$\mathcal{I}_t^s(u) := \int_t^s \int (u\varphi_t + f(u)\varphi_x)\,dx\,dt \\ - \int u(x,s)\varphi(x,s)\,dx + \int u(x,t)\varphi(x,t)\,dx = 0 \tag{6.31}$$

for every test function φ.

In the following we shall assume that all bounded quantities are bounded by the (universal) constant M. Thus φ, φ_x, φ_t, u^δ, etc., are all bounded in absolute value by M. Let $s_1 = t_1$, and let $v_1(x,s)$ denote the weak solution of

$$v_s + f(v)_x = 0, \qquad v(x, s_i) = u^\delta(x, s_i), \tag{6.32}$$

with $i = 1$. For $(s - s_1)$ small, v_1 can be calculated by solving the Riemann problems defined at the jumps of u^δ. The function v_1 can be defined until these interact at some $s_2 > s_1$. If these waves do not interact until $s = t_2$, v_1 can be defined in the whole strip $[t_1, t_2]$, and we set $s_2 = t_2$. If waves in v_1 interact before t_2, then for $s > s_2$ we let v_2 be the solution of (6.32) with $i = 2$. Observe that we use the function u^δ as initial data at time s_2. Continuing in this way, we fill the whole strip $[t_1, t_2]$ with smaller strips $[s_i, s_{i+1})$, in which we have defined v_i. Let v denote the function that equals v_i in the interval $[s_i, s_{i+1})$. We have that $u^\delta(x, s_i) = v(x, s_i)$ for all i. Now

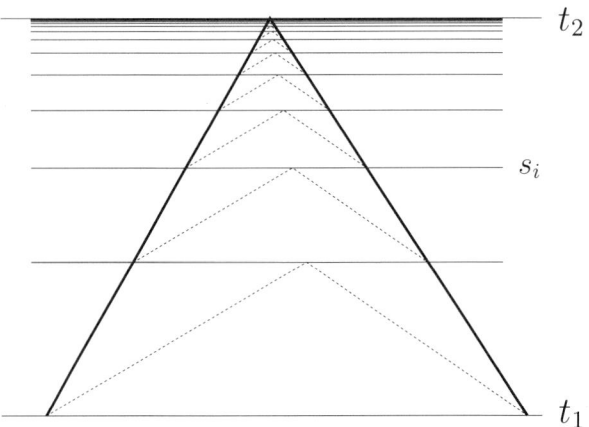

Figure 6.5. Waves of v (dotted lines) and fronts of u^δ (thick lines).

6.2. Convergence

let v^δ be the approximation to v obtained by front tracking in the strip $[s_i, s_{i+1})$ with data as in (6.32). Since no fronts of v^δ will collide in this strip, the fronts in v^δ approximate all waves in the solution of Riemann problems at s_i. Furthermore, we have that $v(x,t) = v^\delta(x,t)$ except when (x,t) is in a rarefaction fan. For such (x,t), $|v - v^\delta| = \mathcal{O}(\delta)$ by the construction of v^δ. We wish to estimate

$$\int |v(x,t) - v^\delta(x,t)| \, dx \tag{6.33}$$

for t between s_i and s_{i+1}. This integral will be a sum of integrals across the rarefaction fans of v. If we have a rarefaction fan with a left state v_l and right state v_r, then the integral across this fan will be a sum of integrals across each step of v^δ. The number of such steps is

$$\frac{|v_l - v_r|}{\mathcal{O}(\delta)},$$

and the width of each step is

$$(t - s_i)\Delta\lambda = (t - s_i)\mathcal{O}(\delta),$$

where $\Delta\lambda$ is the difference in characteristic speed on both sides of a front in v^δ approximating the rarefaction wave in v. Then the integral (6.33) can be estimated as follows:

$$\int |v(x,t) - v^\delta(x,t)| \, dx = \sum_j \frac{|v_l^j - v_r^j|}{\mathcal{O}(\delta)} \mathcal{O}(\delta)(t - s_i)\mathcal{O}(\delta)$$
$$\leq (t - s_i)\,\text{T.V.}\left(u^\delta\right)\mathcal{O}(\delta) \tag{6.34}$$
$$= (t - s_i)\mathcal{O}(\delta),$$

since $\sum_j |v_l^j - v_r^j| = \text{T.V.}\left(u^\delta\right)$.

We also wish to compare u^δ and v^δ in this strip. The difference between these is that some fronts that are tracked in v^δ are ignored in u^δ, since all fronts in v^δ will be first-generation fronts. Consequently, v^δ is different from u^δ in a finite number of wedges emanating from the discontinuities in $u^\delta(\cdot, s_i)$. Then, using Lemma 6.5,

$$\int |v^\delta(x,t) - u^\delta(x,t)| \, dx \leq C(t - s_i) \sum_{\epsilon_j \in \mathcal{G}_{N+1}} |\epsilon_j|$$
$$\leq C(t - s_i)T_{N+1} \tag{6.35}$$
$$= (t - s_i)\mathcal{O}(\delta),$$

when $T \leq 1/(4K)$ and N is chosen as in (6.24). Using (6.34) and (6.35), we obtain

$$\int |u^\delta - v| \, dx = (t - s_i)\mathcal{O}(\delta). \tag{6.36}$$

We can use the result (6.36) to estimate $\mathcal{I}_{s_i}^{s_{i+1}}(u^\delta)$. Since $\mathcal{I}_{s_i}^{s_{i+1}}(v) = 0$, we have that

$$\left|\mathcal{I}_{s_i}^{s_{i+1}}(u^\delta)\right| = \left|\mathcal{I}_{s_i}^{s_{i+1}}(u^\delta) - \mathcal{I}_{s_i}^{s_{i+1}}(v)\right|$$

$$= \left|\int_{s_i}^{s_{i+1}}\int \left((u^\delta - v)\varphi_t + (f(u^\delta) - f(v))\varphi_x\right)\,dx\,dt\right.$$

$$\left. - \int \left(u^\delta(x, s_{i+1}) - v(x, s_{i+1})\right)\varphi(x, s_{i+1})\,dx\right|$$

$$\leq M\left\{\int_{s_i}^{s_{i+1}}\int \left(|u^\delta - v| + |f(u^\delta) - f(v)|\right)\,dx\,dt\right.$$

$$\left. + \int \left|u^\delta(x, s_{i+1}) - v(x, s_{i+1})\right|\,dx\right\}$$

$$\leq M\left\{\int_{s_i}^{s_{i+1}}\int \left(|u^\delta - v| + \mathcal{O}\left(|u^\delta - v|^2\right)\right)\,dx\,dt\right.$$

$$\left. + \int \left|u^\delta(x, s_{i+1}) - v(x, s_{i+1})\right|\,dx\right\}$$

$$\leq \mathcal{O}(\delta)\left((s_{i+1} - s_i) + (s_{i+1} - s_i)^2\right).$$

Using this, it is not difficult to estimate $\mathcal{I}_{t_j}^{t_{j+1}}(u^\delta)$, where t_j and t_{j+1} are consecutive collision times of u^δ, because

$$\left|\mathcal{I}_{t_j}^{t_{j+1}}(u^\delta)\right| \leq \sum_{i=1}^{\infty}\left|\mathcal{I}_{s_i}^{s_{i+1}}(u^\delta)\right| = \sum_{i=1}^{\infty}\mathcal{O}(\delta)\left((s_{i+1} - s_i) + (s_{i+1} - s_i)^2\right). \tag{6.37}$$

Since $t_{j+1} - t_j = \sum_{i=1}^{\infty}(s_{i+1} - s_i)$, the second term on the rightmost part of (6.37) is less than $(t_{j+1} - t_j)^2$, and we get

$$\left|\mathcal{I}_{t_j}^{t_{j+1}}(u^\delta)\right| = \mathcal{O}(\delta)\left((t_{j+1} - t_j) + (t_{j+1} - t_j)^2\right). \tag{6.38}$$

Finally, to show that the limit u is a weak solution, we observe that since u^δ is bounded and $u^\delta \to u$ in L^1_{loc}, $f(u^\delta)$ will converge to $f(u)$ in L^1_{loc}. Thus

$$\lim_{\delta \to 0}\left|\mathcal{I}_0^T(u^\delta)\right| = \left|\mathcal{I}_0^T(u)\right|.$$

By (6.38) we get

$$\left|\mathcal{I}_0^T(u^\delta)\right| = \mathcal{O}(\delta)\sum_j\left((t_{j+1} - t_j) + (t_{j+1} - t_j)^2\right) \leq \mathcal{O}(\delta)\left(T + T^2\right),$$

which shows that $\mathcal{I}_0^T(u) = 0$, and accordingly that u is a weak solution.

We can actually extract some more information about the limit u by examining the approximate solutions u^δ. More precisely, we would like to show that isolated jump discontinuities of u satisfy the Lax entropy condition

$$\lambda_m(u_l) \geq \sigma \geq \lambda_m(u_r) \tag{6.39}$$

for some m between 1 and n, where σ is the speed of the discontinuity, and
$$u_l = \lim_{y \to x-} u(y,t) \quad \text{and} \quad u_r = \lim_{y \to x+} u(y,t).$$

To show this we assume that u has an isolated discontinuity at (x,t), with left and right limits u_l and u_r. We can enclose (x,t) by a trapezoid E_δ with corners defined as follows. Start by finding points
$$x_{\delta,l}^k \to x-, \quad x_{\delta,r}^k \to x+, \quad t_\delta^1 \uparrow t, \quad \text{and} \quad t_\delta^2 \downarrow t,$$
for $k = 1, 2$ as $\delta \to 0$. We let E_δ denote the trapezoid with corners $(x_{\delta,l}^1, t_\delta^1)$, $(x_{\delta,r}^1, t_\delta^1)$, $(x_{\delta,r}^2, t_\delta^2)$, and $(x_{\delta,l}^2, t_\delta^2)$. Recall that convergence in L_{loc}^1 implies pointwise convergence almost everywhere, so we choose these points such that
$$\left. \begin{array}{c} u^\delta(x_{\delta,l}^1, t_\delta^1) \\ u^\delta(x_{\delta,l}^2, t_\delta^2) \end{array} \right\} \to u_l \quad \text{and} \quad \left. \begin{array}{c} u^\delta(x_{\delta,r}^1, t_\delta^1) \\ u^\delta(x_{\delta,r}^2, t_\delta^2) \end{array} \right\} \to u_r$$
as $\delta \to 0$. We can also choose points such that the diagonals of E_δ have slopes not too different from σ; precisely,
$$\left| \frac{x_{\delta,l}^1 - x_{\delta,r}^2}{t_\delta^1 - t_\delta^2} - \sigma \right| \leq \varepsilon(\delta) \quad \text{and} \quad \left| \frac{x_{\delta,r}^1 - x_{\delta,l}^2}{t_\delta^1 - t_\delta^2} - \sigma \right| \leq \varepsilon(\delta), \qquad (6.40)$$
where $\varepsilon(\delta) \to 0$ as $\delta \to 0$. Next for $k = 1, 2$ set
$$M_\delta^k = \frac{\sum |\epsilon_i|}{x_{\delta,r}^k - x_{\delta,l}^k},$$
where the sum is over all rarefaction fronts in the interval $[x_{\delta,l}^k, x_{\delta,r}^k]$. If M_δ^k is unbounded as $\delta \to 0$, then u contains a centered rarefaction wave at (x,t), i.e., a rarefaction wave starting at (x,t). In this case the discontinuity will not be isolated, and hence M_δ^k remains bounded as $\delta \to 0$. Next observe that
$$\frac{\left| u^\delta(x_{\delta,l}^k, t_\delta^k) - u^\delta(x_{\delta,r}^k, t_\delta^k) \right|}{x_{\delta,r}^k - x_{\delta,l}^k} \leq C \frac{\sum |\text{rarefaction fronts}| + \sum |\text{shock fronts}|}{x_{\delta,r}^k - x_{\delta,l}^k}$$
$$= CM_\delta^k + C \frac{\sum |\text{shock fronts}|}{x_{\delta,r}^k - x_{\delta,l}^k}.$$

Here the sums are over fronts crossing the interval $[x_{\delta,l}^k, x_{\delta,r}^k]$. Since the fraction on the left is unbounded as $\delta \to 0$, there must be shock fronts crossing the top and bottom of E_δ for all $\delta > 0$. Furthermore, since the discontinuity is isolated, the total strength of all fronts crossing the left and right sides of E_δ must tend to zero as $\delta \to 0$.

Next we define a *shock line* as a sequence of shock fronts of the same family in u^δ. Assume that a shock line has been defined for $t < t_n$, where t_n is a collision time, and in the interval $[t_{n-1}, t_n)$ consists of the shock front ϵ. In the interval $[t_n, t_{n+1})$, this shock line continues as the front ϵ if ϵ does not collide at t_n. If ϵ collides at t_n, and the approximate solution of the Riemann problem determined by this collision contains an approximate shock front of the same family as ϵ, then the shock line continues as this front. Otherwise, it stops at t_n. Note that we can associate a unique family to each shock line.

From the above reasoning it follows that for all δ there must be shock lines entering E_δ through the bottom that do not exit E_δ through the sides; hence such shock lines must exit E_δ through the top. Assume that the leftmost of these shock lines enters E_δ at $y_{\delta,l}^1$ and leaves E_δ at $y_{\delta,l}^2$. Similarly, the rightmost of the shock lines enters E_δ at $y_{\delta,r}^1$ and leaves E_δ at $y_{\delta,r}^2$. Set

$$v_{\delta,l}^k = u^\delta\left(y_{\delta,l}^k-, t_\delta^k\right) \quad \text{and} \quad v_{\delta,r}^k = u^\delta\left(y_{\delta,r}^k+, t_\delta^k\right).$$

Between $y_{\delta,l}^k$ and $x_{\delta,l}^k$, the function u^δ varies over rarefaction fronts or over shock lines that must enter or leave E_δ through the left or right side. Since the discontinuity is isolated, the total strength of such waves must tend to zero as $\delta \to 0$. Because $u^\delta(x_{\delta,l}^k, t_\delta^k) \to u_l$ as $\delta \to 0$, we have that $v_{\delta,l}^k \to u_l$ as $\delta \to 0$. Similarly, $v_{\delta,r}^k \to u_r$. Since $\varepsilon(\delta) \to 0$, by strict hyperbolicity, the family of all shock lines not crossing the left or right side of E_δ must be the same, say m. The speed of an approximate m-shock front with speed $\tilde{\sigma}$ and left state $v_{\delta,l}^k$ satisfies

$$\lambda_{m-1}\left(v_{\delta,l}^k\right) < \tilde{\sigma} + \mathcal{O}\left(\delta\right) < \lambda_m\left(v_{\delta,l}^k\right). \tag{6.41}$$

Similarly, an approximate m-shock front with right state $v_{\delta,r}^k$ and speed $\hat{\sigma}$ satisfies

$$\lambda_m\left(v_{\delta,r}^k\right) < \hat{\sigma} + \mathcal{O}\left(\delta\right) < \lambda_{m+1}\left(v_{\delta,r}^k\right). \tag{6.42}$$

Then (6.39) follows by noting that $\tilde{\sigma}$ and $\hat{\sigma}$ both tend to σ as $\delta \to 0$, and then letting $\delta \to 0$ in (6.41) and (6.42).

To summarize the results of this chapter we have the following theorem:

Theorem 6.6. *Consider the strictly hyperbolic system of equations*

$$u_t + f(u)_x = 0, \qquad u(x,0) = u_0(x),$$

and assume that $f \in C^2$ is such that each characteristic wave family is either linearly degenerate or genuinely nonlinear. If T.V. (u_0) *is sufficiently small, there exists a global weak solution $u(x,t)$ to this initial value problem. This solution may be constructed by the front-tracking algorithm described in Section 6.1. Furthermore, if u has an isolated jump discontinuity at a point (x,t), then the Lax entropy condition (6.39) holds for some m between 1 and n.*

We have seen that for each $\delta > 0$ there are only a finite number of collisions between the fronts in u^δ for all $t > 0$. Hence there exists a finite

6.2. Convergence

time T_δ such that for $t > T_\delta$, the fronts in u^δ will move apart, and not interact. This has some similarity to the solution of the Riemann problem. One can intuitively make the change of variables $t \mapsto t/\varepsilon$, $x \mapsto x/\varepsilon$ without changing the equation, but the initial data is changed to $u_0(x/\varepsilon)$. Sending $\varepsilon \to 0$, or alternatively $t \to \infty$, we see that u solves the Riemann problem

$$u_t + f(u)_x = 0, \qquad u(x,0) = \begin{cases} u_L & \text{for } x < 0, \\ u_R & \text{for } x \geq 0, \end{cases} \qquad (6.43)$$

where $u_L = \lim_{x \to -\infty} u_0(x)$ and $u_R = \lim_{x \to \infty} u_0(x)$. Thus in some sense, for very large times, u should solve this Riemann problem. Next, we shall show that this (very imprecise statement) is true, but first we need some more information about u^δ.

For $t > T_\delta$, the function u^δ will consist of a finite number of constant states, say u_i^δ, for $i = 0, \ldots, M$. If u_{i-1}^δ is connected with u_i^δ by a wave of a different family from the one connecting u_i^δ to u_{i+1}^δ, we call u_i^δ a *real state*, and we let $\{\bar{u}_i\}_{i=0}^N$ be the set of real states of u^δ. Since the discontinuities of u^δ are moving apart, we must have

$$N \leq n, \qquad (6.44)$$

by strict hyperbolicity. Furthermore, to each pair $(\bar{u}_{i-1}, \bar{u}_i)$ we can associate a family k_i such that $1 \leq k_i < k_{i+1} \leq n$, and we define $k_0 = 0$ and $k_{N+1} = n$. We write the solution of the Riemann problem with left and right data $\bar{u}_0$ and $\bar{u}_N$, respectively, as u, and define ϵ_j, $j = 1, \ldots, n$, by

$$\bar{u}_N = W_n(\epsilon_n) W_{n-1}(\epsilon_{n-1}) \cdots W_1(\epsilon_1) \bar{u}_0,$$

and define the intermediate states

$$u_0 = \bar{u}_0 \quad \text{and} \quad u_j = H_j(\epsilon_j) u_{j-1} \quad \text{for } j = 1, \ldots, n.$$

Now we claim that

$$|u_j - \bar{u}_i| \leq \mathcal{O}(\delta), \quad \text{for } k_{i-1} \leq j \leq k_i. \qquad (6.45)$$

If $N = 1$, this clearly holds, since in this case u^δ consists of two states for $t > T_\delta$, and by construction of u^δ, the pair $(\bar{u}_0, \bar{u}_1)$ is the solution of the same Riemann problem as u is, but possibly with waves of a high generation ignored.

Now assume that (6.45) holds for some $N > 1$. We shall show that it holds for $N+1$ as well. Let v be the solution of the Riemann problem with initial data given by

$$v(x,0) = \begin{cases} \bar{u}_0 & \text{for } x < 0, \\ \bar{u}_N & \text{for } x \geq 0, \end{cases}$$

and let w be the solution of the Riemann problem with initial data

$$w(x,0) = \begin{cases} \bar{u}_N & \text{for } x < 0, \\ \bar{u}_{N+1} & \text{for } x \geq 0. \end{cases}$$

We denote the waves in v and w by ϵ_j^v and ϵ_j^w, respectively. Then by the induction assumption,

$$\left|\bar{\epsilon}_i - \epsilon_{k_i}^v\right| \leq \mathcal{O}(\delta), \qquad \left|\bar{\epsilon}_{N+1} - \epsilon_{k_{N+1}}^w\right| \leq \mathcal{O}(\delta),$$

$$\sum_{i \notin \{k_1, \ldots, k_N\}} |\epsilon_i^v| \leq \mathcal{O}(\delta), \quad \text{and} \quad \sum_{i \neq k_{N+1}} |\epsilon_i^w| \leq \mathcal{O}(\delta),$$

where $\bar{\epsilon}_i$ denotes the strength of the wave separating $\bar{u}_{i-1}$ and $\bar{u}_i$. Notice now that u can be viewed as the interaction of v and w; hence by the interaction estimate,

$$\sum_i |\bar{\epsilon}_i - \epsilon_i^v| \leq \mathcal{O}(\delta) \quad \text{for } i \leq k_N, \text{ and} \quad \left|\epsilon_{k_{N+1}} - \epsilon_{k_{N+1}}^w\right| \leq \mathcal{O}(\delta).$$

Thus (6.45) holds for $N+1$ real states, and therefore for any $N \leq n$. Now we can conclude that for $u = \lim_{\delta \to 0} u^\delta$ the following result holds.

Theorem 6.7. *Assume that $u_L = \lim_{x \to -\infty} u_0(x)$ and $u_R = \lim_{x \to \infty} u_0(x)$ exist. Then as $t \to \infty$, u will consist of a finite number of states $\{u_i\}_{i=0}^N$, where $N \leq n$. These states are the intermediate states in the solution of the Riemann problem (6.43), and they will be separated by the same waves as the corresponding states in the solution of the Riemann problem.*

Proof. By the calculations preceding the lemma, for $t > T_\delta$ we can define a function $\bar{u}_\delta$ that consists of a number of constant states separated by elementary waves, shocks, rarefactions, or contact discontinuities such that these constant states are the intermediate states in the solution of the Riemann problem defined by $\lim_{x \to -\infty} u^\delta(x, t)$ and $\lim_{x \to \infty} u^\delta(x, t)$, and such that for any bounded interval I,

$$\left\|\bar{u}_\delta(\cdot, t) - u^\delta(\cdot, t)\right\|_{L^1(I)} \to 0 \quad \text{as } \delta \to 0.$$

Then for $t > T_\delta$,

$$\|u(\cdot, t) - \bar{u}_\delta(\cdot, t)\|_{L^1(I)} \leq \|u(\cdot, t) - u^\delta(\cdot, t)\|_{L^1(I)} + \|\bar{u}_\delta(\cdot, t) - u^\delta(\cdot, t)\|_{L^1(I)}.$$

Set $t = T_\delta + 1$, and let $\delta \to 0$. Then both terms on the right tend to zero, and $\bar{u}_0 \to u_L$ and $\bar{u}_N \to u_R$. Hence the lemma holds. Note, however, that u does not necessarily equal some $\bar{u}_\delta$ in finite time. □

Remark 6.8. Here is another way to interpret heuristically the asymptotic result for large times. Consider the set

$$\{u^\delta(x, t) \mid x \in \mathbb{R}\}$$

in phase space. There is a certain ordering of that set given by the ordering of x. As $\delta \to 0$, this set will approach some set

$$\{u(x, t) \mid x \in \mathbb{R}\}.$$

Theorem 6.7 states that as $t \to \infty$ this set approaches the set that consists of the states in the solution of the Riemann problem (6.43) with the same order. No statements are made as to how fast this limit is obtained. In particular, if $u_l = u_r = 0$, then $u(x,t) \to 0$ for almost all x as $t \to \infty$.

6.3 Notes

The fundamental result concerning existence of solutions of the general Cauchy problem is due to Glimm [52], where the fundamental approach was given, and where all the basic estimates can be found. Glimm's result for small initial data uses the *random choice method*. The random element is not really essential to the random choice method, as was shown by Liu in [102]. The existence result has been extended for some 2×2 systems, allowing for initial data with large total variation; see [109, 131]. These systems have the rather special property that the solution of the Riemann problem is translation-invariant in phase space.

Our proof of the interaction estimate (6.10) is a modified version of Yong's argument [146].

Front tracking for systems was first used by DiPerna in [43]. In this work a front-tracking process was presented for 2×2 systems, and shown to be well-defined and to converge to a weak solution. Although DiPerna states that "the method is adaptable for numerical calculation," numerical examples of front tracking were first presented by Swartz and Wendroff in [133], in which front tracking was used as a component in a numerical code for solving problems of gas dynamics.

The front tracking presented here contains elements from the front-tracking methods of Bressan [15] and, in particular, of Risebro [123]. In [123] the generation concept was not used. Instead, one "looked ahead" to see whether a buildup of collision times was about to occur. In [5] Baiti and Jenssen showed that one does not really need to use the generation concept nor to look ahead in order to decide which fronts to ignore.

The large time behavior of u was shown to hold for the limit of the Glimm scheme by Liu in [103].

The front-tracking method presented in [123] has been used as a numerical method; see Risebro and Tveito [124, 125] and Langseth [91, 92] for examples of problems in one space dimension. In several space dimensions, front tracking has also been used in conjunction with dimensional splitting with some success for systems; see [68] and [99].

Exercises

6.1 Assume that $f: \mathbb{R}^n \to \mathbb{R}^n$ is three-times differentiable, with bounded derivatives. We study the solution of the system of ordinary differential equations
$$\frac{dx}{dt} = f(x), \quad x(0) = x_0.$$
We write the unique solution as $x(t) = \exp(tf)x_0$.

  a. Show that
  $$\exp(\varepsilon f)x_0 = x_0 + \varepsilon f(x_0) + \frac{\varepsilon^2}{2} df(x_0) f(x_0) + \mathcal{O}(\varepsilon^3).$$

  b. If g is another vector field with the same properties as f, show that
  $$\exp(\varepsilon g)\exp(\varepsilon f)x_0 = x_0 + \varepsilon(f(x_0) + g(x_0))$$
  $$+ \frac{\varepsilon^2}{2}\Big(df(x_0)f(x_0) + dg(x_0)g(x_0)\Big)$$
  $$+ \varepsilon^2 dg(x_0) f(x_0) + \mathcal{O}(\varepsilon^3).$$

  c. The *Lie bracket* of f and g is defined as
  $$[f,g](x) = dg(x)f(x) - df(x)g(x).$$
  Show that
  $$[f,g](x_0) = \lim_{\varepsilon \to 0} \frac{1}{\varepsilon^2}\Big(\exp(\varepsilon g)\exp(\varepsilon f)x_0 - \exp(\varepsilon f)\exp(\varepsilon g)x_0\Big).$$

  d. Indicate how this can be used to give an alternative proof of the interaction estimate (6.10).

6.2 We study the p system with $p(u_1)$ as in Exercise 5.2, and we use the results of Exercise 5.8. Define a front-tracking scheme by introducing a grid in the (μ, τ) plane. We approximate rarefaction waves by choosing intermediate states that are not further apart than δ in (μ, τ). If ϵ is a front with left state (μ_l, τ_l) and right state (μ_r, τ_r), define
$$T(\epsilon) = \begin{cases} |[\![\mu]\!]| - |[\![\tau]\!]| & \text{if } \epsilon \text{ is a 1-wave,} \\ |[\![\tau]\!]| - |[\![\mu]\!]| & \text{if } \epsilon \text{ is a 2-wave,} \end{cases}$$
and define T additively for a sequence of fronts.

  a. Define a front-tracking algorithm based on this, and show that
  $$T_{n+1} \leq T_n,$$
  where T_n denotes the T value of the front-tracking approximation between collision times t_n and t_{n+1}.

b. Find a suitable condition on the initial data so that the front-tracking algorithm produces a convergent subsequence.

c. Show that the limit is a weak solution.

7
Well-Posedness of the Cauchy Problem for Systems

> Ma per seguir virtute e conoscenza.[1]
>
> Dante Alighieri (1265–1321), La Divina Commedia

The goal of this chapter is to show that the limit found by front tracking, that is, the weak solution of the initial value problem

$$u_t + f(u)_x = 0, \quad u(x,0) = u_0(x), \qquad (7.1)$$

is stable in L^1 with respect to perturbations in the initial data. In other words, if $v = v(x,t)$ is another solution found by front tracking, then

$$\|u(\,\cdot\,,t) - v(\,\cdot\,,t)\|_1 \leq C\|u_0 - v_0\|_1$$

for some constant C. Furthermore, we shall show that under some mild extra entropy conditions, any weak solution coincides with the solution constructed by front tracking.

◊ **Example 7.1 (A special system).**

As an example for this chapter we shall consider the special 2×2 system

$$\begin{aligned} u_t + \left(vu^2\right)_x &= 0, \\ v_t + \left(uv^2\right)_x &= 0. \end{aligned} \qquad (7.2)$$

---

[1] Hard to comprehend? It means "[but to] pursue virtue and knowledge."

For simplicity assume that $u > 0$ and $v > 0$. The Jacobian matrix reads

$$\begin{pmatrix} 2uv & u^2 \\ v^2 & 2uv \end{pmatrix}, \tag{7.3}$$

with eigenvalues and eigenvectors

$$\lambda_1 = uv, \qquad r_1 = \begin{pmatrix} -u/v \\ 1 \end{pmatrix},$$

$$\lambda_2 = 3uv, \qquad r_2 = \begin{pmatrix} u/v \\ 1 \end{pmatrix}. \tag{7.4}$$

The system is clearly strictly hyperbolic. Observe that

$$\nabla \lambda_1 \cdot r_1 = 0,$$

and hence the first family is linearly degenerate. The corresponding wave curve $W_1(u_l, v_l) = C_1(u_l, v_l)$ is given by (cf. Theorem 5.7)

$$\frac{du}{dv} = -\frac{u}{v}, \qquad u(v_l) = u_l,$$

or

$$W_1(u_l, v_l) = C_1(u_l, v_l) = \{(u, v) \mid uv = u_l v_l\}.$$

The corresponding eigenvalue λ_1 is constant along each hyperbola.

With the chosen normalization of r_2 we find that

$$\nabla \lambda_2 \cdot r_2 = 6u,$$

and hence the second-wave family is genuinely nonlinear. The rarefaction curves of the second family are solutions of

$$\frac{du}{dv} = \frac{u}{v}, \qquad u(v_l) = u_l,$$

and thus

$$\frac{u}{v} = \frac{u_l}{v_l}.$$

We see that these are straight lines emanating from the origin, and λ_2 increases as u increases. Consequently, R_2 consists of the ray

$$v = u \frac{v_l}{u_l}, \qquad u \geq u_l.$$

The rarefaction speed is given by

$$\lambda_2(u; u_l, v_l) = 3u^2 \frac{v_l}{u_l}.$$

To find the shocks in the second family, we use the Rankine–Hugoniot relation

$$s(u - u_l) = vu^2 - v_l u_l^2,$$
$$s(v - v_l) = v^2 u - v_l^2 u_l,$$

7. Well-Posedness of the Cauchy Problem

which implies

$$\frac{u}{u_l} = \frac{1}{2}\left(\frac{v}{v_l} + \frac{v_l}{v} \pm \left(\frac{v}{v_l} - \frac{v_l}{v}\right)\right) = \begin{cases} v_l/v, \\ v/v_l. \end{cases}$$

(Observe that the solution with $u/u_l = v_l/v$ coincides with the wave curve of the linearly degenerate first family.) The shock part of this curve S_2 consists of the line

$$S_2(u_l, v_l) = \left\{(u, v) \mid v = u\frac{v_l}{u_l}, \quad 0 < u \le u_l\right\}.$$

The shock speed is given by

$$s := \mu_2(u; u_l, v_l) = \left(u^2 + uu_l + u_l^2\right)\frac{v_l}{u_l}.$$

Hence the Hugoniot locus and rarefaction curves coincide for this system. Systems with this property are called *Temple class systems* after Temple [137]. Furthermore, the system is linearly degenerate in the first family and genuinely nonlinear in the second. Summing up, the solution of the Riemann problem for (7.2) is as follows: First the middle state is given by

$$u_m = \sqrt{u_l u_r \frac{v_l}{v_r}}, \quad v_m = \sqrt{v_l v_r \frac{u_l}{u_r}}.$$

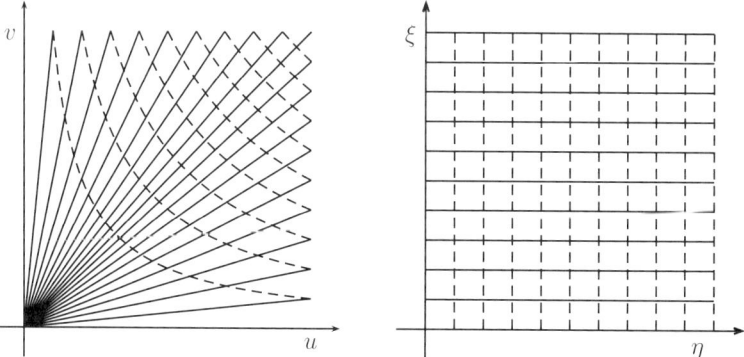

Figure 7.1. The curves W in (u, v) coordinates (left) and (η, ξ) coordinates (right).

If $u_l/v_l \leq u_r/v_r$, the second wave is a rarefaction wave, and the solution can be written as

$$\begin{pmatrix} u \\ v \end{pmatrix}(x,t) = \begin{cases} \begin{pmatrix} u_l \\ v_l \end{pmatrix} & \text{for } x/t \leq u_l v_l, \\ \begin{pmatrix} u_m \\ v_m \end{pmatrix} & \text{for } u_l v_l < x/t \leq 3 u_m v_m, \\ \sqrt{\frac{x}{3t}\frac{v_m}{u_m}} \begin{pmatrix} u_m/v_m \\ 1 \end{pmatrix} & \text{for } 3u_m v_m < x/t \leq 3 u_r v_r, \\ \begin{pmatrix} u_r \\ v_r \end{pmatrix} & \text{for } 3 u_r v_r < x/t. \end{cases} \quad (7.5)$$

In the shock case, that is, when $u_l/v_l > u_r/v_r$, the solution reads

$$\begin{pmatrix} u \\ v \end{pmatrix}(x,t) = \begin{cases} \begin{pmatrix} u_l \\ v_l \end{pmatrix} & \text{for } x/t \leq u_l v_l, \\ \begin{pmatrix} u_m \\ v_m \end{pmatrix} & \text{for } u_l v_l < x/t \leq \mu_2(u_r; u_m, v_m), \\ \begin{pmatrix} u_r \\ v_r \end{pmatrix} & \text{for } \mu_2(u_r; u_m, v_m) < x/t. \end{cases} \quad (7.6)$$

If we set

$$\eta = uv, \quad \xi = \frac{u}{v},$$

and thus

$$u = \sqrt{\eta \xi}, \quad v = \sqrt{\eta/\xi},$$

the solution of the Riemann problem will be especially simple in (η, ξ) coordinates. Given left and right states (η_l, ξ_l), (η_r, ξ_r), the middle state is given by (η_l, ξ_r). Consequently, measured in (η, ξ) coordinates, the total variation of the solution of the Riemann problem equals the total variation of the initial data. This means that we do not need the Glimm functional to show that a front-tracking approximation to the solution of (7.2) has bounded total variation. With this in mind it is easy to show (using the methods of the previous chapters) that there exists a weak solution to the initial value problem for (7.2) whenever the total variation of the initial data is bounded.

We may use these variables to parameterize the wave curves as follows:

$$\begin{pmatrix} u \\ v \end{pmatrix} = \begin{pmatrix} u_l v_l/\eta \\ \eta \end{pmatrix} \text{ (first family)},$$

$$\begin{pmatrix} u \\ v \end{pmatrix} = \begin{pmatrix} u_l \eta/v_l \\ \eta \end{pmatrix} \text{ (second family)}.$$

For future use we note that the rarefaction and shock speeds are as follows:
$$\lambda_1(\eta) = \mu_1(\eta) = \eta,$$
$$\lambda_2(\eta) = 3\eta, \quad \text{and} \quad \mu_2(\eta_l, \eta_r) = (\eta_l + \sqrt{\eta_l \eta_r} + \eta_r).$$

$\diamond$

As a reminder we now summarize some properties of the front-tracking approximation for a fixed δ.

1. For all positive times t, $u^\delta(x,t)$ has finitely many discontinuities, each having position $x_i(t)$. These discontinuities can be of two types: shock fronts or approximate rarefaction fronts. Furthermore, only finitely many interactions between discontinuities occur for $t \geq 0$.

2. Along each shock front, the left and right states
$$u_{l,r} = u^\delta(x_i \mp, t) \tag{7.7}$$
are related by
$$u_r = S_{\hat{\imath}}(\epsilon_i) u_l + e_i,$$
where ϵ_i is the strength of the shock and $\hat{\imath}$ is the family of the shock. The "error" e_i is a vector of small magnitude. Furthermore, the speed of the shock, $\dot{x}$, satisfies
$$|\dot{x} - \mu_{\hat{\imath}}(u_l, u_r)| \leq \mathcal{O}(1)\delta, \tag{7.8}$$
where $\mu_{\hat{\imath}}(u_l, u_r)$ is the $\hat{\imath}$th eigenvalue of the averaged matrix
$$M(u_l, u_r) = \int_0^1 df((1-\alpha)u_l + \alpha u_r)\, d\alpha;$$
cf. (5.64)–(5.65).

3. Along each rarefaction front, the values u_l and u_r are related by
$$u_r = R_{\hat{\imath}}(\epsilon_i) u_l + e_i. \tag{7.9}$$
Also,
$$|\dot{x} - \lambda_{\hat{\imath}}(u_r)| \leq \mathcal{O}(1)\delta \quad \text{and} \quad |\dot{x} - \lambda_{\hat{\imath}}(u_l)| \leq \mathcal{O}(1)\delta \tag{7.10}$$
where $\lambda_{\hat{\imath}}(u)$ is the $\hat{\imath}$th eigenvalue of $df(u)$.

4. The total magnitude of all errors is small:
$$\sum_i |e_i| \leq \delta. \tag{7.11}$$

Also, recall that for a suitable constant C_0 the Glimm functional
$$G(u^\delta(\,\cdot\,,t)) = T(u^\delta(\,\cdot\,,t)) + C_0 Q(u^\delta(\,\cdot\,,t))$$

is nonincreasing for each collision of fronts, where T and Q are defined by (6.19) and (6.18), respectively, and that the interaction potential

$$Q\left(u^\delta(\,\cdot\,,t)\right)$$

is strictly decreasing for each collision of fronts.

7.1 Stability

> Details are always vulgar.
>
> Oscar Wilde, The Picture of Dorian Gray (1891)

Now let v^δ be another front-tracking solution with initial condition v_0. To compare u^δ and v^δ in the L^1-norm, i.e., to estimate $\|u^\delta - v^\delta\|_1$, we introduce the vector $q = q(x,t) = (q_1,\dots,q_n)$ by

$$v^\delta(x,t) = H_n\left(q_n\right) H_{n-1}\left(q_{n-1}\right) \cdots H_1\left(q_1\right) u^\delta(x,t) \qquad (7.12)$$

and the intermediate states w_i,

$$w_0 = u^\delta(x,t), \quad w_i = H_i\left(q_i\right) w_{i-1}, \quad \text{for } 1 \le i \le n, \qquad (7.13)$$

with velocities

$$\mu_i = \mu_i(w_{i-1}, w_i). \qquad (7.14)$$

As in Chapter 5, $H_k(\epsilon)u$ denotes the kth Hugoniot curve through u, parameterized such that

$$\frac{d}{d\epsilon} H_k\left(\epsilon\right) u \Big|_{\epsilon=0} = r_k(u).$$

Note that in the definition of q we use *both* parts of this curve, not only the part where $\epsilon < 0$. The vector q represents a "solution" of the Riemann problem with left state u^δ and right state v^δ using only shocks. (For $\epsilon > 0$ these will be weak solutions; that is, they satisfy the Rankine–Hugoniot condition. However, they will not be Lax shocks.)

Later in this section we shall use the fact that genuine nonlinearity implies that $\mu_k\left(u, H_k(\epsilon)u\right)$ will be increasing in ϵ, i.e.,

$$\frac{d}{d\epsilon} \mu_k\left(u, H_k(\epsilon)u\right) \ge c > 0,$$

for some constant c depending only on f.

As our model problem showed, the L^1 distance is more difficult to control than the "q-distance." However, it turns out that even the q-distance is not quite enough, and we need to introduce a weighted form. We let $\mathcal{D}\left(u^\delta\right)$ and $\mathcal{D}\left(v^\delta\right)$ denote the sets of all discontinuities in u and v, respectively, and define the functional $\Phi\left(u^{\delta}, v^{\delta}\right)$ as

$$\Phi(u^{\delta}, v^{\delta}) = \sum_{k=1}^{n} \int_{-\infty}^{\infty} |q_k(x)| \, W_k(x) \, dx. \tag{7.15}$$

Here the weights W_k are defined as

$$W_k = 1 + \kappa_1 A_k + \kappa_2 \left(Q\left(u^{\delta}\right) + Q\left(v^{\delta}\right)\right), \tag{7.16}$$

where $Q\left(u^{\delta}\right)$ and $Q\left(v^{\delta}\right)$ are the interaction potentials of u^{δ} and v^{δ}, respectively; cf. (6.18). The quantity A_k is the *total strength of all waves in* u^{δ} *or* v^{δ} *that approach the k-wave* $q_k(x)$. More precisely, if the kth field is linearly degenerate, then

$$A_k(x) = \sum_{\substack{i, x_i < x \\ \hat{i} > k}} |\epsilon_i| + \sum_{\substack{i, x > x_i \\ \hat{i} < k}} |\epsilon_i|. \tag{7.17}$$

The summation is over all discontinuities $x_i \in \mathcal{D}\left(u^{\delta}\right) \cup \mathcal{D}(v^{\delta})$. If the kth field is genuinely nonlinear, we must also account for waves of the same family approaching each other, and define

$$A_k(x) = \sum_{\substack{i, x_i < x \\ \hat{i} > k}} |\epsilon_i| + \sum_{\substack{i, x > x_i \\ \hat{i} < k}} |\epsilon_i|$$

$$+ \begin{cases} \displaystyle\sum_{\substack{i \in \mathcal{D}(u^{\delta}) \\ \hat{i}=k, x_i < x}} |\epsilon_i| + \sum_{\substack{i \in \mathcal{D}(v^{\delta}) \\ \hat{i}=k, x < x_i}} |\epsilon_i| & \text{if } q_k(x) < 0, \\[2em] \displaystyle\sum_{\substack{i \in \mathcal{D}(v^{\delta}) \\ \hat{i}=k, x_i < x}} |\epsilon_i| + \sum_{\substack{i \in \mathcal{D}(u^{\delta}) \\ \hat{i}=k, x < x_i}} |\epsilon_i| & \text{if } q_k(x) > 0. \end{cases} \tag{7.18}$$

In plain words, a q_k shock is approached by k-waves in u^{δ} from the left, and k-waves in v^{δ} from the right. Similarly, a q_k rarefaction wave is approached by k-waves in v^{δ} from the left and k-waves in u^{δ} from the right.

Once the values of the constants κ_1 and κ_2 are determined, we will assume that the total variations of u^{δ} and v^{δ} are so small that

$$1 \leq W_k(x) \leq 2. \tag{7.19}$$

In this case we see that Φ is equivalent to the L^1 norm; i.e., there exists a finite constant C_1 such that

$$\frac{1}{C_1}\|u^{\delta} - v^{\delta}\|_1 \leq \Phi\left(u^{\delta}, v^{\delta}\right) \leq C_1 \|u^{\delta} - v^{\delta}\|_1. \tag{7.20}$$

We can also define, with obvious modifications, $\Phi(u^{\delta_1}(t), v^{\delta_2}(t))$ with two different parameters δ_1 and δ_2. Our first goal will be to show that

$$\Phi\left(u^{\delta_1}(t), v^{\delta_2}(t)\right) - \Phi\left(u^{\delta_1}(s), v^{\delta_2}(s)\right) \leq C_2(t-s)(\delta_1 \vee \delta_2), \tag{7.21}$$

for all $0 \le t \le s$. Once this inequality is in place we can show that the sequence of front-tracking approximations is a Cauchy sequence in L^1 for

$$\begin{aligned}\left\|u^{\delta_1}(t) - u^{\delta_2}(t)\right\|_1 &\le C_1 \Phi\left(u^{\delta_1}(t), u^{\delta_2}(t)\right) \\ &\le C_1 \Phi\left(u^{\delta_1}(0), u^{\delta_2}(0)\right) + C_1 C_2 t(\delta_1 \vee \delta_2) \\ &\le C_1^2 \left\|u^{\delta_1}(0) - u^{\delta_2}(0)\right\|_1 + C_1 C_2 t(\delta_1 \vee \delta_2).\end{aligned}$$

Letting δ_1 and δ_2 tend to zero, we have the convergence of the whole sequence, and not only a subsequence.

The first step in order to prove (7.21) is to choose κ_2 so large that the weights W_k do not increase when fronts in u^{δ_1} or v^{δ_2} collide. This is possible, since the total variations of both u^{δ_1} and v^{δ_2} are uniformly small; hence the terms $\kappa_1 A_k$ are uniformly bounded, and by the interaction estimate, Q decreases for all collisions. This ensures the inequalities (7.19).

Then we must examine how Φ changes between collisions. Observe that $\Phi(t)$ is piecewise linear and continuous in t. Let

$$\mathcal{D} = \mathcal{D}\left(u^{\delta_1}\right) \cup \mathcal{D}\left(v^{\delta_2}\right).$$

We differentiate Φ and find that

$$\begin{aligned}&\frac{d}{dt}\Phi\left(u^{\delta_1}, v^{\delta_2}\right) \\ &= \sum_{i \in \mathcal{D}} \sum_{k=1}^{n} \{|q_k(x_i-)|\, W_k(x_i-) - |q_k(x_i+)|\, W_k(x_i+)\} \dot{x}_i \\ &= \sum_{i \in \mathcal{D}} \sum_{k=1}^{n} \left\{ \left|q_k^{i,+}\right| W_k^{i,+}\left(\mu_k^{i,+} - \dot{x}_i\right) - \left|q_k^{i,-}\right| W_k^{i,-}\left(\mu_k^{i,-} - \dot{x}_i\right) \right\}, \\ &=: \sum_{i \in \mathcal{D}} \sum_{k=1}^{n} E_{i,k}, \end{aligned} \qquad (7.22)$$

where

$$\begin{aligned}\mu_k^{i,\pm} &= \mu_k(x_i\pm), & \mu_k(x) &= \mu_k(\omega_{k-1}(x), \omega_k(x)), \\ q_k^{i,\pm} &= q_k(x_i\pm), & \text{and} \quad W_k^{i,\pm} &= W_k(x_i\pm).\end{aligned}$$

The second equality in (7.22) is obtained by adding terms

$$\left|q_k^{i,-}\right| W_k^{i,-} \mu_k^{i,-} - \left|q_k^{(i-1),+}\right| W_k^{(i-1),+} \mu_k^{(i-1),+} = 0,$$

and observing that there are only a finite number of terms in the sum in (7.22).

◇ **Example 7.2 (Example 7.1 (cont'd.)).**

Let us check how this works for our special system. The two front-tracking approximations are denoted by u and v, and for simplicity we omit the superscript δ. These are made by approximating a rarefaction wave between $\eta_l = n\delta$ and $\eta_r = m\delta$, $m > n$, by a series of discontinuities with speed $3j\delta$, $j = n, \ldots, m-1$. In other words, we use the characteristic speed to the left of the discontinuity. The functions u and v are well-defined by standard techniques.

Since we managed this far without the interaction potential, we define the weights also without these (they are needed only to bound the weights, anyway). Hence for the example we use

$$W_k(x) = 1 + \kappa A_k(x). \tag{7.23}$$

Now we shall estimate

$$\frac{d}{dt}\Phi(u,v) = \sum_{i \in \mathcal{D}}(E_{i,1} + E_{i,2}). \tag{7.24}$$

To this end we consider a fixed discontinuity at x (to simplify the notation we do not use a subscript on this discontinuity) in one of the functions, say v. This discontinuity gives a contribution to the right-hand side of (7.24), denoted by $E_1 + E_2$, symmetry where

$$E_j = W_j^+ \, |q_j^+| \, (\mu_j^+ - \dot{x}) - W_j^- \, |q_j^-| \, (\mu_j^- - \dot{x}), \quad j = 1, 2.$$

For this 2×2 system we have

$$A_1(x) = \sum_{x_i < x,\, \hat{\imath}=2} |\epsilon_i|,$$

$$A_2(x) = \sum_{x_i > x,\, \hat{\imath}=1} |\epsilon_i|$$

$$+ \begin{cases} \displaystyle\sum_{\substack{\hat{\imath}=2,\, x_i < x \\ x_i \in \mathcal{D}(u)}} |\epsilon_i| + \sum_{\substack{\hat{\imath}=2,\, x_i > x \\ x_i \in \mathcal{D}(v)}} |\epsilon_i| & \text{if } q_2 < 0, \\[2ex] \displaystyle\sum_{\substack{\hat{\imath}=2,\, x_i < x \\ x_i \in \mathcal{D}(v)}} |\epsilon_i| + \sum_{\substack{\hat{\imath}=2,\, x_i > x \\ x_i \in \mathcal{D}(u)}} |\epsilon_i| & \text{if } q_2 > 0. \end{cases}$$

To estimate $E_1 + E_2$ we study several cases.

Case 1. Assume first that the jump at x is a *contact discontinuity*, that is, of the first family, in which case

$$A_1^+ = A_1^-,$$

and consequently,

$$W_1^+ = W_1^-. \tag{7.25}$$

Furthermore,

$$q_1^+ = q_1^- + \epsilon \quad \text{and} \quad \mu_1^+ = \mu_1^- = \dot{x} - q_2^-.$$

Then
$$E_1 = W_1^+ |q_1^+| (\mu_1^+ - \dot{x}) - W_1^- |q_1^-| (\mu_1^- - \dot{x})$$
$$= W_1^- \{|q_1^- + \epsilon| - |q_1^-|\} (-q_2^-)$$
$$\leq W_1^- |q_2^-| |\epsilon|. \qquad (7.26)$$

For the weights of the second family we find that
$$A_2^+ = A_2^- - |\epsilon|, \quad W_2^+ = W_2^- - \kappa |\epsilon|, \quad q_2^+ = q_2^-, \quad \mu_2^- = \mu_2^+.$$

To estimate $\mu_2^- - \dot{x}$ we exploit that μ_2^- is a discontinuity of the second family, while $\dot{x}$ is a contact discontinuity of the first family. Thus we can estimate from below their difference by the smallest difference in speeds between waves in the first- and second-wave families. We find that $\mu_2^- - \dot{x} \geq c = \min_{u,v} \{\eta\} > 0$. Hence
$$E_2 = W_2^+ |q_2^+| (\mu_2^+ - \dot{x}) - W_2^- |q_2^-| (\mu_2^- - \dot{x})$$
$$= |q_2^-| (\mu_2^- - \dot{x}) (-\kappa |\epsilon|)$$
$$\leq -\kappa c |q_2^-| |\epsilon|. \qquad (7.27)$$

Then
$$E_1 + E_2 = |q_2^-| |\epsilon| (W_1^- - \kappa c) \leq 0 \qquad (7.28)$$

if $\kappa c \geq \sup_x W_1(x)$. (Throughout this argument we will choose larger and larger κ.) This inequality, (7.28), is the desired estimate in the case where x is a contact discontinuity.

Case 2. The case where x is a *genuinely nonlinear wave*, that is, belongs to the second family, is more complicated. There are two distinct cases, that of an (approximate) rarefaction wave and that of a shock wave. First we treat the term E_1, which is common to the two cases. Here
$$A_1^+ = A_1^- + |\epsilon|, \quad W_1^+ = W_1^- + \kappa |\epsilon|, \quad q_1^+ = q_1^-,$$
$$\mu_1^+ = \mu_1^-, \quad \text{and} \quad \mu_1^- - \dot{x} < -c.$$

Consequently,
$$E_1 = W_1^+ |q_1^+| (\mu_1^+ - \dot{x}) - W_1^- |q_1^-| (\mu_1^- - \dot{x})$$
$$= \kappa |\epsilon| |q_1^-| (\mu_1^- - \dot{x})$$
$$\leq -\kappa c |q_1^-| |\epsilon| \leq 0. \qquad (7.29)$$

We split the estimate for E_2 into several cases.

Case 2a (rarefaction wave). First we consider the case where x is an approximate rarefaction wave. By the construction of v we have
$$\epsilon = \delta > 0 \quad \text{and} \quad q_2^+ = q_2^- + \epsilon.$$

7.1. Stability

The speeds appearing in E_2 are given by

$$\mu_2^+ = 2\eta_u + q_2^- + \epsilon + \sqrt{\eta_u\left(\eta_u + q_2^- + \epsilon\right)},$$
$$\mu_2^- = 2\eta_u + q_2^- + \sqrt{\eta_u\left(\eta_u + q_2^-\right)},$$
$$\dot{x} = 3\left(\eta_u + q_2^-\right).$$

We define the auxiliary speed

$$\tilde{\mu} = \mu_2\left(v^-, v^+\right) = 2\eta_u + 2q_2^- + \epsilon + \sqrt{\left(\eta_u + q_2^-\right)\left(\eta_u + q_2^- + \epsilon\right)}.$$

It is easily seen that

$$0 \le \epsilon \le \tilde{\mu} - \dot{x} \le 2\epsilon.$$

We have several subcases. First we assume that $q_2^- > 0$, in which case $q_2^+ > 0$ as well; see Figure 7.2. In this case $A_2^+ = A_2^- + |\epsilon|$. Hence

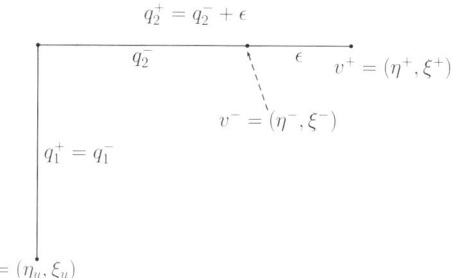

Figure 7.2. $q_2^- > 0$.

$$E_2 = W_2^+ \left|q_2^+\right|\left(\mu_2^+ - \dot{x}\right) - \left(W_2^+ - \kappa\left|\epsilon\right|\right)\left|q_2^-\right|\left(\mu_2^- - \dot{x}\right)$$
$$= W_2^+ \left\{\left(q_2^- + \epsilon\right)\left(\mu_2^+ - \tilde{\mu}\right) - q_2^-\left(\mu_2^- - \tilde{\mu}\right)\right\}$$
$$+ W_2^+ \left(q_2^+ - q_2^-\right)\left(\tilde{\mu} - \dot{x}\right) + \kappa\left|\epsilon\right|\left|q_2^-\right|\left(\mu_2^- - \dot{x}\right).$$

We need to estimate the term $\left\{\left(q_2^- + \epsilon\right)\left(\mu_2^+ - \tilde{\mu}\right) - q_2^-\left(\mu_2^- - \tilde{\mu}\right)\right\}$. This estimate is contained in Lemma 7.4 in the general case, and it is verified directly for this model right after the proof of Lemma 7.4. We obtain

$$\left|\left(q_2^- + \epsilon\right)\left(\mu_2^+ - \tilde{\mu}\right) - q_2^-\left(\mu_2^- - \tilde{\mu}\right)\right| \le \mathcal{O}\left(1\right)\left|\epsilon\right|\left|q_2^-\right|\left(\left|q_2^-\right| + \left|\epsilon\right|\right),$$

and thus

$$E_2 \le \mathcal{O}\left(1\right)\left|\epsilon\right|\left|q_2^-\right|\left(\left|q_2^-\right| + \left|\epsilon\right|\right) + W_2^+\left|\epsilon\right|\left|\tilde{\mu} - \dot{x}\right| + \kappa\left|\epsilon\right|\left|q_2^-\right|\left(\mu_2^- - \dot{x}\right)$$
$$\le \mathcal{O}\left(1\right)\left|\epsilon\right|\left|q_2^-\right|\left(\left|q_2^-\right| + \left|\epsilon\right|\right) + 2W_2^+\left|\epsilon\right|^2 + \kappa\left|\epsilon\right|\left|q_2^-\right|\left(\mu_2^- - \dot{x}\right).$$

We estimate $\mu_2^- - \dot{x} \le -q_2^- \le 0$, and hence

$$E_2 \le \left|\epsilon\right|\left|q_2^-\right|^2\left(\mathcal{O}\left(1\right) - \kappa\right) + \mathcal{O}\left(1\right)\left|\epsilon\right|^2\left|q_2^-\right| + \mathcal{O}\left(1\right)\left|\epsilon\right|^2 \le M\left|\epsilon\right|\delta,$$

for some constant M if we choose κ big enough. We have used that W_2^+ is bounded. Therefore,
$$E_1 + E_2 \leq M |\epsilon| \delta.$$

Now for the case where $q_2^- < 0$. Here we have two further subcases, $q_2^+ < 0$ and $q_2^+ > 0$. First we assume that $q_2^+ < 0$, and thus both q_2^- and q_2^+ are negative. Note that
$$|q_2^+| = |q_2^-| - |\epsilon|,$$
$$0 \leq -q_2^- \leq \mu_2^- - \dot{x} \leq -2q_2^-, \quad \text{and} \quad A_2^+ = A_2^- - |\epsilon|.$$

Thus
$$\begin{aligned}
E_2 &= \left(W_2^- - \kappa |\epsilon|\right) |q_2^+| \left(\mu_2^+ - \dot{x}\right) - W_2^- |q_2^-| \left(\mu_2^- - \dot{x}\right) \\
&= W_2^- \left\{(q_2^+ - \epsilon)(\mu_2^- - \tilde{\mu}) - q_2^+ (\mu_2^+ - \tilde{\mu})\right\} \\
&\quad - W_2^- |\epsilon| (\tilde{\mu} - \dot{x}) - \kappa |\epsilon| |q_2^+| \left(\mu_2^+ - \dot{x}\right) \\
&\leq \mathcal{O}(1) |\epsilon| |q_2^+| (|q_2^+| + |\epsilon|) + \mathcal{O}(1) |\epsilon|^2 - \kappa |\epsilon| |q_2^+|^2 \\
&\leq |\epsilon| |q_2^-|^2 \left(\mathcal{O}(1) - \kappa\right) + \mathcal{O}(1) |\epsilon|^2 \\
&\leq M |\epsilon| \delta,
\end{aligned}$$
where we have used Lemma 7.4 (with $\varepsilon = \epsilon$, $\varepsilon' = q_2^+$) and chosen κ sufficiently large. Thus we conclude that $E_1 + E_2 \leq M |\epsilon| \delta$ in this case as well.

Now for the last case in which $\epsilon > 0$, namely $q_2^- < 0 < q_2^+$. Since $q_2^+ = q_2^- + \epsilon$, we have
$$|q_2^+| \leq \delta, \qquad |q_2^-| \leq \delta.$$
Furthermore, $A_2^+ = A_2^-$, and thus $W_2^+ = W_2^-$. We see that
$$0 \leq -q_2^- \leq \mu_2^- - \dot{x} \leq -2q_2^-, \qquad \mu_2^+ - \dot{x} \leq 2\epsilon - q_2^-,$$
and hence
$$\begin{aligned}
E_2 &= W_2^+ \left\{q_2^+ \left(\mu_2^+ - \dot{x}\right) + |q_2^-| \left(\mu_2^- - \dot{x}\right)\right\} \\
&\leq W_2^+ \left\{q_2^+ (2|\epsilon| + q_2^-) + |q_2^-| \, 2 \, |q_2^-|\right\} \\
&\leq M |\epsilon| \delta,
\end{aligned}$$
for some constant M.

Case 2b (shock wave). When x is a shock front, we have $\epsilon < 0$. In this case
$$\dot{x} = \tilde{\mu} = \mu_2(v^-, v^+) = 2\eta_u + 2q_2^- + \epsilon + \sqrt{(\eta_u + q_2^-)(\eta_u + q_2^- + \epsilon)}.$$

We first consider the case where $q_2^- < 0$. Then
$$q_2^+ = q_2^- + \epsilon < 0, \quad |q_2^+| = |q_2^-| + |\epsilon|, \quad \text{and} \quad A_2^+ = A_2^- - |\epsilon|,$$

and we obtain
$$\begin{aligned}
E_2 &= \left(W_2^- - \kappa\,|\epsilon|\right)\left|q_2^+\right|\left(\mu_2^+ - \dot{x}\right) - W_2^-\left|q_2^-\right|\left(\mu_2^- - \dot{x}\right) \\
&= -W_2^-\left((q_2^- + \epsilon)(\mu_2^+ - \dot{x}) - q_2^-(\mu_2^- - \dot{x})\right) \\
&\quad - \kappa\,|\epsilon|\left(\left|q_2^-\right| + |\epsilon|\right)(\mu_2^+ - \dot{x}) \\
&\leq \mathcal{O}(1)\,|\epsilon|\left|q_2^-\right|\left(\left|q_2^-\right| + |\epsilon|\right) - \kappa\,|\epsilon|\left(\left|q_2^-\right| + |\epsilon|\right)\left|q_2^-\right| \\
&\leq |\epsilon|\left|q_2^-\right|\left(\left|q_2^-\right| + |\epsilon|\right)(\mathcal{O}(1) - \kappa) \leq 0.
\end{aligned}$$

Lemma 7.4 (with $\varepsilon' = \epsilon$, $\varepsilon = q_2^-$) implies
$$\left|(q_2^- + \epsilon)\left(\mu_2^+ - \dot{x}\right) - q_2^-\left(\mu_2^- - \dot{x}\right)\right| \leq \mathcal{O}(1)\,|\epsilon|\left|q_2^-\right|\left(\left|q_2^-\right| + |\epsilon|\right).$$

Furthermore,
$$\begin{aligned}
\mu_2^+ - \dot{x} &= -q_2^- + \sqrt{\eta_u\left(\eta_u + q_2^- + \epsilon\right)} \\
&\quad - \sqrt{\left(\eta_u + q_2^-\right)\left(\eta_u + q_2^- + \epsilon\right)} \\
&= -q_2^-\left(1 + \frac{\sqrt{\eta_u + q_2^+}}{\sqrt{\eta_u} + \sqrt{\eta_u + q_2^+}}\right) \\
&\geq -q_2^- = \left|q_2^-\right|.
\end{aligned}$$

If $q_2^- > 0$, then there are two further cases to be considered, depending on the sign of q_2^+. We first consider the case where $q_2^+ < 0$, and thus $q_2^+ < 0 < q_2^-$. Now $A_2^+ = A_2^-$. Furthermore,
$$\mu_2^- - \dot{x} \geq -2q_2^- \geq 0,$$
$$\mu_2^+ - \dot{x} = -q_2^-\left(1 + \frac{\sqrt{\eta_u + q_2^+}}{\sqrt{\eta_u} + \sqrt{\eta_u + q_2^+}}\right) < -\left|q_2^-\right|.$$

Thus
$$\mu_2^+ < \dot{x} < \mu_2^-,$$
and we easily obtain
$$E_2 = W_2^-\left\{\left|q_2^+\right|\left(\mu_2^+ - \dot{x}\right) - \left|q_2^-\right|\left(\mu_2^- - \dot{x}\right)\right\} < 0.$$

This leaves the final case where $q_2^\pm > 0$. In this case we have that $A_2^+ = A_2^- + |\epsilon|$. We still have
$$\mu_2^- - \dot{x} = -q_2^+\left(1 + \frac{\sqrt{\eta_u + q_2^-}}{\sqrt{\eta_u} + \sqrt{\eta_u + q_2^+}}\right) \leq -q_2^+ < 0,$$

and thus
$$\left|\dot{x} - \mu_2^-\right| \geq q_2^+.$$

Furthermore, by Lemma 7.4, we have that
$$\left|(q_2^- + \epsilon)(\mu_2^+ - \dot{x}) - q_2^-(\mu_2^- - \dot{x})\right| \leq \mathcal{O}(1)\left|q_2^+\right||\epsilon|\left(\left|q_2^+\right| + |\epsilon|\right).$$

Then we calculate
$$\begin{aligned}
E_2 &= W_2^+ \left|q_2^+\right|(\mu_2^+ - \dot{x}) - \left(W_2^+ - \kappa|\epsilon|\right)\left|q_2^-\right|(\mu_2^- - \dot{x}) \\
&= W_2^+ \left((q_2^- + \epsilon)(\mu_2^+ - \dot{x}) - q_2^-(\mu_2^- - \dot{x})\right) + \kappa|\epsilon|\left|q_2^-\right|(\mu_2^- - \dot{x}) \\
&\leq W_2^+ \left|q_2^+(\mu_2^+ - \dot{x}) - q_2^-(\mu_2^- - \dot{x})\right| - \kappa|\epsilon|\left|\mu_2^- - \dot{x}\right|\left|q_2^-\right| \\
&\leq \mathcal{O}(1)|\epsilon|\left|q_2^-\right|\left(\left|q_2^+\right| + |\epsilon|\right) - \kappa|\epsilon|\left|q_2^-\right|\left|q_2^+\right| \\
&\leq \mathcal{O}(1)|\epsilon|^2 + |\epsilon|\left|q_2^-\right|^2(\mathcal{O}(1) - \kappa) \\
&\leq M|\epsilon|\delta
\end{aligned}$$

if κ is sufficiently large. This is the last case.

Now we have shown that in all cases,
$$E_1 + E_2 \leq M|\epsilon|\delta.$$

Summing over all discontinuities in u and v we conclude that
$$\frac{d}{dt}\Phi(u,v) \leq C'\delta,$$

for some finite constant C' independent of δ. ◇

We shall now show that
$$\sum_{k=1}^{n} E_{i,k} \leq \mathcal{O}(1)|\epsilon_i|(\delta_1 \vee \delta_2) + \mathcal{O}(1)|e_i|, \tag{7.30}$$

and this estimate is easily seen to imply (7.21). To prove (7.30) we shall need some preliminary results:

Lemma 7.3. *Assume that the vectors* $\epsilon = (\epsilon_1, \ldots, \epsilon_n)$, $\epsilon' = (\epsilon'_1, \ldots, \epsilon'_n)$, *and* $\epsilon'' = (\epsilon''_1, \ldots, \epsilon''_n)$ *satisfy*
$$H(\epsilon)u = H(\epsilon'')H(\epsilon')u$$

for some vector u, where
$$H(\epsilon) = H_n(\epsilon_n)H_{n-1}(\epsilon_{n-1})\cdots H_1(\epsilon_1).$$

Then
$$\sum_{k=1}^{n}|\epsilon_k - \epsilon'_k - \epsilon''_k| = \mathcal{O}(1)\left(\sum_j |\epsilon'_j \epsilon''_j|\left(|\epsilon'_j| + |\epsilon''_j|\right) + \sum_{\substack{k,\ell \\ k \neq \ell}} |\epsilon'_j \epsilon''_\ell|\right). \tag{7.31}$$

If the scalar ϵ and the vector $\epsilon' = (\epsilon'_1, \ldots, \epsilon'_n)$ satisfy
$$R_\ell(\epsilon)u = H(\epsilon')u$$

7.1. Stability

where R_ℓ denotes the ℓth rarefaction curve, then

$$|\epsilon - \epsilon'_\ell| + \sum_{k \neq \ell} |\epsilon'_k| = \mathcal{O}(1) |\epsilon| \left(|\epsilon'_\ell| (|\epsilon| + |\epsilon'_\ell|) + \sum_{k \neq \ell} |\epsilon'_k| \right). \qquad (7.32)$$

Proof. The proof of this lemma is a straightforward modification of the proof of the interaction estimate (6.14). □

Lemma 7.4. *Let $\bar{\omega} \in \Omega$ be sufficiently small, and let ε and ε' be real numbers. Define*

$$\begin{aligned} \omega &= H_k(\varepsilon)\bar{\omega}, & \mu &= \mu_k(\bar{\omega}, \omega), \\ \omega' &= H_k(\varepsilon')\omega, & \mu' &= \mu_k(\omega, \omega'), \\ \omega'' &= H_k(\varepsilon + \varepsilon')\bar{\omega}, & \mu'' &= \mu_k(\bar{\omega}, \omega''). \end{aligned}$$

Then one has

$$|(\varepsilon + \varepsilon')(\mu'' - \mu') - \varepsilon(\mu - \mu')| \leq \mathcal{O}(1) |\varepsilon \varepsilon'| (|\varepsilon| + |\varepsilon'|). \qquad (7.33)$$

Proof. The proof of this is again in the spirit of the proof of the interaction estimate, equation (6.10). Let the function Ψ be defined as

$$\Psi(\varepsilon, \varepsilon') = (\varepsilon + \varepsilon') \mu'' - \varepsilon \mu - \varepsilon' \mu'.$$

Then Ψ is at least twice differentiable, and satisfies

$$\Psi(\varepsilon, 0) = \Psi(0, \varepsilon') = 0, \qquad \frac{\partial^2 \Psi}{\partial \varepsilon \partial \varepsilon'}(0, 0) = 0.$$

Consequently,

$$\Psi(\varepsilon, \varepsilon') = \int_0^\varepsilon \int_0^{\varepsilon'} \frac{\partial^2 \Psi}{\partial \varepsilon \partial \varepsilon'}(r, s) \, ds \, dr = \mathcal{O}(1) \int_0^{|\varepsilon|} \int_0^{|\varepsilon'|} (|r| + |s|) \, dr \, ds.$$

From this the lemma follows. □

◇ **Example 7.5 (Lemma 7.4 for Example 7.1).**

If $k = 2$, let $\bar{\omega}$, ω', and ω'' denote the η coordinate, since only this will influence the speeds. Then a straightforward calculation yields

$$|(\varepsilon + \varepsilon')(\mu'' - \mu') - \varepsilon(\mu - \mu')|$$
$$= |\varepsilon| |\varepsilon'| (|\varepsilon| + |\varepsilon'|)$$
$$\times \frac{\sqrt{\bar{\omega}} + \sqrt{\omega'} + \sqrt{\omega''}}{\bar{\omega}\left(\sqrt{\omega'} + \sqrt{\omega''}\right) + \omega'\left(\sqrt{\bar{\omega}} + \sqrt{\omega''}\right) + \omega''\left(\sqrt{\bar{\omega}} + \sqrt{\omega'}\right) + 2\sqrt{\bar{\omega}\omega'\omega''}}$$
$$\leq \frac{|\varepsilon| |\varepsilon'| (|\varepsilon| + |\varepsilon'|)}{\min\{\bar{\omega}, \omega', \omega''\}}$$

verifying the lemma in this case. ◇

If the kth characteristic field is genuinely nonlinear, then the characteristic speed $\lambda_k(H_k(\epsilon)\omega)$ is increasing in ϵ, and we can even choose the parameterization such that

$$\lambda_k(H_k(\epsilon)\omega) - \lambda_k(\omega) = \epsilon,$$

for all sufficiently small ϵ and ω. This also implies that $\mu_k(\omega, H_k(\epsilon)\omega)$ is strictly increasing in ϵ. However, the Hugoniot locus through the point ω does not in general coincide with the Hugoniot locus through the point $H_k(q)\omega$. Therefore, it is not so straightforward comparing speeds defined on different Hugoniot loci. When proving (7.30) we shall need to do this, and we repeatedly use the following lemma:

Lemma 7.6. *For some state ω define*

$$\Psi(q) = \mu_k(H_k(q)\omega, H_k(\epsilon)H_k(q)\omega) - \mu_k(\omega, H_k(\epsilon + q)\omega).$$

Then Ψ is at least twice differentiable for all $k = 1, \ldots, n$. Furthermore, if the kth characteristic field is genuinely nonlinear, then for sufficiently small $|q|$ and $|\epsilon|$,

$$\Psi'(q) \geq c > 0, \tag{7.34}$$

where c depends only on f for all sufficiently small $|\omega|$.

Proof. Let the vector ϵ' be defined by $\mathcal{H}(\epsilon')\omega = H_k(\epsilon)H_k(q)\omega$. Then by Lemma 7.3

$$|\epsilon'_k - (q + \epsilon)| + \sum_{i \neq k} |\epsilon'_i| \leq \mathcal{O}(1)|q\epsilon|(|\epsilon| + |q|).$$

Consequently,

$$H_k(\epsilon + q)\omega = H_k(\epsilon)H_k(q)\omega + \mathcal{O}(1)|q\epsilon|(|\epsilon| + |q|).$$

Using this we find that

$$\left|\frac{H_k(\epsilon)H_k(q)\omega - H_k(\epsilon)\omega}{q}\right| = \left|\frac{H_k(\epsilon + q)\omega - H_k(\epsilon)\omega}{q}\right| + \mathcal{O}(1)|\epsilon|(|\epsilon| + |q|).$$

Therefore,

$$\frac{d}{dq}\{H_k(\epsilon)H_k(q)\omega\}\bigg|_{q=0} = \frac{d}{d\epsilon}\{H_k(\epsilon)\omega\} + \mathcal{O}(1)|\epsilon|^2. \tag{7.35}$$

Hence, we compute

$$\Psi'(0) = \nabla_1 \mu_k(\omega, H_k(\epsilon)\omega) \cdot r_k(\omega)$$
$$- \nabla_2 \mu_k(\omega, H_k(\epsilon)\omega) \cdot \left(\frac{d}{d\epsilon}\{H_k(\epsilon)\omega\} - \frac{d}{dq}\{H_k(\epsilon)H_k(q)\omega\}\bigg|_{q=0}\right)$$
$$= \nabla_1 \mu_k(\omega, H_k(\epsilon)\omega) \cdot r_k(\omega) + \mathcal{O}(1)|\epsilon|^2$$
$$\geq c' > 0,$$

for sufficiently small $|\epsilon|$. The value of the constant c' (and its existence) depends on the genuine nonlinearity of the system and hence on f. Since Ψ' is continuous for small $|q|$, the lemma follows. $\square$

We shall prove (7.30) in the case where the front at x_i is a front in v^{δ_2}; the case where it is a front in u^{δ_1} is completely analogous. We therefore fix i, and study the relation between q_k^- and q_k^+. Since the front is going to be fixed from now on, we drop the subscript i. For simplicity we write $\delta = \delta_2$. Assume the the family of the front x is ℓ and the front has strength ϵ. The situation is as in Figure 7.3. A key observation is that we can regard the

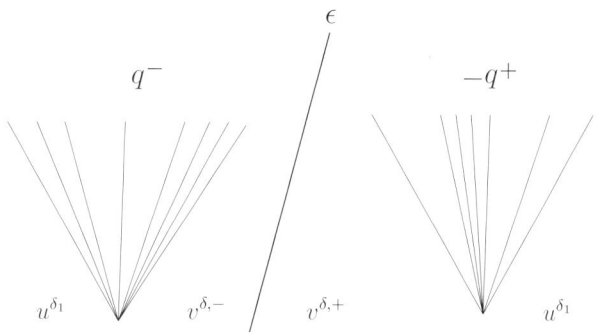

Figure 7.3. The setting in the proof of (7.30).

waves q_k^+ as being the result of an interaction between the waves q_k^- and ϵ; similarly, the waves $-q_k^-$ are the result of an interaction between ϵ and $-q_k^+$.

Regarding the weights, from (7.16) and (7.18) we find that

$$W_k^+ - W_k^- = \begin{cases} \kappa_1 |\epsilon| & \text{if } k < \ell, \\ -\kappa_1 |\epsilon| & \text{if } k > \ell, \end{cases} \quad (7.36)$$

while for $k = \ell$ we obtain

$$W_\ell^+ - W_\ell^- = \begin{cases} \kappa_1 |\epsilon| & \text{if } \min\{q_\ell^-, q_\ell^+\} > 0, \\ -\kappa_1 |\epsilon| & \text{if } \max\{q_\ell^-, q_\ell^+\} < 0, \\ \mathcal{O}(1) & \text{if } q_\ell^- q_\ell^+ < 0. \end{cases} \quad (7.37)$$

The proof of (7.30) is a study of cases. We split the estimate into two subgroups, depending on whether the front at x is an approximate rarefaction wave or a shock. Within each of subgroups we discuss three subcases depending on the signs of $q_\ell^\pm$. In all cases we discuss the terms E_k ($k \neq \ell$) and E_ℓ separately. For $k \neq \ell$ we write E_k (remember that we dropped the subscript i) as

$$E_k = \left(|q_k^+| - |q_k^-|\right) W_k^+ \left(\mu_k^+ - \dot{x}\right)$$

$$+ \left|q_k^-\right| \left(W_k^+ - W_k^-\right) \left(\mu_k^+ - \dot{x}\right) + \left|q_k^-\right| W_k^- \left(\mu_k^+ - \mu_k^-\right). \tag{7.38}$$

By the strict hyperbolicity of the system, we have that

$$\mu_k^+ - \dot{x} \leq -c < 0, \quad \text{for } k < \ell,$$
$$\mu_k^+ - \dot{x} \geq c > 0, \quad \text{for } k > \ell,$$

where c is some fixed constant depending on the system. Thus we always have that

$$\left(W_k^+ - W_k^-\right) \left(\mu_k^+ - \dot{x}\right) \leq -c\kappa_1 \left|\epsilon\right|, \quad k \neq \ell. \tag{7.39}$$

We begin with the case where the front at x is an approximate rarefaction wave ($\epsilon > 0$). In this case

$$R_\ell(\epsilon) v^{\delta,-} + e = H\left(q^+\right) u^{\delta_1} = H\left(q^+\right) H\left(-q^-\right) v^{\delta,-} = H\left(\tilde{q}\right) v^{\delta,-}$$

for some vector $\tilde{q}$. Hence

$$H\left(-q^-\right) v^{\delta,-} = H\left(-q^+\right) H\left(\tilde{q}\right) v^{\delta,-}, \tag{7.40}$$
$$R_\ell(\epsilon) v^{\delta,-} + e = H\left(\tilde{q}\right) v^{\delta,-}. \tag{7.41}$$

From (7.31) and (7.40) we obtain

$$\sum_k \left|q_k^+ - q_k^- - \tilde{q}_k\right|$$
$$= \mathcal{O}\left(1\right) \left(\sum_k \left|q_k^+ \tilde{q}_k\right| \left(\left|q_k^+\right| + \left|\tilde{q}_k\right|\right) + \sum_{\substack{k,j \\ k \neq j}} \left|q_k^+ \tilde{q}_j\right|\right), \tag{7.42}$$

and from (7.32) and (7.41) we obtain

$$\left|\tilde{q}_\ell - \epsilon\right| + \sum_{k \neq \ell} \left|\tilde{q}_k\right| = \mathcal{O}\left(1\right) \left|\epsilon\right| \left(\left|\tilde{q}_\ell\right| \left(\left|\tilde{q}_\ell\right| + \left|\epsilon\right|\right) + \sum_{k \neq \ell} \left|\tilde{q}_k\right|\right) + \mathcal{O}\left(1\right) \left|e\right|.$$

This implies that

$$\left|\tilde{q}_\ell - \epsilon\right| \leq \mathcal{O}\left(1\right) \left|\epsilon\right| + \mathcal{O}\left(1\right) \left|e\right|,$$
$$\sum_{k \neq \ell} \left|\tilde{q}_k\right| \leq \mathcal{O}\left(1\right) \left|\epsilon\right| + \mathcal{O}\left(1\right) \left|e\right|. \tag{7.43}$$

Furthermore, since ϵ is an approximate rarefaction, $0 \leq \epsilon \leq \delta$. Therefore, we can replace $\tilde{q}_\ell$ with ϵ and $\tilde{q}_k$ ($k \neq \ell$) with zero on the right-hand side of (7.42), making an error of $\mathcal{O}\left(1\right) \delta$. Indeed,

$$\left|q_\ell^+ - q_\ell^- - \epsilon\right| + \sum_{k \neq \ell} \left|q_k^+ - q_k^-\right|$$
$$\leq \sum_k \left|q_k^+ - q_k^- - \epsilon\right| + \left|\tilde{q}_\ell - \epsilon\right| + \sum_{k \neq \ell} \left|\tilde{q}_k\right|$$

$$\leq \mathcal{O}(1) \left(\sum_k |q_k^+ \tilde{q}_k| \left(|q_k^+| + |\tilde{q}_k|\right) + \sum_{\substack{k,j \\ k \neq j}} |q_k^+ \tilde{q}_j| \right)$$

$$+ \mathcal{O}(1) |\epsilon| \left(|\tilde{q}_\ell| (|\tilde{q}_\ell| + |\epsilon|) + \sum_{k \neq \ell} |\tilde{q}_k| \right) + \mathcal{O}(1) |e|.$$

Using (7.43) and the fact that $\epsilon \leq \delta$ we conclude that

$$\left| q_\ell^+ - q_\ell^- - \epsilon \right| + \sum_{k \neq \ell} \left| q_k^+ - q_k^- \right|$$
$$= \mathcal{O}(1) |\epsilon| \left(\delta + |q_\ell^+| \left(|q_\ell^+| + |\epsilon|\right) + \sum_{k \neq \ell} |q_k^+| \right) + \mathcal{O}(1) |e|. \tag{7.44}$$

Similarly,

$$\left| q_\ell^+ - q_\ell^- - \epsilon \right| + \sum_{k \neq \ell} \left| q_k^+ - q_k^- \right|$$
$$= \mathcal{O}(1) |\epsilon| \left(\delta + |q_\ell^-| \left(|q_\ell^-| + |\epsilon|\right) + \sum_{k \neq \ell} |q_k^-| \right) + \mathcal{O}(1) |e|. \tag{7.45}$$

Since in this case $0 \leq \epsilon \leq \delta$, and the total variation is small, we can assume that the right-hand sides of (7.44)–(7.45) are smaller than $\epsilon + \mathcal{O}(1)|e|$. Also, the error e is small; cf. (7.11). Then

$$0 < q_\ell^+ - q_\ell^- < 2\epsilon + \mathcal{O}(1) |e| \leq 2\delta + \mathcal{O}(1) |e|. \tag{7.46}$$

We can also use the estimates (7.44) and (7.45) to make a simplifying assumption throughout the rest of our calculations. Since the total variation of $u - v$ is uniformly bounded, we can assume that the right-hand sides of (7.44) and (7.45) are bounded by

$$\frac{1}{2} |\epsilon| + \mathcal{O}(1) |e|.$$

In particular, we then find that

$$\epsilon - \frac{1}{2} |\epsilon| - \mathcal{O}(1) |e| \leq q_\ell^+ - q_\ell^- \leq \epsilon + \frac{1}{2} |\epsilon| + \mathcal{O}(1) |e|.$$

Hence if $\epsilon > 0$, from the left inequality we find that

$$q_\ell^+ > q_\ell^-$$

or

$$|\epsilon| \leq \mathcal{O}(1) |e|,$$

and if $\epsilon < 0$, from the right inequality above,

$$q_\ell^+ < q_\ell^-$$

or
$$|\epsilon| \leq \mathcal{O}(1)|e|.$$

If $\epsilon > 0$ and $q_\ell^- \geq q_\ell^+$ or $\epsilon < 0$ and $q_\ell^+ \geq q_\ell^-$, then $|\epsilon| \leq \mathcal{O}(1)|e|$. In this case we find for $k \neq \ell$, or $k = \ell$ and $q_\ell^- q_\ell^+ > 0$, that

$$\begin{aligned}E_k &= \{|q_k^-|\,(W_k^- - W_k^+) + W_k^+\,(|q_k^-| - |q_k^+|)\}\,\dot{x} \\ &\leq \{|q_k^-|\,\kappa_1\,|\epsilon| + |W_k^+|\,(|\epsilon|/2 + \mathcal{O}(1)\,|e|)\}\,|\dot{x}| \\ &\leq \mathcal{O}(1)\,|e|.\end{aligned} \qquad (7.47)$$

If $k = \ell$ and $q_\ell^- q_\ell^+ < 0$, then for $\epsilon > 0$ we have that $q_\ell^+ - q_\ell^- \geq \mathcal{O}(1)|e|$, so if $q_\ell^+ < q_\ell^-$, we must have that

$$|q_\ell^+| \leq \mathcal{O}(1)\,|e| \quad \text{and} \quad q_\ell^- \leq \mathcal{O}(1)\,|e|.$$

Similarly, if $\epsilon < 0$ and $q_\ell^+ > q_\ell^-$, we obtain

$$q_\ell^+ < \mathcal{O}(1)\,|e| \quad \text{and} \quad |q_\ell^-| \leq \mathcal{O}(1)\,|e|.$$

Then we find that

$$E_\ell = \{|q_\ell^-|\,W_\ell^- - |q_\ell^+|\,W_\ell^+\}\,\dot{x} \leq \mathcal{O}(1)\,|e|. \qquad (7.48)$$

These observations imply that if $|\epsilon| = \mathcal{O}(1)|e|$, we have that

$$\sum_k E_k = \mathcal{O}(1)\,|e|,$$

which is what we want to show. Thus in the following we can assume that either

$$\epsilon > 0 \quad \text{and} \quad q_\ell^+ > q_\ell^-,$$

or

$$\epsilon < 0 \quad \text{and} \quad q_\ell^+ < q_\ell^-. \qquad (7.49)$$

Now follows a discussion of several different cases, depending on whether the front is an approximate rarefaction wave or a shock wave, and on the signs of q_ℓ^- and q_ℓ^+.

Case R1: $0 < q_\ell^- < q_\ell^+$, $\epsilon > 0$.
For $k \neq \ell$ we recall (7.38) that

$$\begin{aligned}E_k &= (|q_k^+| - |q_k^-|)\,W_k^+\,(\mu_k^+ - \dot{x}) \\ &\quad + |q_k^-|\,(W_k^+ - W_k^-)\,(\mu_k^+ - \dot{x}) + |q_k^-|\,W_k^-\,(\mu_k^+ - \mu_k^-).\end{aligned} \qquad (7.50)$$

The second term in (7.50) is less than or equal to (cf. (7.39))

$$-c\kappa_1\,|q_k^-|\,|\epsilon|.$$

Furthermore, by (7.45),
$$\left|q_k^+\right| - \left|q_k^-\right| \leq \mathcal{O}\left(1\right)|\epsilon|\left(\delta + \left|q_\ell^-\right|\left(\left|q_\ell^-\right| + |\epsilon|\right) + \sum_{k \neq \ell}\left|q_k^-\right|\right) + \mathcal{O}\left(1\right)|e|.$$

By the continuity of μ_k,
$$\left|\mu_k^+ - \mu_k^-\right| = \mathcal{O}\left(1\right)\left(|\epsilon| + |e|\right).$$

Hence from (7.38), we find that
$$E_k \leq \mathcal{O}\left(1\right)|\epsilon|\left(\delta + \left|q_\ell^-\right|\left(\left|q_\ell^-\right| + |\epsilon|\right) + \sum_{\tilde{k} \neq \ell}\left|q_{\tilde{k}}^-\right|\right) + \mathcal{O}\left(1\right)|e| - c\kappa_1 \left|q_k^-\right| |\epsilon|$$
$$\leq \mathcal{O}\left(1\right)|\epsilon|\left(\delta + \sum_{\tilde{k} \neq \ell}\left|q_{\tilde{k}}^-\right|\right) + \mathcal{O}\left(1\right)|e|$$
$$- c\kappa_1 |\epsilon|\left|q_k^-\right| + \mathcal{O}\left(1\right)|\epsilon|\left|q_\ell^-\right|\left(\left|q_\ell^-\right| + |\epsilon|\right). \tag{7.51}$$

For $k = \ell$ the situation is more complicated. We define states and speeds
$$\begin{aligned}\tilde{\omega}_\ell &= H_\ell\left(q_\ell^- + \epsilon\right)\omega_{\ell-1}^-, & \tilde{\mu}_\ell &= \mu_\ell\left(\omega_{\ell-1}^-, \tilde{\omega}_\ell\right), \\ \omega_\ell^\star &= H_\ell\left(\epsilon\right)\omega_\ell^-, & \mu_\ell^\star &= \mu_\ell\left(\omega_\ell^-, \omega_\ell^\star\right);\end{aligned} \tag{7.52}$$

see Figure 7.4. Remember that

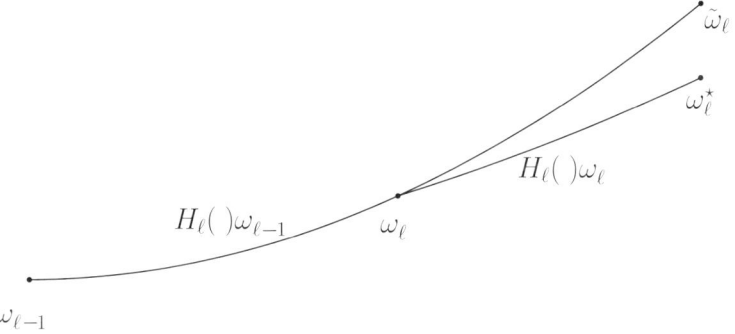

Figure 7.4. The situation for $0 < q_\ell^- < q_\ell^+$, $\epsilon > 0$, and $k = \ell$.

$$\mu_\ell^\pm = \mu_\ell\left(\omega_{\ell-1}^\pm, \omega_\ell^\pm\right).$$

Now by Lemma 7.4, with $\omega = \omega_{\ell-1}^-$, $\varepsilon = q_\ell^-$, and $\varepsilon'' = q_\ell^- + \epsilon$,
$$\left|\left(q_\ell^- + \epsilon\right)\left(\tilde{\mu}_\ell - \mu_\ell^\star\right) - q_\ell^-\left(\mu_\ell^- - \mu_\ell^\star\right)\right| = \mathcal{O}\left(1\right)\left|q_\ell^-\right||\epsilon|\left(\left|q_\ell^-\right| + |\epsilon|\right). \tag{7.53}$$

We also find that (cf. (7.10) and the fact that $\mu_\ell(u, u) = \lambda_\ell(u)$)
$$\left|\mu_\ell^\star - \dot{x}\right| \leq \left|\mu_\ell\left(\omega_\ell^-, \omega_\ell^\star\right) - \mu_\ell\left(v^{\delta,-}, v^{\delta,-}\right)\right| + \mathcal{O}\left(1\right)\delta$$

$$= \left|\mu_\ell\left(\omega_\ell^-, H_\ell(\epsilon)\omega_\ell^-\right) - \mu_\ell\left(\omega_n^-, \omega_n^-\right)\right| + \mathcal{O}\left(1\right)\delta$$
$$\leq \left|\mu_\ell\left(\omega_\ell^-, H_\ell(\epsilon)\omega_\ell^-\right) - \mu_\ell(\omega_\ell^-, \omega_\ell^-)\right|$$
$$+ \left|\mu_\ell(\omega_\ell^-, \omega_\ell^-) - \mu_\ell(\omega_\ell^-, \omega_{\ell+1}^-)\right|$$
$$+ \left|\mu_\ell(\omega_\ell^-, \omega_{\ell+1}^-) - \mu_\ell(\omega_{\ell+1}^-, \omega_{\ell+2}^-)\right| + \cdots$$
$$+ \left|\mu_\ell(\omega_{n-1}^-, \omega_n^-) - \mu_\ell(\omega_n^-, \omega_n^-)\right| + \mathcal{O}\left(1\right)\delta$$
$$\leq \mathcal{O}\left(1\right)\left(|\epsilon| + \left|\omega_\ell^- - \omega_{\ell+1}^-\right| + \cdots + \left|\omega_{n-1}^-, \omega_n^-\right|\right) + \mathcal{O}\left(1\right)\delta$$
$$\leq \mathcal{O}\left(1\right)\left(|\delta| + |q_\ell^-| + \sum_{k > \ell}|q_k^-|\right). \tag{7.54}$$

Furthermore,

$$\left|\mu_\ell^+ - \tilde{\mu}_\ell\right| = \left|\mu_\ell\left(\omega_{\ell-1}^+, H_\ell\left(q_\ell^+\right)\omega_{\ell-1}^+\right) - \mu_\ell\left(\omega_{\ell-1}^-, H_\ell\left(q_\ell^- + \epsilon\right)\omega_{\ell-1}^-\right)\right|$$
$$\leq \left|\mu_\ell\left(\omega_{\ell-1}^+, H_\ell\left(q_\ell^+\right)\omega_{\ell-1}^+\right) - \mu_\ell\left(H_\ell\left(q_\ell^+\right)\omega_{\ell-1}^+, \omega_{\ell-1}^-\right)\right|$$
$$+ \left|\mu_\ell\left(H_\ell\left(q_\ell^+\right)\omega_{\ell-1}^+, \omega_{\ell-1}^-\right) - \mu_\ell\left(\omega_{\ell-1}^+, H_\ell\left(q_\ell^- + \epsilon\right)\omega_{\ell-1}^-\right)\right|$$
$$\leq \mathcal{O}\left(1\right)\left(\left|\omega_{\ell-1}^+ - \omega_{\ell-1}^-\right| + \left|H_\ell\left(q_\ell^+\right)\omega_{\ell-1}^+ - H_\ell\left(q_\ell^- + \epsilon\right)\omega_{\ell-1}^-\right|\right)$$
$$\leq \mathcal{O}\left(1\right)\left(\left|\omega_{\ell-1}^+ - \omega_{\ell-1}^-\right| + \left|H_\ell\left(q_\ell^+\right)\omega_{\ell-1}^+ - H_\ell\left(q_\ell^- + \epsilon\right)\omega_{\ell-1}^+\right|\right.$$
$$\left. + \left|H_\ell\left(q_\ell^- + \epsilon\right)\omega_{\ell-1}^+ - H_\ell\left(q_\ell^- + \epsilon\right)\omega_{\ell-1}^+\right|\right)$$
$$\leq \mathcal{O}\left(1\right)\left(\left|\omega_{\ell-1}^+ - \omega_{\ell-1}^-\right| + \left|q_\ell^+ - q_\ell^- - \epsilon\right|\right)$$
$$\leq \mathcal{O}\left(1\right)\left(\left|q_{\ell-2}^+ - q_{\ell-2}^-\right| + \cdots + \left|q_1^+ - q_1^-\right| + \left|q_\ell^+ - q_\ell^- - \epsilon\right|\right)$$
$$= \mathcal{O}\left(1\right)\epsilon\left(\delta + |q_\ell^-|\left(|q_\ell^-| + |\epsilon|\right) + \sum_{k \neq \ell}|q_k^-|\right) + \mathcal{O}\left(1\right)|e|. \tag{7.55}$$

Since the ℓth field is genuinely nonlinear, then by Lemma 7.6

$$\mu_\ell^\star - \tilde{\mu}_\ell \geq c\left|q_\ell^-\right| \tag{7.56}$$

for some constant $c > 0$ depending only on the system. Remember that in this case

$$W_\ell^+ = W_\ell^- + \kappa_1 |\epsilon|.$$

Moreover, ϵ, q_ℓ^+, and q_ℓ^- are positive. Using the above inequalities, we compute

$$E_\ell = W_\ell^+ q_\ell^+ \left(\mu_\ell^+ - \dot{x}\right) - W_\ell^- q_\ell^- \left(\mu_\ell^- - \dot{x}\right)$$
$$= \left(W_\ell^- + \kappa_1 |\epsilon|\right) q_\ell^+ \left(\mu_\ell^+ - \dot{x}\right) - W_\ell^- q_\ell^- \left(\mu_\ell^- - \dot{x}\right)$$
$$= \kappa_1 \epsilon q_\ell^+ \left(\mu_\ell^+ - \dot{x}\right) + W_\ell^- \left\{q_\ell^+ \left(\mu_\ell^+ - \dot{x}\right) - q_\ell^- \left(\mu_\ell^- - \dot{x}\right)\right\}$$
$$= \kappa_1 \epsilon \left\{\left(q_\ell^- + \epsilon\right)\left(\tilde{\mu}_\ell - \mu_\ell^\star\right) + q_\ell^+ \left(\mu_\ell^+ - \dot{x}\right) - \left(q_\ell^- + \epsilon\right)\left(\tilde{\mu}_\ell - \mu_\ell^\star\right)\right\}$$
$$+ W_\ell^- \left\{q_\ell^+ \left(\mu_\ell^+ - \dot{x}\right) - q_\ell^- \left(\mu_\ell^- - \dot{x}\right)\right\}$$
$$= \kappa_1 \epsilon \left(q_\ell^- + \epsilon\right)\left(\tilde{\mu}_\ell - \mu_\ell^\star\right)$$
$$+ \kappa_1 \epsilon \left\{\left(q_\ell^- + \epsilon\right)\left(\mu_\ell^+ - \dot{x} - \left(\tilde{\mu}_\ell - \mu_\ell^\star\right)\right) + \left(q_\ell^+ - q_\ell^- - \epsilon\right)\left(\mu_\ell^+ - \dot{x}\right)\right\}$$

$$
\begin{aligned}
&+ W_\ell^- \left\{ q_\ell^+ \left(\mu_\ell^+ - \dot{x} \right) - q_\ell^- \left(\mu_\ell^- - \dot{x} \right) \right\} \\
&\leq \kappa_1 \epsilon \left(q_\ell^- + \epsilon \right) \left(\tilde{\mu}_\ell - \mu_\ell^\star \right) + \kappa_1 \epsilon \left(q_\ell^- + \epsilon \right) \left(|\mu_\ell^+ - \tilde{\mu}_\ell| + |\mu_\ell^\star - \dot{x}| \right) \\
&\quad + \kappa_1 \epsilon \left| q_\ell^+ - q_\ell^- - \epsilon \right| |\mu_\ell^+ - \dot{x}| + W_\ell^- \left\{ q_\ell^+ \left(\mu_\ell^+ - \dot{x} \right) - q_\ell^- \left(\mu_\ell^- - \dot{x} \right) \right\} \\
&\leq -c \kappa_1 q_\ell^- \epsilon \left(q_\ell^- + \epsilon \right) \\
&\quad + \kappa_1 \epsilon \left(q_\ell^- + \epsilon \right) \left(\mathcal{O}(1) \epsilon \left(\delta + q_\ell^- \left(q_\ell^- + \epsilon \right) + \sum_{k \neq \ell} |q_k^-| \right) \right. \\
&\quad \left. + \delta + \mathcal{O}(1) \sum_{k > \ell} |q_k^-| + \mathcal{O}(1) |e| \right) \\
&\quad + \mathcal{O}(1) \kappa_1 \epsilon^2 \left(\delta + q_\ell^- \left(q_\ell^- + \epsilon \right) + \sum_{k \neq \ell} |q_k^-| \right) + \mathcal{O}(1) |e| \\
&\quad + W_\ell^- \left\{ q_\ell^+ \left(\mu_\ell^+ - \dot{x} \right) - q_\ell^- \left(\mu_\ell^- - \dot{x} \right) \right\} \\
&\leq -c \kappa_1 q_\ell^- \epsilon \left(q_\ell^- + \epsilon \right) + \mathcal{O}(1) \kappa_1 \epsilon \left(\delta + q_\ell^- \left(q_\ell^- + \epsilon \right) + \sum_{k \neq \ell} |q_k^-| \right) + \mathcal{O}(1) |e| \\
&\quad + W_\ell^- \left\{ \left| \left(q_\ell^- + \epsilon \right) \left(\tilde{\mu}_\ell - \mu_\ell^\star \right) - q_\ell^- \left(\mu_\ell^- - \mu_\ell^\star \right) \right| + \left| q_\ell^+ - q_\ell^- - \epsilon \right| |\mu_\ell^+ - \dot{x}| \right. \\
&\quad \left. \epsilon |\mu_\ell^\star - \dot{x}| + \left(q_\ell^- + \epsilon \right) |\mu_\ell^+ - \tilde{\mu}_\ell| \right\} \\
&\leq -c \kappa_1 |q_\ell^-| |\epsilon| \left(|q_\ell^-| + |\epsilon| \right) + \mathcal{O}(1) |\epsilon| \left(\delta + |q_\ell^-| \left(|q_\ell^-| + |\epsilon| \right) + \sum_{k \neq \ell} |q_k^-| \right) \\
&\quad + \mathcal{O}(1) |e| \\
&\leq \mathcal{O}(1) |\epsilon| \left(\delta + \sum_{k \neq \ell} |q_k^-| \right) + \mathcal{O}(1) |e| + |\epsilon| |q_\ell^-| \left(|q_\ell^-| + |\epsilon| \right) \left(\mathcal{O}(1) - c \kappa_1 \right).
\end{aligned}
$$
(7.57)

Adding (7.57) and (7.51) we obtain
$$
\begin{aligned}
\sum_k E_k &= E_\ell + \sum_{k \neq \ell} E_k \\
&\leq \mathcal{O}(1) \epsilon \delta + \mathcal{O}(1) |e| + \epsilon \sum_{k \neq \ell} |q_k^-| \left(\mathcal{O}(1) - c \kappa_1 \right) \\
&\quad + \epsilon |q_\ell^-| \left(|q_\ell^-| + \epsilon \right) \left(\mathcal{O}(1) - c \kappa_1 \right) \\
&\leq \mathcal{O}(1) \epsilon \delta + \mathcal{O}(1) |e|,
\end{aligned}
$$
(7.58)

which holds for sufficiently large κ_1. This implies (7.30) in Case R1.

Case R2: $q_\ell^- < q_\ell^+ < 0$, $\epsilon > 0$.
Writing E_k as in (7.38), and using (7.44) (instead of (7.45) as in the previous case), we find for $k \neq \ell$ that

7. Well-Posedness of the Cauchy Problem

$$E_k \leq \mathcal{O}(1) \left(\delta + |q_\ell^+| \left(|q_\ell^+| + |\epsilon| \right) + \sum_{k \neq \ell} |q_k^+| \right) + \mathcal{O}(1) |\epsilon| - c\kappa_1 |q_k^+| |\epsilon|. \tag{7.59}$$

For $k = \ell$ the situation is similar to the previous case. We define auxiliary states and speeds

$$\begin{aligned} \tilde{w}_\ell &= H_\ell \left(q_\ell^+ - \epsilon \right) w_{\ell-1}^+ & \tilde{\mu}_\ell &= \mu_\ell \left(w_{\ell-1}^+, \tilde{w}_\ell \right), \\ w_\ell^\star &= H_\ell(-\epsilon) w_\ell^+ & \mu_\ell^\star &= \mu_\ell \left(w_\ell^+, w_\ell^\star \right); \end{aligned} \tag{7.60}$$

see Figure 7.5. Remember that

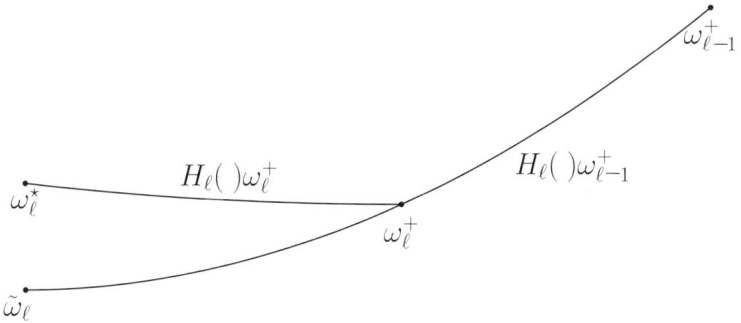

Figure 7.5. The situation for $q_\ell^- < q_\ell^+ < 0$, $\epsilon > 0$, and $k = \ell$.

$$w_\ell^+ = H_\ell \left(q_\ell^+ \right) w_{\ell-1}^+ \quad \text{and} \quad \mu_\ell^+ = \mu_\ell \left(w_{\ell-1}^+, w_\ell^+ \right).$$

In this case we use (7.33) with $\bar{\omega} = w_\ell^+$, $\varepsilon = q_\ell^+$, and $\varepsilon' = -\epsilon$. This gives

$$\left| \left(q_\ell^+ - \epsilon \right) \left(\tilde{\mu}_\ell - \mu_\ell^\star \right) - q_\ell^+ \left(\mu_\ell^+ - \mu_\ell^\star \right) \right| = \mathcal{O}(1) |q_\ell^+| |\epsilon| \left(|q_\ell^+| + |\epsilon| \right). \tag{7.61}$$

As in (7.54), we find that

$$\begin{aligned} |\mu_\ell^\star - \dot{x}| &\leq \left| \mu_\ell \left(w_\ell^+, w_\ell^\star \right) - \mu_l \left(v^{\delta,+}, v^{\delta,+} \right) \right| + \mathcal{O}(1) \delta \\ &= \left| \mu_\ell \left(w_\ell^+, H_\ell(-\epsilon) w_\ell^+ \right) - \mu_\ell \left(v^{\delta,+}, v^{\delta,+} \right) \right| + \mathcal{O}(1) \delta \\ &\leq \mathcal{O}(1) \delta + \mathcal{O}(1) \left(|w_\ell^+ - w_n^+| + \epsilon \right) \\ &\leq \mathcal{O}(1) \delta + \mathcal{O}(1) \sum_{k \neq \ell} |q_k^+|. \end{aligned} \tag{7.62}$$

We also obtain the analogue of (7.55), namely,

$$\begin{aligned} |\mu_\ell^- - \tilde{\mu}_\ell| &= \left| \mu_\ell \left(w_{\ell-1}^-, H_\ell \left(q_\ell^- \right) w_{\ell-1}^- \right) - \mu_\ell \left(w_{\ell-1}^+, H_\ell \left(q_\ell^+ - \epsilon \right) w_{\ell-1}^+ \right) \right| \\ &= \mathcal{O}(1) \left(|w_{\ell-1}^- - w_{\ell-1}^+| + |q_\ell^+ - q_\ell^- - \epsilon| \right) \\ &= \mathcal{O}(1) \left(\delta + |q_\ell^+| \left(|q_\ell^+| + |\epsilon| \right) + \sum_{k \neq \ell} |q_k^+| \right) + \mathcal{O}(1) |\epsilon|. \tag{7.63} \end{aligned}$$

7.1. Stability

By genuine nonlinearity, using Lemma 7.6, we find that
$$\tilde{\mu}_\ell - \mu_\ell^\star > c \left| q_\ell^+ \right|, \tag{7.64}$$
for some constant c. Now
$$W_\ell^- = W_\ell^+ + \kappa_1 |\epsilon|.$$
Using the above estimates (7.61)–(7.64), we compute
$$\begin{aligned}
E_\ell &= W_\ell^+ \left| q_\ell^+ \right| (\mu_\ell^+ - \dot{x}) - (W_\ell^+ + \kappa_1 \epsilon) \left| q_\ell^- \right| (\mu_\ell^- - \dot{x}) \\
&= -\kappa_1 \epsilon \left| q_\ell^- \right| (\mu_\ell^- - \dot{x}) - W_\ell^+ \left\{ q_\ell^+ (\mu_\ell^+ - \dot{x}) - q_\ell^- (\mu_\ell^- - \dot{x}) \right\} \\
&\leq -\kappa_1 \epsilon \left(\left| q_\ell^+ \right| + \epsilon \right) (\tilde{\mu}_\ell - \mu_\ell^\star) + \kappa_1 \epsilon \left(\left| q_\ell^+ \right| + \epsilon \right) \left(\left| \mu_\ell^- - \tilde{\mu}_\ell \right| + \left| \dot{x} - \mu_\ell^\star \right| \right) \\
&\quad + \kappa_1 \epsilon \left| q_\ell^+ - q_\ell^- - \epsilon \right| \left| \mu_\ell^- - \dot{x} \right| - W_\ell^+ \left\{ q_\ell^+ (\mu_\ell^+ - \dot{x}) - q_\ell^- (\mu_\ell^- - \dot{x}) \right\} \\
&\leq -c\epsilon \kappa_1 \left| q_\ell^+ \right| \left(\left| q_\ell^+ \right| + \epsilon \right) \\
&\quad + \mathcal{O}(1) \kappa_1 \epsilon \left(\left| q_\ell^+ \right| + \epsilon \right) \left(\delta + \left| q_\ell^+ \right| \left(\left| q_\ell^+ \right| + \epsilon \right) + \sum_{k \neq \ell} \left| q_k^+ \right| + |\epsilon| \right) \\
&\quad - W_\ell^+ \left\{ q_\ell^+ (\mu_\ell^+ - \dot{x}) - q_\ell^- (\mu_\ell^- - \dot{x}) \right\} \\
&\leq -c\epsilon \kappa_1 \left| q_\ell^+ \right| \left(\left| q_\ell^+ \right| + \epsilon \right) \\
&\quad + \mathcal{O}(1) \kappa_1 \epsilon \left(\left| q_\ell^+ \right| + \epsilon \right) \left(\delta + \left| q_\ell^+ \right| \left(\left| q_\ell^+ \right| + \epsilon \right) + \sum_{k \neq \ell} \left| q_k^+ \right| + |\epsilon| \right) \\
&\quad + W_\ell^+ \Big\{ \left| \left| q_\ell^+ \right| (\mu_\ell^+ - \mu_\ell^\star) - \left(\left| q_\ell^+ \right| + \epsilon \right) (\tilde{\mu}_\ell - \mu_\ell^\star) \right| \\
&\quad + \left| q_\ell^+ - q_\ell^- - \epsilon \right| \left| \mu_\ell^- - \dot{x} \right| + \epsilon \left| \mu_\ell^\star - \dot{x} \right| + \left(\left| q_\ell^+ \right| + \epsilon \right) \left| \mu_\ell^- - \tilde{\mu}_\ell \right| \Big\} \\
&\leq -c\epsilon \kappa_1 \left| q_\ell^+ \right| \left(\left| q_\ell^+ \right| + \epsilon \right) + \mathcal{O}(1) \epsilon \left(\delta + \left| q_\ell^+ \right| \left(\left| q_\ell^+ \right| + \epsilon \right) + \sum_{k \neq \ell} \left| q_k^+ \right| \right) \\
&\quad + \mathcal{O}(1) |\epsilon| \\
&\leq \mathcal{O}(1) \epsilon \left(\delta + \sum_{k \neq \ell} \left| q_k^+ \right| \right) + \mathcal{O}(1) |\epsilon| + \epsilon \left| q_\ell^+ \right| \left(\left| q_\ell^+ \right| + \epsilon \right) \left(\mathcal{O}(1) - c\kappa_1 \right).
\end{aligned} \tag{7.65}$$

Finally,
$$\begin{aligned}
\sum_k E_k &= E_\ell + \sum_{k \neq \ell} E_k \\
&\leq \mathcal{O}(1) \epsilon \delta + \mathcal{O}(1) |\epsilon| + \epsilon \sum_{k \neq \ell} \left| q_k^+ \right| \left(\mathcal{O}(1) - c\kappa_1 \right) \\
&\quad + \epsilon \left| q_\ell^+ \right| \left(\left| q_\ell^+ \right| + \epsilon \right) \left(\mathcal{O}(1) - c\kappa_1 \right) \\
&\leq \mathcal{O}(1) \epsilon \delta + \mathcal{O}(1) |\epsilon|
\end{aligned} \tag{7.66}$$

by choosing κ_1 larger if necessary. Hence (7.30) holds in this case as well.

Case R3: $q_\ell^- < 0 < q_\ell^+$, $\epsilon > 0$.
Since the front at x is a rarefaction front, both estimates (7.51) and (7.59) hold. Moreover, we have that
$$q_\ell^+ - q_\ell^- - |q_\ell^+| + |q_\ell^-| < 2\epsilon \le 2\delta.$$
Then from $AD + BC \le (A+B)(D+C)$ for positive A, B, C, and D, we obtain

$$\begin{aligned}
E_\ell &= W_\ell^+ \left|q_\ell^+\right| \left(\mu_\ell^+ - \dot{x}\right) - W_\ell^- \left|q_\ell^-\right| \left(\mu_\ell^- - \dot{x}\right) \\
&\le \mathcal{O}\left(1\right) \left(\left|q_\ell^+\right| + \left|q_\ell^-\right|\right) \left(\left|\mu_\ell^+ - \dot{x}\right| + \left|\mu_\ell^- - \dot{x}\right|\right) \\
&\le \mathcal{O}\left(1\right) \epsilon \left(\left|\mu_\ell^+ - \dot{x}\right| + \left|\mu_\ell^- - \dot{x}\right|\right) \\
&= \mathcal{O}\left(1\right) \epsilon \left(\left|\mu_\ell\left(\omega_{\ell-1}^+, \omega_\ell^+\right) - \mu_\ell\left(v^{\delta,+}, v^{\delta,+}\right)\right| \right. \\
&\qquad\qquad\quad \left. + \left|\mu_\ell\left(\omega_{\ell-1}^-, \omega_\ell^-\right) - \mu_\ell\left(v^{\delta,-}, v^{\delta,-}\right)\right|\right) \\
&= \mathcal{O}\left(1\right) \epsilon \left(\delta + \left|q_\ell^+\right| + \left|q_\ell^-\right| + \sum_{k>\ell} \left|q_k^+\right| + \sum_{k<\ell} \left|q_k^-\right|\right) \\
&= \mathcal{O}\left(1\right) \epsilon \left(\delta + \sum_{k>\ell} \left|q_k^+\right| + \sum_{k<\ell} \left|q_k^-\right|\right). \quad (7.67)
\end{aligned}$$

Using (7.51) for $k < \ell$ and (7.59) for $k > \ell$, and choosing κ_1 sufficiently large, we obtain (7.30).

Now we shall study the cases where the front at x is a shock front. Also, here we prove (7.30) in three cases depending on q_ℓ^- and q_ℓ^+. If the front at x is a shock front, then by the construction of the front-tracking approximation we have

$$H_\ell(\epsilon) v^{\delta,-} = v^{\delta,+} + e,$$

or

$$H_\ell(\epsilon) H\left(q^-\right) u^{\delta_1} = H\left(q^+\right) u^{\delta_1} + e,$$

where $q^\pm = \left(q_1^\pm, \ldots, q_n^\pm\right)$, and e is the error of the front at x. Then we can use (7.31) and continuity of the mapping H to find that

$$\begin{aligned}
\left|q_\ell^+ - q_\ell^- - \epsilon\right| &+ \sum_{k \ne \ell} \left|q_k^+ - q_k^-\right| \\
&= \mathcal{O}\left(1\right) |\epsilon| \left(\left|q_\ell^-\right|\left(\left|q_\ell^-\right| + |\epsilon|\right) + \sum_{k \ne \ell}\left|q_k^-\right|\right) + \mathcal{O}\left(1\right) |e|.
\end{aligned} \quad (7.68)$$

We also have that

$$u^{\delta_1} = H\left(-q^+\right) v^{\delta,+} = H\left(-q^+\right) \left(H_\ell(\epsilon) v^{\delta,-} + e\right),$$

or

$$H\left(-q^-\right) v^{\delta,-} = H\left(-q^+\right) H_\ell(\epsilon) v^{\delta,-} + \mathcal{O}\left(1\right) |e|,$$

by the continuity of H. From this we obtain

$$\left|q_\ell^+ - q_\ell^- - \epsilon\right| + \sum_{k \neq \ell} \left|q_k^+ - q_k^-\right|$$
$$= \mathcal{O}(1) |\epsilon| \left(|q_\ell^+| \left(|q_\ell^+| + |\epsilon|\right) + \sum_{k \neq \ell} |q_k^+| \right) + \mathcal{O}(1) |e|. \tag{7.69}$$

Case S1: $0 < q_\ell^+ < q_\ell^-$, $\epsilon < 0$.
If $k \neq \ell$, then we can write E_k as (7.38) and use the arguments leading to (7.51) and the estimate (7.69) to obtain

$$E_k \leq \mathcal{O}(1) |\epsilon| \left(|q_\ell^+| \left(|q_\ell^+| + |\epsilon|\right) + \sum_{\tilde{k} \neq \ell} |q_{\tilde{k}}^+| \right) + \mathcal{O}(1) |e| - c\kappa_1 |q_k^+| |\epsilon|. \tag{7.70}$$

For $k = \ell$ we define the auxiliary states and speeds as in (7.60); see Figure 7.6. Then the estimate (7.61) holds. Also, using (7.69) we find that

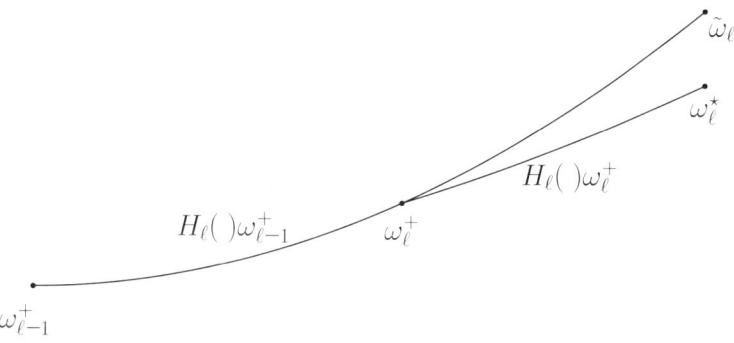

Figure 7.6. The situation for $0 < q_\ell^+ < q_\ell^-$, $\epsilon < 0$, and $k = \ell$.

$$\left|\mu_\ell^- - \tilde{\mu}_\ell\right| = \mathcal{O}(1) \left(|\omega_{\ell-1}^- - \omega_{\ell-1}^+| + |q_\ell^+ - q_\ell^- - \epsilon| \right)$$
$$= \mathcal{O}(1) |\epsilon| \left(|q_\ell^+| \left(|q_\ell^+| + |\epsilon|\right) + \sum_{k \neq \ell} |q_k^+| \right) + \mathcal{O}(1) |e|. \tag{7.71}$$

Moreover,

$$\left|\mu_\ell^\star - \dot{x}\right| = \left|\mu_\ell \left(\omega_\ell^+, \omega_\ell^\star\right) - \mu_l \left(v^{\delta,-}, v^{\delta,+}\right)\right|$$
$$\leq \left|\mu_\ell \left(\omega_\ell^+, H_\ell(-\epsilon)\omega_\ell^+\right) - \mu_\ell \left(v^{\delta,+}, H_\ell(-\epsilon) v^{\delta,+}\right)\right| + \mathcal{O}(1) |e|$$
$$= \mathcal{O}(1) \left(|\omega_\ell^+ - \omega_n^+|\right) + \mathcal{O}(1) |e|$$
$$= \mathcal{O}(1) \left(\sum_{k > \ell} |q_k^+| + |e| \right). \tag{7.72}$$

262   7. Well-Posedness of the Cauchy Problem

By Lemma 7.6, we have
$$\mu_\ell^\star - \tilde{\mu}_\ell > cq_\ell^+. \tag{7.73}$$

In this case
$$W_\ell^+ = W_\ell^- + \kappa_1 |\epsilon|, \tag{7.74}$$

and
$$\epsilon < 0 < q_\ell^+ < q_\ell^-.$$

We estimate
$$\begin{aligned}
E_\ell &= W_\ell^+ q_\ell^+ \left(\mu_\ell^+ - \dot{x}\right) - \left(W_\ell^+ - \kappa_1 |\epsilon|\right) q_\ell^- \left(\mu_\ell^- - \dot{x}\right) \\
&= \kappa_1 |\epsilon| q_\ell^- \left(\mu_\ell^- - \dot{x}\right) + W_\ell^+ \left\{ q_\ell^+ \left(\mu_\ell^+ - \dot{x}\right) - q_\ell^- \left(\mu_\ell^- - \dot{x}\right) \right\} \\
&= \kappa_1 |\epsilon| \left\{ \left(q_\ell^+ + |\epsilon|\right) \left(\mu_\ell^- - \mu_\ell^\star\right) + q_\ell^- \left(\mu_\ell^- - \dot{x}\right) - \left(q_\ell^+ + |\epsilon|\right) \left(\mu_\ell^- - \mu_\ell^\star\right) \right\} \\
&\quad + W_\ell^+ \left\{ q_\ell^+ \left(\mu_\ell^+ - \dot{x}\right) - q_\ell^- \left(\mu_\ell^- - \dot{x}\right) \right\} \\
&= \kappa_1 |\epsilon| \left(q_\ell^+ + |\epsilon|\right) \left(\tilde{\mu}_\ell - \mu_\ell^\star\right) \\
&\quad + \kappa_1 |\epsilon| \left\{ \left(q_\ell^+ + |\epsilon|\right) \left(\left(\mu_\ell^- - \dot{x}\right) - \left(\tilde{\mu}_\ell - \mu_\ell^\star\right)\right) \right. \\
&\qquad \left. + \left(q_\ell^- - q_\ell^+ - |\epsilon|\right) \left(\mu_\ell^- - \dot{x}\right) \right\} \\
&\quad + W_\ell^+ \left\{ q_\ell^+ \left(\mu_\ell^+ - \dot{x}\right) - q_\ell^- \left(\mu_\ell^- - \dot{x}\right) \right\} \\
&\leq \kappa_1 |\epsilon| \left(q_\ell^+ + |\epsilon|\right) \left(\tilde{\mu}_\ell - \mu_\ell^\star\right) \\
&\quad + \kappa_1 |\epsilon| \left(q_\ell^+ + |\epsilon|\right) \left(|\mu_\ell^- - \tilde{\mu}_\ell| + |\mu_\ell^\star - \dot{x}|\right) \\
&\quad + \kappa_1 |\epsilon| \left|q_\ell^+ - q_\ell^- - \epsilon\right| \left|\mu_\ell^- - \dot{x}\right| \\
&\quad + W_\ell^+ \left\{ q_\ell^+ \left(\mu_\ell^+ - \dot{x}\right) - q_\ell^- \left(\mu_\ell^- - \dot{x}\right) \right\} \\
&\leq -c\kappa_1 \left(q_\ell^+ + |\epsilon|\right) |\epsilon| q_\ell^+ \\
&\quad + \mathcal{O}(1) \kappa_1 |\epsilon|^2 \left(q_\ell^+ \left(q_\ell^+ + |\epsilon|\right) + \sum_{k \neq \ell} |q_k^+| \right) + \mathcal{O}(1) |\epsilon| \\
&\quad + \mathcal{O}(1) \kappa_1 |\epsilon| \left(\sum_{k > \ell} |q_k^+| \right) + \mathcal{O}(1) |\epsilon| \\
&\quad + W_\ell^+ \left\{ \left| q_\ell^+ \left(\mu_\ell^+ - \mu_\ell^\star\right) - \left(q_\ell^+ - \epsilon\right) \left(\tilde{\mu}_\ell - \mu_\ell^\star\right) \right| \right. \\
&\qquad \left. + \left|q_\ell^+ - q_\ell^- - \epsilon\right| \left|\mu_\ell^- - \dot{x}\right| + |\epsilon| \left|\mu_\ell^\star - \dot{x}\right| + \left(q_\ell^+ + |\epsilon|\right) \left|\mu_\ell^- - \tilde{\mu}_\ell\right| \right\} \\
&\leq -c\kappa_1 \left(q_\ell^+ + |\epsilon|\right) |\epsilon| q_\ell^+ + \mathcal{O}(1) |\epsilon| \left(q_\ell^+ \left(q_\ell^+ + |\epsilon|\right) + \sum_{k \neq \ell} |q_k^+| \right) \\
&\quad + \mathcal{O}(1) |\epsilon| \\
&\leq \mathcal{O}(1) \sum_{k \neq \ell} |q_k^+| + |\epsilon| \, |q_\ell^+| \left(q_\ell^+ + |\epsilon|\right) \left(\mathcal{O}(1) - c\kappa_1\right) + \mathcal{O}(1) |\epsilon|. \tag{7.75}
\end{aligned}$$

7.1. Stability

As before, setting κ_1 sufficiently large, (7.75) and (7.70) imply

$$\sum_k E_k = E_\ell + \sum_{k\neq\ell} E_k \leq \mathcal{O}(1)|\epsilon|, \qquad (7.76)$$

which is (7.30).

Case S2: $q_\ell^+ < q_\ell^- < 0$, $\epsilon < 0$.
In this case we proceed as in Case S1, but using (7.68) instead of (7.69). For $k \neq \ell$ this gives the estimate

$$E_k \leq \mathcal{O}(1)|\epsilon|\left(|q_\ell^-|(|q_\ell^-|+|\epsilon|) + \sum_{\tilde{k}\neq\ell}|q_{\tilde{k}}^-|\right) + \mathcal{O}(1)|\epsilon| - c\kappa_1 |q_k^-||\epsilon|. \quad (7.77)$$

We now define the intermediate states $\tilde{w}_\ell$, $w_\ell^\star$ and the speeds $\tilde{\mu}_\ell$ and $\mu_\ell^\star$ as in (7.52); see Figure 7.7. Then the estimate (7.53) holds. As in Case R1,

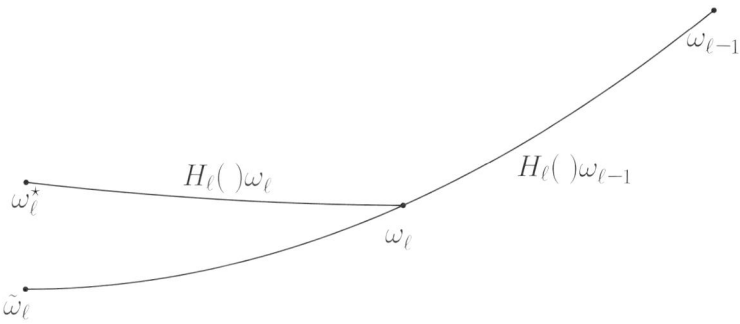

Figure 7.7. The situation for $q_\ell^+ < q_\ell^- < 0$, $\epsilon < 0$, and $k = \ell$.

we compute

$$|\mu_\ell^+ - \tilde{\mu}_\ell| = \mathcal{O}(1)\left(|w_{\ell-1}^- - w_{\ell-1}^+| + |q_\ell^+ - q_\ell^- - \epsilon|\right)$$
$$= \mathcal{O}(1)|\epsilon|\left(|q_\ell^-|(|q_\ell^-|+|\epsilon|) + \sum_{k\neq\ell}|q_k^-|\right) + \mathcal{O}(1)|\epsilon|, \quad (7.78)$$

and

$$|\mu_\ell^\star - \dot{x}| \leq |\mu_\ell(w_\ell^-, H_\ell(\epsilon)w_\ell^-) - \mu_\ell(v^{\delta,-}, H_\ell(\epsilon)v^{\delta,-})| + \mathcal{O}(1)|\epsilon|$$
$$\leq \mathcal{O}(1)\delta + \mathcal{O}(1)|w_\ell^- - w_0^-| + \mathcal{O}(1)|\epsilon|$$
$$\leq \mathcal{O}(1)|\epsilon| + \mathcal{O}(1)\sum_{k<\ell}|q_k^-|. \qquad (7.79)$$

In this case, genuine nonlinearity and Lemma 7.6 imply that

$$\tilde{\mu}_\ell - \mu_\ell^\star > c|q_\ell^-|, \qquad (7.80)$$

264   7. Well-Posedness of the Cauchy Problem

with $c > 0$. Moreover, now
$$W_\ell^+ = W_\ell^- - \kappa_1 |\epsilon|.$$

Now we can use the (by now) familiar technique of estimating F_ℓ:

$$\begin{aligned}
E_\ell &= \left(W_\ell^- - \kappa_1 |\epsilon|\right) |q_\ell^+| \left(\mu_\ell^+ - \dot{x}\right) - W_\ell^- |q_\ell^-| \left(\mu_\ell^- - \dot{x}\right) \\
&\leq -\kappa_1 |\epsilon| \left(|q_\ell^-| + |\epsilon|\right) \left(\tilde{\mu}_\ell - \mu_\ell^\star\right) \\
&\quad + \kappa_1 |\epsilon| \left(|q_\ell^-| + |\epsilon|\right) \left(|\mu_\ell^+ - \tilde{\mu}_\ell| + |\mu_\ell^\star - \dot{x}|\right) \\
&\quad + \kappa_1 |\epsilon| \left|q_\ell^+ - q_\ell^- - \epsilon\right| |\mu_\ell^+ - \dot{x}| \\
&\quad + W_\ell^- \left\{|q_\ell^+| (\mu_\ell^+ - \dot{x}) - |q_\ell^-| (\mu_\ell^- - \dot{x})\right\} \\
&\leq -c\kappa_1 |q_\ell^-| |\epsilon| \left(|q_\ell^-| + |\epsilon|\right) \\
&\quad + \mathcal{O}(1) \kappa_1 |\epsilon| \left(|q_\ell^-| + |\epsilon|\right) \left(|q_\ell^-|\left(|q_\ell^-| + |\epsilon|\right) + \sum_{k \neq \ell} |q_k^-|\right) \\
&\quad + W_\ell^- \Big\{ \left|q_\ell^- \left(\mu_\ell^- - \mu_\ell^\star\right) - (q_\ell^- + \epsilon)(\tilde{\mu}_\ell - \mu_\ell^\star)\right| \\
&\quad\quad + \left|q_\ell^+ - q_\ell^- - \epsilon\right| |\mu_\ell^+ - \dot{x}| \\
&\quad\quad + |\epsilon| |\mu_\ell^\star - \dot{x}| + \left(|q_\ell^-| + |\epsilon|\right) |\mu_\ell^+ - \tilde{\mu}_\ell|\Big\} + \mathcal{O}(1) |\epsilon| \\
&\leq -c\kappa_1 |q_\ell^-| |\epsilon| \left(|q_\ell^-| + |\epsilon|\right) + \mathcal{O}(1) \left(|q_\ell^-|\left(|q_\ell^-| + |\epsilon|\right) + \sum_{k \neq \ell} |q_k^-|\right) \\
&\quad + \mathcal{O}(1) |\epsilon| \\
&\leq \mathcal{O}(1) \sum_{k \neq \ell} |q_k^-| + |\epsilon| |q_\ell^-| \left(|q_\ell^-| + |\epsilon|\right) \left(\mathcal{O}(1) - c\kappa_1\right) + \mathcal{O}(1) |\epsilon|. \quad (7.81)
\end{aligned}$$

Combining (7.81) and (7.77) we obtain

$$\sum_k E_k = E_\ell + \sum_{k \neq \ell} E_k \leq \mathcal{O}(1) |\epsilon|, \qquad (7.82)$$

which is (7.30).

Case S3: $q_\ell^+ < 0 < q_\ell^-$, $\epsilon < 0$.

For $k \neq \ell$, the estimate (7.77) remains valid.

Next we consider the case $k = \ell$. The $\mathcal{O}(1)$ that multiplies $|\epsilon|$ in (7.69) (or (7.69)) is proportional to the total variation of the initial data. Hence we can assume that this is arbitrarily small by choosing T.V.(u_0) sufficiently small. Since all terms $q_j^\pm$ are bounded, we can and will assume that

$$\left|q_\ell^+ - q_\ell^- - \epsilon\right| \leq \frac{1}{2} |\epsilon| + \mathcal{O}(1) |\epsilon|. \qquad (7.83)$$

Without loss of generality we may assume that $|q_\ell^+| \geq |q_\ell^-|$. This implies that

$$\left|q_\ell^+ - q_\ell^- - \epsilon\right| \geq \left|q_\ell^- - q_\ell^+\right| - |\epsilon| = q_\ell^- - q_\ell^+ + \epsilon \geq 2q_\ell^- + \epsilon. \qquad (7.84)$$

7.1. Stability

Thus

$$2q_\ell^- + \epsilon \le \frac{1}{2}|\epsilon| + \mathcal{O}(1)|e|, \tag{7.85}$$

or

$$q_\ell^- + \epsilon \le -\frac{1}{4}|\epsilon| + \mathcal{O}(1)|e|, \tag{7.86}$$

which can be rewritten as

$$|q_\ell^- + \epsilon - \mathcal{O}(1)|e|| \ge \frac{1}{4}|\epsilon|. \tag{7.87}$$

From this we conclude that

$$|q_\ell^- + \epsilon| \ge \frac{1}{4}|\epsilon| - \mathcal{O}(1)|e|. \tag{7.88}$$

We define the auxiliary states $\tilde{\omega}_\ell$, $\omega_\ell^\star$ and the speeds $\tilde{\mu}_\ell$ and $\mu_\ell^\star$ as in (7.52); see Figure 7.8. Then estimates (7.78) and (7.79) hold.

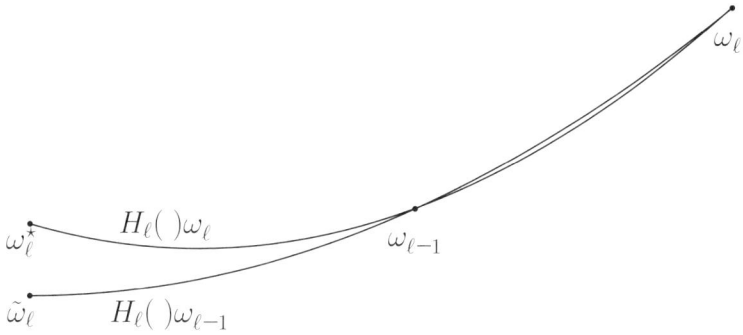

Figure 7.8. The situation for $q_\ell^+ < 0 < q_\ell^-$, $\epsilon < 0$, and $k = \ell$.

By Lemma 7.6 we have that

$$\tilde{\mu}_\ell - \mu_\ell^\star \le 0, \tag{7.89}$$
$$\mu_\ell^- - \mu_\ell^\star \ge c|q_\ell^- + \epsilon|, \tag{7.90}$$

for a positive constant c. Recalling that $W_\ell^- \ge 1$, and using (7.89), (7.90), and the estimates (7.78) and (7.79) (which remain valid in this case), we compute

$$E_\ell = W_\ell^+ |q_\ell^+| (\mu_\ell^+ - \dot{x}) - W_\ell^- |q_\ell^-| (\mu_\ell^- - \dot{x})$$
$$\le W_\ell^+ |q_\ell^+| (\tilde{\mu}_\ell - \mu_\ell^\star) - W_\ell^- |q_\ell^-| (\mu_\ell^- - \mu_\ell^\star)$$
$$\quad + W_\ell^+ |q_\ell^+| (|\mu_\ell^+ - \tilde{\mu}_\ell| + |\mu_\ell^\star - \dot{x}|) + W_\ell^- |q_\ell^-| |\mu_\ell^\star - \dot{x}|$$
$$\le -|q_\ell^-| c |q_\ell^- + \epsilon| + \mathcal{O}(1) |\epsilon| \left(q_\ell^- (q_\ell^- + |\epsilon|) + \sum_{\tilde{k} \ne \ell} |q_{\tilde{k}}^-| \right) + \mathcal{O}(1) |e|$$

$$\leq \frac{-c}{4}\left|q_\ell^-\right||\epsilon| + \mathcal{O}\left(1\right)|\epsilon|\left(q_\ell^-\left(q_\ell^- + |\epsilon|\right) + \sum_{\tilde{k}\neq\ell}\left|q_{\tilde{k}}^-\right|\right) + \mathcal{O}\left(1\right)|e|. \quad (7.91)$$

Now (7.77) and (7.91) are used to balance the terms containing the factor $\sum_{k\neq\ell}|q_k^-|$. The remaining term,

$$\left|q_\ell^-\right||\epsilon|\left(-\frac{1}{4}c + \mathcal{O}\left(1\right)\left(q_\ell^- + |\epsilon|\right)\right),$$

can be made negative by choosing T.V. (u_0) (and hence $\mathcal{O}\left(1\right)$) sufficiently small.

Hence also in this case (7.30) holds.

Finally, if q_ℓ^- or q_ℓ^+ is zero, (7.30) can easily be shown to be a limit of one of the previous cases.

Summing up, we have proved the following theorem:

Theorem 7.7. *Let u^{δ_1} and v^{δ_2} be front-tracking approximations, with accuracies defined by δ_1, δ_2,*

$$G\left(u^{\delta_1}(t)\right) < M, \quad \text{and} \quad G\left(v^{\delta_2}(t)\right) < M, \quad \text{for } t \geq 0. \quad (7.92)$$

For sufficiently small M there exist constants κ_1, κ_2, and C_2 such that the functional Φ defined by (7.15) and (7.16) satisfies (7.21). Furthermore, there exists a constant C (independent of δ_1 and δ_2) such that

$$\left\|u^{\delta_1}(t) - v^{\delta_2}(t)\right\|_1 \leq C\left\|u^{\delta_1}(0) - v^{\delta_2}(0)\right\|_1 + Ct\left(\delta_1 \vee \delta_2\right). \quad (7.93)$$

To state the next theorem we need the following definition. Let the domain $\mathcal{D}$ be defined as the L^1 closure of the set

$$\mathcal{D}_0 = \left\{u \in L^1(\mathbb{R}; \mathbb{R}^n) \mid u \text{ is piecewise constant and } G(u) < M\right\}; \quad (7.94)$$

that is, $\mathcal{D} = \overline{\mathcal{D}}_0$. Since the total variation is small, we will assume that all possible values of u are in a (small) neighborhood $\Omega \subset \mathbb{R}^n$.

Theorem 7.8. *Let $f_j \in C^2(\mathbb{R}^n)$, $j = 1, \ldots, n$. Consider the strictly hyperbolic equation $u_t + f(u)_x = 0$. Assume that each wave family is either genuinely nonlinear or linearly degenerate. For all initial data u_0 in $\mathcal{D}$, defined by (7.94), any sequence of front-tracking approximations u^δ converges to a unique limit u as $\delta \to 0$. Furthermore, let u and v denote solutions*

$$u_t + f(u)_x = 0,$$

with initial data u_0 and v_0, respectively, obtained as a limit of a front-tracking approximation. Then

$$\|u(t) - v(t)\|_1 \leq C\|u_0 - v_0\|_1. \quad (7.95)$$

Proof. First we use (7.93) to conclude that any front-tracking approximation u^δ has a unique limit u as $\delta \to 0$. Then we take the limit $\delta \to 0$ in (7.93) to conclude that (7.95) holds. □

Note that this also gives an error estimate for front tracking for systems. If we denote the limit of the sequence $\{u^\delta\}$ by u and $v^{\delta_2} = u^\delta$, then by letting $\delta_2 \to 0$ in (7.93)

$$\left\|u^\delta(\,\cdot\,,t) - u(\,\cdot\,,t)\right\|_1 \leq C\left(\left\|u_0^\delta - u_0\right\|_1 + \delta t\right) = \mathcal{O}\left(1\right)\delta$$

for some finite constant C. Hence front tracking for systems is a first-order method.

7.2 Uniqueness

Let $\mathcal{S}_t$ denote the map that maps initial data u_0 into the solution u of

$$u_t + f(u)_x = 0, \quad u|_{t=0} = u_0$$

at time t, that is, $u = \mathcal{S}_t u_0$. In Chapter 6 we showed the existence of the semigroup $\mathcal{S}_t$, and in the previous section its stability for initial data in the class $\mathcal{D}$ as limits of approximate solutions obtained by front tracking. Thus we know that it satisfies

$$\mathcal{S}_0 u = u, \qquad \mathcal{S}_t \mathcal{S}_s u = \mathcal{S}_{t+s} u,$$
$$\|\mathcal{S}_t u - \mathcal{S}_s v\|_1 \leq L\left(|t - s| + \|u - v\|_1\right)$$

for all $t, s \geq 0$ and u, v in $\mathcal{D}$.

In this section we prove uniqueness of solutions that have initial data in $\mathcal{D}$.

We want to demonstrate that any other solution u coincides with this semigroup. To do this we will basically need three assumptions. The first is that u is a weak solution, the second is that it satisfies Lax's entropy conditions across discontinuities, and the third is that it has locally bounded variation on a certain family of curves. Concretely, we define an *entropy solution* to

$$u_t + f(u)_x = 0, \quad u|_{t=0} = u_0,$$

to be a bounded measurable function $u = u(x,t)$ of bounded total variation satisfying the following conditions:

A The function $u = u(x,t)$ is a weak solution of the Cauchy problem (7.1) taking values in $\mathcal{D}$, i.e.,

$$\int_0^T \int_\mathbb{R} (u\varphi_t + f(u)\varphi_x)\, dx\, dt + \int_\mathbb{R} \varphi(x,0)u_0(x)\, dx = 0 \qquad (7.96)$$

for all test functions φ whose support is contained in the strip $[0, T)$.

B Assume that u has a jump discontinuity at some point (x,t), i.e., there exist states $u_{l,r} \in \Omega$ and speed σ such that if we let

$$U(y,s) = \begin{cases} u_l & \text{for } y < x + \sigma(s-t), \\ u_r & \text{for } y \geq x + \sigma(s-t), \end{cases} \quad (7.97)$$

then

$$\lim_{\rho \to 0} \frac{1}{\rho^2} \int_{t-\rho}^{t+\rho} \int_{x-\rho}^{x+\rho} |u(y,s) - U(y,s)| \, dy \, ds = 0. \quad (7.98)$$

Furthermore, there exists k such that

$$\lambda_k(u_l) \geq \sigma \geq \lambda_k(u_r). \quad (7.99)$$

C There exists a $\theta > 0$ such that for all Lipschitz functions γ with Lipschitz constant not exceeding θ, the total variation of $u(x, \gamma(x))$ is locally bounded.

Remark 7.9. One can prove, see Exercise 7.1, that the front-tracking solution constructed in the previous chapter is an entropy solution of the conservation law.

There is a direct argument showing that any weak solution, whether it is a limit of a front-tracking approximation or not, satisfies a Lipschitz continuity in time of the spatial L^1-norm, as long as the solution has a uniform bound on the total variation. We present that argument here.

Theorem 7.10. *Let $u_0 \in \mathcal{D}$, and let u denote any weak solution of (7.1) such that* T.V.$(u(t)) \leq C$. *Then*

$$\|u(\,\cdot\,,t) - u(\,\cdot\,,s)\|_1 \leq C \, \|f\|_{\mathrm{Lip}} \, |t-s|, \quad s,t \geq 0. \quad (7.100)$$

Proof. Let $0 < s < t < T$, and let α_h be a smooth approximation to the characteristic function of the interval $[s,t]$, so that

$$\lim_{h \to 0} \alpha_h = \chi_{[s,t]}.$$

Furthermore, define

$$\varphi_h(y,\tau) = \alpha_h(\tau)\phi(y),$$

where ϕ is any smooth function with compact support. If we insert this into the weak formulation

$$\int_0^T \int_{\mathbb{R}} (u\varphi_{h,t} + f(u)\varphi_{h,x}) \, dx \, dt + \int_{\mathbb{R}} \varphi_h(x,0)u(x,0) \, dx = 0, \quad (7.101)$$

and let $h \to 0$, we obtain

$$\int \phi(y) \left(u(y,t) - u(y,s)\right) dy + \int_s^t \int \phi_y f(u) \, dy \, ds = 0.$$

From this we obtain

$$\|u(\,\cdot\,,t) - u(\,\cdot\,,s)\|_1 = \sup_{|\phi|\leq 1} \int \phi(y)\,(u(y,t) - u(y,s))\,dy$$

$$= -\sup_{|\phi|\leq 1} \int_s^t \int \phi(y)_y f(u)\,dy\,ds$$

$$\leq \int_s^t \text{T.V.}(f(u))\,ds$$

$$\leq C\,\|f\|_{\text{Lip}}(t-s),$$

which proves the claim. Here we first used Exercise A.1, Theorem A.2, subsequently the definition (A.1) for T.V.(f), and finally the Lipschitz continuity of f and the bound on the total variation on u. □

Remark 7.11. This argument provides an alternative to the proof of the Lipschitz continuity in Theorem 2.14 in the scalar case.

Before we can compare an arbitrary entropy solution to the semigroup solution, we need some preliminary results. Firstly, Theorem 7.10 says that any function $u(\,\cdot\,,t)$ taking values in $\mathcal{D}$ and satisfying **A** is L^1 Lipschitz continuous:

$$\|u(\,\cdot\,,t) - u(\,\cdot\,,s)\|_1 \leq L(t-s),$$

for $t \geq s$.

Furthermore, by the structure theorem for functions of bounded variation [148, Theorem 5.9.6], u is continuous almost everywhere. For the sake of definiteness, we shall assume that all functions in $\mathcal{D}$ are right continuous. Also, there exists a set $\mathcal{N}$ of zero Lebesgue measure in the interval $[0,T]$ such that for $t \in [0,T]\setminus\mathcal{N}$, the function $u(\,\cdot\,,t)$ is either continuous at x or has a jump discontinuity there. Intuitively, the set $\mathcal{N}$ can be thought of as the set of times when collisions of discontinuities occur.

Lemma 7.12. *If (7.96)–(7.98) hold, then*

$$u_l = \lim_{y\to x-} u(y,t), \quad u_r = \lim_{y\to x+} u(y,t),$$

$$\text{and} \quad \sigma(u_l - u_r) = f(u_l) - f(u_r).$$

Proof. Let P_λ denote the parallelogram

$$P_\lambda = \{(y,s) \mid |t-s| \leq \lambda,\ |y - x - \sigma(s-t)| \leq \lambda\}.$$

Integrating the conservation law over P_λ, we obtain

$$\left(\int_{x-\lambda+\lambda\sigma}^{x+\lambda+\lambda\sigma} u(y,t+\lambda)\,dy - \int_{x-\lambda-\lambda\sigma}^{x+\lambda-\lambda\sigma} u(y,t-\lambda)\,dy\right)$$

$$+ \int_{t-\lambda}^{t+\lambda} (f(u) - \sigma u)(x + \lambda + \sigma(s-t), s)\,ds$$

$$-\int_{t-\lambda}^{t+\lambda}(f(u)-\sigma u)(x-\lambda+\sigma(s-t),s)\,ds=0.$$

If we furthermore integrate this with respect to λ from $\lambda=0$ to $\lambda=\rho$, and divide by ρ^2, we obtain

$$\frac{1}{\rho^2}\left(\int_0^\rho\int_{x-\lambda+\lambda\sigma}^{x+\lambda+\lambda\sigma}u(y,t+\lambda)\,dy\,d\lambda-\int_0^\rho\int_{x-\lambda-\lambda\sigma}^{x+\lambda-\lambda\sigma}u(y,t-\lambda)\,dy\,d\lambda\right)$$

$$+\frac{1}{\rho^2}\left(\int_0^\rho\int_{t-\lambda}^{t+\lambda}(f(u)-\sigma u)(x+\lambda+\sigma(s-t),s)\,ds\,d\lambda\right.$$

$$\left.-\int_0^\rho\int_{t-\sigma}^{t+\sigma}(f(u)-\sigma u)(x-\lambda+\sigma(s-t),s)\,ds\,d\lambda\right)=0.$$

Now let $\rho\to 0$. Then

$$\frac{1}{\rho^2}\int_0^\rho\int_{x-\lambda+\lambda\sigma}^{x+\lambda+\lambda\sigma}u(y,t+\lambda)\,dy\,d\lambda\to\frac{1}{2}(u_l+u_r),$$

$$\frac{1}{\rho^2}\int_0^\rho\int_{x-\lambda-\lambda\sigma}^{x+\lambda-\lambda\sigma}u(y,t-\lambda)\,dy\,d\lambda\to\frac{1}{2}(u_l+u_r),$$

$$\frac{1}{\rho^2}\int_0^\rho\int_{t-\lambda}^{t+\lambda}(f(u)-\sigma u)(x+\lambda+\sigma(s-t),s)\,ds\,d\lambda\to f(u_r)-\sigma u_r,$$

$$\frac{1}{\rho^2}\int_0^\rho\int_{t-\lambda}^{t+\lambda}(f(u)-\sigma u)(x-\lambda+\sigma(s-t),s)\,ds\,d\lambda\to f(u_l)-\sigma u_l.$$

Hence

$$\frac{1}{2}(u_l+u_r)-\frac{1}{2}(u_l+u_r)+(f(u_r)-\sigma u_r)-(f(u_l)-\sigma u_l)=0.$$

This concludes the proof of the lemma. □

The next lemma states that if u satisfies **C**, then the discontinuities cannot cluster too tightly together.

Lemma 7.13. *Assume that $u\colon[0,T]\to\mathcal{D}$ satisfies* **C**. *Let $t\in[0,T]$ and $\varepsilon>0$. Then the set*

$$B_{t,\varepsilon}=\left\{x\in\mathbb{R}\mid\limsup_{s\to t+,\,y\to x}|u(x,t)-u(y,s)|>\varepsilon\right\}\qquad(7.102)$$

has no limit points.

Proof. Assume that $B_{t,\varepsilon}$ has a limit point denoted by x_0. Then there is a monotone sequence $\{x_i\}_{i=1}^\infty$ in $B_{t,\varepsilon}$ converging to x_0. Without loss of generality we assume that the sequence is decreasing. Since $u(x,t)$ is right continuous, we can find a point z_i in $\langle x_i,x_{i-1}\rangle$ such that

$$|u(z_i,t)-u(x_i,t)|\le\frac{\varepsilon}{2}.$$

Now choose $s_i > t$ and $y_i \in \langle z_{i+1}, z_i \rangle$ such that
$$|u(y_i, s_i) - u(x_i, t)| \geq \varepsilon, \quad |s_i - t| \leq \theta \max\{|y_i - z_i|, |y_i - z_{i+1}|\}.$$
We define a curve $\gamma(x)$ for $x \in [x_0, x_1]$ passing through all the points (z_i, t) and (y_i, s_i) by
$$\gamma(x) = \begin{cases} t & \text{for } x = x_0 \text{ or } x \geq z_1, \\ s_i - (x - y_i)\frac{s_i - t}{z_i - y_i} & \text{for } x \in [y_i, z_i], \\ t + (x - z_{i+1})\frac{s_i - t}{y_i - z_{i+1}} & \text{for } x \in [z_{i+1}, y_i]. \end{cases} \quad (7.103)$$
Then γ is Lipschitz continuous with Lipschitz constant θ, and we have that
$$|u(y_i, s_i) - u(z_i, t)| \geq \frac{\varepsilon}{2}$$
for all $i \in \mathbb{N}$. This means that the total variation of $u(x, \gamma(x))$ is infinite, violating **C**, concluding the proof of the lemma. $\square$

In the following, we let $\sigma^\star$ be a number strictly larger than the absolute value of any characteristic speed, and we also demand that $\sigma^\star \geq 1/\theta$, where θ is the constant in **C**. The next lemma says that if u satisfies **C**, then discontinuities cannot propagate faster than $\sigma^\star$. Precisely, we have the following result.

Lemma 7.14. *Assume that $u\colon [0,T] \to \mathcal{D}$ satisfies **C**. Then for $(x,t) \in \langle 0, T \rangle \times \mathbb{R}$,*
$$\lim_{\substack{s \to t^+,\, y \to x\pm \\ |x-y| > \sigma^\star(s-t)}} u(y, s) = u(x\pm, t). \quad (7.104)$$

Proof. We assume that the lemma does not hold. Then, for some (x_0, t) there exist decreasing sequences $s_j \to t$ and $y_j \to x_0$ such that
$$|y_j - x_0| \geq \sigma^\star(s_j - t), \quad |u(y_j, s_j) - u(x_0, t)| \geq \varepsilon$$
for some $\varepsilon > 0$ and $j \in \mathbb{N}$. Now let
$$z_0 = y_1 + \frac{s_1 - t}{\theta},$$
where as before θ is defined by **C**. Now we define a subsequence of $\{(y_j, s_j)\}$ as follows. Set $j_1 = 1$ and for $i \geq 1$ define
$$\begin{cases} z_i = y_{j_i} - \frac{s_{j_i} - t}{\theta}, \\ j_{i+1} = \min\{k \mid s_k \leq t - \theta(y_k - z_i)\}. \end{cases}$$
Then
$$y_{j_i} \in \langle z_{i+1}, z_i \rangle \quad \text{and} \quad |s_{j_i} - t| \leq \theta \max\{|y_{j_i} - z_i|, |y_{j_i} - z_{i+1}|\}$$
for all i. Let γ be the curve defined in (7.103). Since we have that $z_i \to x_0$, we have that
$$|u(z_i, t) - u(x_0, t)| \leq \frac{\varepsilon}{2}$$

for sufficiently large i. Consequently,

$$|u(z_i, t) - u(y_{j_i}, s_{j_i})| \geq \frac{\varepsilon}{2},$$

and the total variation of $u(x, \gamma(x))$ is infinite, contradicting **C**. □

The next lemma concerns properties of the semigroup $\mathcal{S}_t$. We assume that u is a continuous function $u \colon [0, T] \to \mathcal{D}$, and wish to estimate $\mathcal{S}_T u(0) - u(T)$. Let h be a small number such that $Nh = T$. Then we can calculate

$$\|\mathcal{S}_T u(0) - u(T)\|_1 \leq \sum_{i=1}^{N} \|\mathcal{S}_{T-(i-1)h} u((i-1)h) - \mathcal{S}_{T-ih} u(ih)\|_1$$

$$\leq L \sum_{i=1}^{N} \left\| \frac{1}{h} \left(u(ih) - \mathcal{S}_h u((i-1)h) \right) \right\|_1 h.$$

Letting h tend to zero we formally obtain the following lemma:

Lemma 7.15. *Assume that $u \colon [0, T] \to \mathcal{D}$ is Lipschitz continuous in the L^1-norm. Then for every interval $[a, b]$ we have*

$$\|\mathcal{S}_T u(0) - u(T)\|_{L^1([a+\sigma^* T, b-\sigma^* T]; \mathbb{R}^n)}$$
$$\leq \mathcal{O}(1) \int_0^T \left\{ \liminf_{h \to 0+} \frac{1}{h} \|\mathcal{S}_h u(t) - u(t+h)\|_{L^1([a+\sigma^*(t+h), b-\sigma^*(t+h)]; \mathbb{R}^n)} \right\} dt.$$

Proof. For ease of notation we set

$$\|\cdot\| = \|\cdot\|_{L^1([a+\sigma^*(t+h), b-\sigma^*(t+h)]; \mathbb{R}^n)}.$$

Observe that by finite speed of propagation, we can define $u(x, 0)$ to be zero outside of $[a, b]$, and the Lipschitz continuity of the semigroup will look identical written in the norm $\|\cdot\|$ to how it looked before. Let

$$\phi(t) = \liminf_{h \to 0+} \frac{1}{h} \|u(t+h) - \mathcal{S}_h u(t)\|.$$

Note that ϕ is measurable, and for all $h > 0$, the function

$$\phi_h(t) = \frac{1}{h} \|u(t+h) - \mathcal{S}_h u(t)\|$$

is continuous. Hence we have that

$$\phi(t) = \lim_{\varepsilon \to 0+} \inf_{h \in \mathbb{Q} \cap [0, \varepsilon]} \phi_h(t),$$

and therefore ϕ is Borel measurable. Define functions

$$\Psi(t) = \|\mathcal{S}_{T-t} u(t) - \mathcal{S}_T u(0)\|,$$

$$\psi(t) = \Psi(t) - L \int_0^t \phi(s) \, ds. \tag{7.105}$$

The function ψ is a Lipschitz function, and hence

$$\psi(T) = \int_0^T \psi'(s)\,ds. \tag{7.106}$$

Furthermore, Rademacher's theorem[2] implies that there exists a null set $\mathcal{N}_1 \subseteq [0,T]$ such that Ψ and ψ are differentiable outside $\mathcal{N}_1$. Furthermore, using that Lebesgue measurable functions are approximately continuous almost everywhere (see [47, p. 47]) we conclude that there exists another null set $\mathcal{N}_2$ such that ϕ is continuous outside $\mathcal{N}_2$. Let $\mathcal{N} = \mathcal{N}_1 \cup \mathcal{N}_2$. Outside $\mathcal{N}$ we have

$$\psi'(t) = \lim_{h \to 0} \frac{1}{h}\big(\Psi(t+h) - \Psi(t)\big) - L\psi(t). \tag{7.107}$$

Using properties of the semigroup we infer

$$\begin{aligned}\Psi(t+h) - \Psi(t) &= \|\mathcal{S}_{T-t-h}u(t+h) - \mathcal{S}_T u(0)\| - \|\mathcal{S}_{T-t}u(t) - \mathcal{S}_T u(0)\| \\ &\leq \|\mathcal{S}_{T-t-h}u(t+h) - \mathcal{S}_{T-t}u(t)\| \\ &= \|\mathcal{S}_{T-t-h}u(t+h) - \mathcal{S}_{T-t-h}\mathcal{S}_h u(t)\| \\ &\leq L\,\|u(t+h) - \mathcal{S}_h u(t)\|,\end{aligned}$$

which implies

$$\lim_{h \to 0} \frac{1}{h}\big(\Psi(t+h) - \Psi(t)\big) \leq L \liminf_{h \to 0} \frac{1}{h}\|u(t+h) - \mathcal{S}_h u(t)\| = L\phi(t).$$

Thus $\psi' \leq 0$ almost everywhere, and we conclude that

$$\psi(T) \leq 0, \tag{7.108}$$

which proves the lemma. $\square$

The next two lemmas are technical results valid for functions satisfying (7.97) and (7.98).

Lemma 7.16. *Assume that $u\colon [0,T] \to \mathcal{D}$ is Lipschitz continuous, and that for some (x,t) equations (7.97) and (7.98) hold. Then for all positive α we have*

$$\lim_{\rho \to 0+}\sup_{|h|\leq \rho} \int_0^\alpha |u(x + \lambda h + \rho y, t+h) - u_r|\,dy = 0, \tag{7.109}$$

$$\lim_{\rho \to 0+}\sup_{|h|\leq \rho} \int_{-\alpha}^0 |u(x + \lambda h + \rho y, t+h) - u_l|\,dy = 0. \tag{7.110}$$

---

[2] Rademacher's theorem states that a Lipschitz function is differentiable almost everywhere; see [47, p. 81].

274   7. Well-Posedness of the Cauchy Problem

Proof. We assume that the limit in (7.109) is not zero. Then there exist sequences $\rho_i \to 0$ and $|h_i| < \rho_i$ and a $\delta > 0$ such that

$$\int_0^\alpha |u(x + \lambda h_i + \rho_i y, t + h_i) - u_r| \, dy > \delta \tag{7.111}$$

for all i. Without loss of generality we assume that $h_i > 0$, and let

$$v(z, h) = u(x + \lambda h + z, t + h).$$

Then the map $h \mapsto v(\,\cdot\,, h)$ is Lipschitz continuous with respect to the L^1 norm, since

$$\|v(\,\cdot\,, h) - v(\,\cdot\,, \eta)\|_1 = \int |u(z, t + h) - u(\lambda(\eta - h) + z, t + \eta)| \, dz$$

$$\leq \int |u(z, t + h) - u(z, t + \eta)| \, dz$$

$$+ \int |u(z, t + \eta) - u(\lambda(\eta - h) + z, t + \eta)| \, dz$$

$$\leq M |h - \eta| + \lambda |\eta - h| \, \text{T.V.} \, (u(t + \eta))$$

$$\leq \widetilde{M} |\eta - h|.$$

From (7.111) we obtain

$$\int_0^{\alpha \rho_i} |u(x + \lambda h + z, t + h) - u_r| \, dz$$

$$\geq \int_0^{\alpha \rho_i} |u(x + \lambda h_i + z, t + h_i) - u_r| \, dz$$

$$- \int_0^{\alpha \rho_i} |u(x + \lambda h_i + z, t + h_i) - u(x + \lambda h + z, t + h)| \, dz$$

$$\geq \delta \rho_i - \widetilde{M} |h_i - h|.$$

We can (safely) assume that $\delta/\widetilde{M} < 1$ (if this is not so, then (7.111) will hold for smaller δ as well). We integrate the last inequality with respect to h, for h in $[-\rho_i, \rho_i]$. Since $[h_i - \rho_i \delta/\widetilde{M}, h_i] \subset [-\rho_i, \rho_i]$, we obtain

$$\int_{-\rho_i}^{\rho_i} \int_0^{\alpha \rho_i} |u(x + \lambda h + z, t + h) - u_r| \, dz \, dh$$

$$\geq \int_{h_i - \rho_i \delta/\widetilde{M}}^{h_i} (\delta \rho_i - \widetilde{M}(h_i - h)) \, dh$$

$$= (\delta^2 \rho_i^2)/(2\widetilde{M}).$$

Comparing this with (7.97) and (7.98) yields a contradiction. The limit (7.111) is proved similarly. □

7.2. Uniqueness 275

For the next lemma, recall that a (signed) Radon measure is a (signed) regular Borel measure[3] that is finite on compact sets.

Lemma 7.17. *Assume that w is in $L^1(\langle a,b \rangle ; \mathbb{R}^n)$ such that for some Radon measure μ, we have that*

$$\left| \int_{x_1}^{x_2} w(x)\,dx \right| \leq \mu([x_1, x_2]) \quad \text{for all } a < x_1 < x_2 < b. \tag{7.112}$$

Then

$$\int_a^b |w(x)|\,dx \leq \mu(\langle a,b \rangle). \tag{7.113}$$

Proof. First observe that the assumptions of the lemma also hold if the closed interval on the right-hand side of (7.112) is replaced by an open interval. We have that

$$\left| \int_{x_1}^{x_2} w(x)\,dx \right| = \lim_{\varepsilon \to 0} \left| \int_{x_1+\varepsilon}^{x_2-\varepsilon} w(x)\,dx \right|$$
$$\leq \lim_{\varepsilon \to 0} \mu([x_1+\varepsilon, x_2-\varepsilon]) = \mu(\langle x_1, x_2 \rangle).$$

Secondly, since w is in L^1, it can be approximated by piecewise constant functions. Let v be a piecewise constant function with discontinuities located at $a = x_0 < x_1 < \cdots < x_N = b$, and

$$\int_a^b |w(x) - v(x)|\,dx \leq \varepsilon.$$

Then we have

$$\int_a^b |w(x)|\,dx \leq \int_a^b |w(x) - v(x)| + \int_a^b |v(x)|\,dx$$
$$\leq \varepsilon + \sum_i \int_{x_{i-1}}^{x_i} |v(x)|\,dx$$
$$= \varepsilon + \sum_i \left| \int_{x_{i-1}}^{x_i} v(x)\,dx \right|$$
$$\leq \varepsilon + \sum_i \left| \int_{x_{i-1}}^{x_i} (v(x) - w(x))\,dx \right| + \sum_i \left| \int_{x_{i-1}}^{x_i} w(x)\,dx \right|$$
$$\leq \varepsilon + \int_a^b |v(x) - w(x)|\,dx + \sum_i \mu(\langle x_{i-1}, x_i \rangle)$$
$$\leq 2\varepsilon + \mu(\langle a,b \rangle).$$

---

[3] A Borel measure μ is regular if it is outer regular on all Borel sets (i.e., $\mu(B) = \inf\{\mu(A) \mid A \supseteq B,\ A \text{ open}\}$ for all Borel sets B) and inner regular on all open sets (i.e., $\mu(U) = \sup\{\mu(K) \mid K \subset U,\ K \text{ compact}\}$ for all open sets U).

Since ε can be made arbitrarily small, this proves the lemma. □

Next we need two results that state how well the semigroup is approximated firstly by the solution of a Riemann problem with states that are close to the initial state for the semigroup, and secondly by the solution of the linearized equation. To define this precisely, let w_0 be a function in $\mathcal{D}$, fix a point x on the real line (which will remain fixed throughout the next lemma and its proof), and let $w(y,t)$ be the solution of the Riemann problem

$$w_t + f(w)_y = 0, \qquad w(y,0) = \begin{cases} w_0(x-) & \text{for } y < 0, \\ w_0(x+) & \text{for } y \geq 0. \end{cases}$$

(If w_0 is continuous at x, then $w(y,t) = w_0(x)$ is constant.) Define $\tilde{A} = df(w_0(x+))$, and let $\tilde{u}$ be the solution of the linearized equation

$$\tilde{u}_t + \tilde{A}\tilde{u}_y = 0, \qquad \tilde{u}(y,0) = w_0(y). \tag{7.114}$$

Furthermore, define $\hat{u}(y,t)$ by

$$\hat{u}(y,t) = \begin{cases} w(y-x,t) & \text{for } |y-x| \leq \sigma^* t, \\ w_0(y) & \text{otherwise.} \end{cases} \tag{7.115}$$

Then we can state the following lemma.

Lemma 7.18. *Let $w_0 \in \mathcal{D}$. Then we have*

$$\frac{1}{h}\int_{x-\rho+h\sigma^*}^{x+\rho-h\sigma^*} |(S_h w_0)(y) - \hat{u}(y,h)|\, dy$$

$$= \mathcal{O}(1)\, \text{T.V.}_{\langle x-\rho,x\rangle \cup \langle x, x+\rho\rangle}(w_0), \tag{7.116}$$

$$\frac{1}{h}\int_{x-\rho+h\sigma^*}^{x+\rho-h\sigma^*} |(S_h w_0)(y) - \tilde{u}(y,h)|\, dy$$

$$= \mathcal{O}(1)\left(\text{T.V.}_{\langle x-\rho,x+\rho\rangle}(w_0)\right)^2, \tag{7.117}$$

for all x and all positive h and ρ such that $x - \rho + h\sigma^* < x + \rho - h\sigma^*$.

Proof. We first prove (7.117). In the proof of this we shall need the following general result:

Let $\bar{v}$ be the solution of $\bar{v}_t + f(\bar{v})_y = 0$ with Riemann initial data

$$\bar{v}(y,0) = \begin{cases} u_l & \text{for } y < 0, \\ u_r & \text{for } y \geq 0, \end{cases}$$

for some states $u_{l,r} \in \Omega$. We have that this Riemann problem is solved by waves separating constant states $u_l = v_0, v_1, \ldots, v_n = u_r$. Let u^c be a constant in Ω and set $A^c = df(u^c)$. Assume that u_l and u_r satisfy

$$A^c(u_l - u_r) = \lambda_k^c(u_l - u_r);$$

7.2. Uniqueness

i.e., λ_k^c is the kth eigenvalue and $u_l - u_r$ is the kth eigenvector of A^c. Let $\tilde{v}$ be defined by

$$\tilde{v}(y,t) = \begin{cases} u_l & \text{for } y < \lambda_k^c t, \\ u_r & \text{for } y \geq \lambda_k^c t \end{cases}$$

($\tilde{v}$ solves $u_t + A^c u_y = 0$ with a single jump at $y = 0$ from u_l to u_r as initial data). We wish to estimate

$$I = \frac{1}{t} \int_{-\sigma^* t}^{\sigma^* t} |\bar{v}(y,t) - \tilde{v}(y,t)| \, dy.$$

Note that since $\bar{v}$ and $\tilde{v}$ are equal outside the range of integration, the limits in the integral can be replaced by $\mp\infty$.

Due to the hyperbolicity of the system, the vectors $\{r_j(u)\}_{j=1}^n$ form a basis in $\mathbb{R}^n$, and hence we can find unique numbers $\bar{\varepsilon}_j^{l,r}$ such that

$$u_r - u_l = \sum_{j=1}^n \bar{\varepsilon}_j^l r_j(u_l) = \sum_{j=n}^1 \bar{\varepsilon}_j^r r_j(u_r). \tag{7.118}$$

From $u_r - u_l = \varepsilon^c r_k(u^c)$ for some ε^c it follows that

$$\bar{\varepsilon}_i^l = l_i(u_l) \cdot \sum_{j=1}^n \bar{\varepsilon}_j^l r_j(u_l)$$
$$= l_i(u_l) \cdot (u_l - u_r)$$
$$= (l_i(u_l) - l_i(u^c)) \cdot (u_l - u_r) + l_i(u^c) \cdot (u_l - u_r)$$
$$= (l_i(u_l) - l_i(u^c)) \cdot (u_l - u_r) + \varepsilon^c l_i(u^c) \cdot r_k(u^c)$$
$$= (l_i(u_l) - l_i(u^c)) \cdot (u_l - u_r), \quad i \neq k.$$

Thus we conclude (using an identical argument for the right state) that

$$\begin{aligned} |\bar{\varepsilon}_i^l| &\leq C \, |u_l - u_r| \, |u_l - u^c|, \quad i \neq k, \\ |\bar{\varepsilon}_i^r| &\leq C \, |u_l - u_r| \, |u_r - u^c|, \quad i \neq k. \end{aligned} \tag{7.119}$$

Let ε_i denote the strength of the ith wave in $\bar{v}$. Then, by construction of the solution of the Riemann problem, for $i < k$ we have that

$$|\varepsilon_i - \bar{\varepsilon}_i^l| \leq C \left(|v_{i-1} - u_l|^2 + |v_i - u_l|^2 \right) \leq C \, |u_l - u_r|^2,$$

while for $i > k$ we find that

$$|\varepsilon_i - \bar{\varepsilon}_i^r| \leq C \, |u_l - u_r|^2,$$

for some constant C. Assume that the k-wave in $\bar{v}$ moves with speed in the interval $[\underline{\lambda}_k, \bar{\lambda}^k]$; i.e., if the k-wave is a shock, then $\underline{\lambda}_k = \bar{\lambda}^k = \mu_k(v_{k-1}, v_k)$, and if the wave is a rarefaction wave, then $\underline{\lambda}_k = \lambda_k(v_{k-1})$ and $\bar{\lambda}_k = \lambda_k(v_k)$.

Set $\underline{s} = \min(\underline{\lambda}_k, \tilde{\lambda}_k)$ and $\bar{s} = \max(\bar{\lambda}_k, \tilde{\lambda}_k)$. Then we can write I as

$$I = \frac{1}{t}\left(\int_{-\infty}^{\underline{s}} |u_l - \bar{v}(y,t)|\, dy\right.$$
$$\left. + \int_{\underline{s}}^{\bar{s}} |\hat{v}(y,t) - \bar{v}(y,t)|\, dy + \int_{\bar{s}}^{\infty} |u_r - \bar{v}(y,t)|\, dy\right)$$
$$= I_1 + I_2 + I_3.$$

Next we note that the first integral above can be estimated as

$$I_1 \le C\sum_{i=1}^{k-1} |v_i - u_l| \le C\sum_{i=1}^{k-1} |\varepsilon_i| \le C\left(\sum_{i=1}^{k-1} |\bar{\varepsilon}_i^l| + |u_r - u_l|^2\right),$$

and similarly,

$$I_3 \le C\left(\sum_{i=k+1}^{n} |\bar{\varepsilon}_i^r| + |u_l - u_r|^2\right).$$

Using (7.119) we obtain

$$I_1 + I_3 \le C|u_l - u_r|(|u_l - u^c| + |u_r - u^c| + |u_l - u_r|)$$
$$\le C|u_l - u_r|(|u_l - u^c| + |u_r - u^c|), \qquad (7.120)$$

for some constant C. It remains to estimate I_2. We first assume that the k-wave in $\bar{v}$ is a shock wave and that $\lambda_k^c > \mu_k(v_{k-1}, v_k)$. Then

$$I_3 = (\lambda_k^c - \mu_k(v_{k-1}, v_k))|u_l - v_k|$$
$$\le C|u_l - v_k|(|u^c - v_{k-1}| + |u^c - v_k|)$$
$$\le C|u_l - u_r|(|u_l - u^c| + |u_r - u^c| + |v_k - u_r| + |v_{k-1} - u_l|),$$
$$\le C|u_l - u_r|(|u_l - u^c| + |u_r - u^c|) + C|u_l - u_r|(|u_l - u^c| + |u_r - u^c|))$$
$$\le C|u_l - u_r|(|u_l - u^c| + |u_r - u^c|) \qquad (7.121)$$

by the above estimates for $|v_k - u_r|$ and $|v_{k-1} - u_l|$. If $\lambda_k^c \le \mu_k(v_{k-1}, v_k)$ or the k-wave is a rarefaction wave, we similarly establish (7.121). Combining this with (7.120) we find that

$$I \le C|u_l - u_r|(|u_l - u^c| + |u_r - u^c|). \qquad (7.122)$$

Having established this preliminary estimate we turn to the proof of (7.117). Let $\bar{\omega}_0$ be a piecewise constant approximation to ω_0 such that

$$\bar{\omega}_0(x\pm) = \omega_0(x\pm), \qquad \int_{x-\rho}^{x+\rho} |\bar{\omega}_0(y) - \omega_0(y)|\, dy \le \epsilon,$$
$$\text{T.V.}_{\langle x-\rho, x+\rho\rangle}(\bar{\omega}_0) \le \text{T.V.}_{\langle x-\rho, x+\rho\rangle}(\omega_0). \qquad (7.123)$$

Furthermore, let v be the solution of the linear hyperbolic problem

$$v_t + \tilde{A}v_y = 0, \quad v(y,0) = \bar{\omega}_0(y),$$

where again $\tilde{A} = df(w_0(x+))$. Let the eigenvalues and the right and left eigenvectors of $\tilde{A}$ be denoted by $\tilde{\lambda}_k$, $\tilde{r}_k$, and $\tilde{l}_k$, respectively, for $k = 1, \ldots, n$, normalized so that

$$\left|\tilde{l}_k\right| = 1, \quad \tilde{l}_k \cdot \tilde{r}_j = \begin{cases} 0 & \text{for } j \neq k, \\ 1 & \text{for } j = k. \end{cases} \tag{7.124}$$

Then it is not too difficult to verify (see Exercise 10) that $v(y, t)$ is given by

$$v(y, t) = \sum_k \left(\tilde{l}_k \cdot \bar{\omega}_0(y - \tilde{\lambda}_k t)\right) \tilde{r}_k. \tag{7.125}$$

We can also construct v by front tracking. Since the eigenvalues are constant and the initial data piecewise constant, front tracking will give the exact solution. Hence v will be piecewise constant with a finite number of jumps occurring at $x_i(t)$, where we have that

$$\frac{d}{dt} x_i(t) = \tilde{\lambda}_k,$$

$$\left(\tilde{A} - \tilde{\lambda}_k I\right) \left(v(x_i(t)+, t) - v(x_i(t)-, t)\right) = 0,$$

for all t where we do not have a collision of fronts, that is, for all but a finite number of t's. Now we apply the estimate (7.122) to each individual front x_i. Then we obtain, introducing $v_i^\pm = v(x_i(t)\pm, t)$,

$$\int_{x-\rho+\sigma^\star \varepsilon}^{x+\rho-\sigma^\star \varepsilon} |(\mathcal{S}_\varepsilon v(\,\cdot\,, \tau))(y) - v(y, \tau + \varepsilon)|\, dy$$

$$\leq \varepsilon \mathcal{O}(1) \sum_i \left|v_i^+ - v_i^-\right| \left(\left|v_i^+ - w_0(x+)\right| + \left|v_i^- - w_0(x+)\right|\right)$$

$$\leq \varepsilon \mathcal{O}(1) \, \text{T.V.}_{\langle x-\rho, x+\rho\rangle}(\bar{\omega}_0) \sum_i \left|v_i^+ - v_i^-\right|$$

$$\leq \varepsilon \mathcal{O}(1) \left(\text{T.V.}_{\langle x-\rho, x+\rho\rangle}(w_0)\right)^2. \tag{7.126}$$

Recall that $\tilde{A} = df(w_0(x+))$ and that $\tilde{u}$ was defined by (7.114), that is,

$$\tilde{u}_t + \tilde{A}\tilde{u}_y = 0, \qquad \tilde{u}(y, 0) = w_0(y). \tag{7.127}$$

In analogy to formula (7.125) we have that $\tilde{u}$ satisfies

$$\tilde{u}(y, t) = \sum_k \left(\tilde{l}_k \cdot w_0(y - \tilde{\lambda}_k t)\right) \tilde{r}_k. \tag{7.128}$$

Regarding the difference between $\tilde{u}$ and v, we find that

$$\int_{x-\rho+\sigma^\star h}^{x+\rho-\sigma^\star h} |v(y, h) - \tilde{u}(y, h)|\, dy \tag{7.129}$$

$$= \int_{x-\rho+\sigma^\star h}^{x+\rho-\sigma^\star h} \left|\sum_k \left(\tilde{l}_k \cdot (\bar{\omega}_0 - w_0)(y - \tilde{\lambda}_k h)\right) \tilde{r}_k\right| dy$$

$$\leq \mathcal{O}\left(1\right) \int_{x-\rho}^{x+\rho} |\bar{w}_0(y) - w_0(y)|\, dy$$

$$\leq \mathcal{O}\left(1\right)\epsilon. \tag{7.130}$$

By the Lipschitz continuity of the semigroup we have that

$$\int_{x-\rho+\sigma^\star h}^{x+\rho-\sigma^\star h} |\mathcal{S}_h\bar{w}_0(y) - \mathcal{S}_h w_0(y)|\, dy \leq L\int_{x-\rho}^{x+\rho} |\bar{w}_0(y) - w_0(y)|\, dy \leq L\epsilon. \tag{7.131}$$

Furthermore, by Lemma 7.15 with $a = x - \rho$, $b = x + \rho$, $T = h$, and $t = 0$, and using (7.126), we obtain

$$\frac{1}{h}\int_{x-\rho+\sigma^\star h}^{x+\rho-\sigma^\star h} |(\mathcal{S}_h\bar{w}_0)(y) - v(y,h)|\, dy$$

$$\leq \frac{\mathcal{O}\left(1\right)}{h}\int_0^h \liminf_{\varepsilon\to 0+} \frac{1}{\varepsilon}\int_{x-\rho+\sigma^\star\varepsilon}^{x+\rho-\sigma^\star\varepsilon} |(\mathcal{S}_\varepsilon v\left(\,\cdot\,,\tau\right))(y) - v(y,\tau+\varepsilon)|\, dy\, d\tau$$

$$\leq \mathcal{O}\left(1\right)\left(\text{T.V.}_{\langle x-\rho,x+\rho\rangle}(w_0)\right)^2. \tag{7.132}$$

Consequently, using (7.132), (7.131), and (7.130) we find that

$$\frac{1}{h}\int_{x-\rho+h\sigma^\star}^{x+\rho-h\sigma^\star} |(\mathcal{S}_h w_0)(y) - \tilde{u}(y,h)|\, dy$$

$$\leq \mathcal{O}\left(1\right)\left(\text{T.V.}_{\langle x-\rho,x+\rho\rangle}(w_0)\right)^2 + \frac{L\epsilon}{h} + \mathcal{O}\left(1\right)\frac{\epsilon}{h}.$$

Since ϵ is arbitrary, this proves (7.117).

Now we turn to the proof of (7.116). First we define z to be the function

$$z(y,t) = \begin{cases} u_l & \text{for } y < \lambda t, \\ u_r & \text{for } y \geq \lambda t, \end{cases}$$

where $|\lambda| \leq \sigma^\star$. Recall that $\bar{v}(y,t)$ denotes the solution of $\bar{v}_t + f(\bar{v})_y = 0$ with Riemann initial data

$$\bar{v}(y,0) = \begin{cases} u_l & \text{for } y < 0, \\ u_r & \text{for } y \geq 0. \end{cases}$$

Then trivially we have that

$$\int_{-\sigma^\star t}^{\sigma^\star t} |z(y,t) - \bar{v}(y,t)|\, dy \leq t\mathcal{O}\left(1\right)|u_l - u_r|. \tag{7.133}$$

Let $\bar{w}_0$ be as (7.123) but replacing the TV bound by

7.2. Uniqueness

$$\text{T.V.}_{\langle x-\rho,x\rangle\cup\langle x,x+\rho\rangle}(\bar{\omega}_0) \leq \text{T.V.}_{\langle x-\rho,x\rangle\cup\langle x,x+\rho\rangle}(\omega_0).$$

Recall that $\hat{u}(y,t)$ was defined in (7.115) by

$$\hat{u}(y,t) = \begin{cases} \omega(y-x,t) & \text{for } |y-x| \leq \sigma^* t, \\ \omega_0(y) & \text{otherwise.} \end{cases}$$

Let J_h be the set

$$J_h = \{y \mid h\sigma^* < |y-x| < \rho - h\sigma^*\},$$

and let $\hat{v}$ be the function defined by

$$\hat{v}(y,t) = \begin{cases} \hat{u}(y,t) & \text{for } |x-y| \leq \sigma^* t, \\ \bar{\omega}_0(y) & \text{otherwise.} \end{cases}$$

Then we have that

$$\int_{x-\rho+\sigma^* h}^{x+\rho-\sigma^* h} |\hat{v}(y,h) - \hat{u}(y,h)|\, dy \leq \int_{J_h} |\bar{\omega}_0(y) - \omega_0(y)|\, dy \leq \epsilon. \tag{7.134}$$

Note that the bound (7.131) remains valid. We need a replacement for (7.126). In this case we wish to estimate

$$I = \int_{x-\rho+\sigma^* \varepsilon}^{x+\rho-\sigma^* \varepsilon} |(\mathcal{S}_\varepsilon v(\,\cdot\,,\tau))(y) - \bar{v}(y,\tau+\varepsilon)|\, dy.$$

For $|x-y| > \sigma^* t$, the function $\bar{v}(y,t)$ is discontinuous across lines located at x_i. In addition it may be discontinuous across the lines $|x-y| = \sigma^* t$. Inside the region $|x-y| \leq \sigma^* t$, v is an exact entropy solution, coinciding with the semigroup solution. Using this and (7.133) we find that

$$I = \left(\int_{x-\rho+\sigma^*\varepsilon}^{x-\sigma^*\tau} + \int_{x+\sigma^*\tau}^{x+\rho-\sigma^*\varepsilon}\right) |(\mathcal{S}_{\tau+\varepsilon}\bar{\omega}_0)(y) - \bar{\omega}_0(y)|\, dy$$
$$+ \int_{x-\sigma^*\tau}^{x+\sigma^*\tau} |(\mathcal{S}_\varepsilon \hat{u}(\,\cdot\,,\tau))(y) - \hat{u}(y,\tau+\varepsilon)|\, dy$$

282   7. Well-Posedness of the Cauchy Problem

$$\leq \varepsilon\mathcal{O}(1)\left(\sum_{|x_i-x|<\sigma^*\tau}|\bar{\omega}_0(x_i+)-\bar{\omega}_0(x_i-)|\right)$$

$$+L\left(\int_{x-2\sigma^*\tau}^{x}|\bar{\omega}_0(y)-u_l|\,dy+\int_{x}^{x+2\sigma^*\tau}|\bar{\omega}_0(y)-u_r|\,dy\right)$$

$$\leq \varepsilon\mathcal{O}(1)\,\text{T.V.}_{\langle x-\rho,x\rangle\cup\langle x,x+\rho\rangle}(\omega_0). \tag{7.135}$$

Now using Lemma 7.15 we find that

$$\frac{1}{h}\int_{x-\rho+\sigma^*h}^{x+\rho-\sigma^*h}|(\mathcal{S}_h\bar{\omega}_0)(y)-\bar{v}(y,t)|\,dy$$

$$\leq \frac{\mathcal{O}(1)}{h}\int_0^h\liminf_{\varepsilon\to 0+}\frac{1}{\varepsilon}\int_{x-\rho+\sigma^*\varepsilon}^{x+\rho-\sigma^*\varepsilon}|(\mathcal{S}_\varepsilon\bar{v}(\cdot,\tau))(y)-\bar{v}(y,\tau+\varepsilon)|\,dy\,d\tau$$

$$\leq \mathcal{O}(1)\,\text{T.V.}_{\langle x-\rho,x\rangle\cup\langle x,x+\rho\rangle}(\omega_0). \tag{7.136}$$

As before, since ε is arbitrary, (7.131), (7.134), and (7.136) imply (7.116). □

Remark 7.19. Note that if ω_0 is continuous at x, then Lemma 7.18 and (7.117) say that the linearized equation gives a good local approximation of the action of the semigroup. If ω_0 has a discontinuity at x, then

$$\text{T.V.}_{\langle x-\rho,x+\rho\rangle}(\omega_0)$$

does not become small as ρ tends to zero; hence we must resort to (7.116) in this case. Since the total variation of any function in $\mathcal{D}$ is small, (7.117) is a much stronger estimate than (7.116).

Now that the preliminary technicalities are out of the way, we can set about proving that an entropy solution coincides with the semigroup.

Let u be an entropy solution. To prove that $u(\cdot,t)=\mathcal{S}_t u_0$, it suffices to show, applying Lemma 7.15, that

$$\liminf_{h\to 0}\frac{1}{h}\|\mathcal{S}_h u(\cdot,t)-u(\cdot,t+h)\|_{L^1([a,b])}=0, \tag{7.137}$$

for all $a<b$, and for all $t\in[0,T]\setminus\mathcal{N}$.

Assume therefore that $t\notin\mathcal{N}$. Then by the structure theorem, see [148, Theorem 5.9.6], there exists a null set $\mathcal{N}\subset[0,T]$ such that outside that set u is either continuous or has a jump discontinuity (as a function of x). Therefore, we split the argument into two cases, one where u has a jump discontinuity, and the case where u is continuous or has a small jump in the sense that it is not in the set $B_{t,\varepsilon}$.

Consider first a point (x,t) where u has jump discontinuity.[4] By condition **B** there exist $u_{l,r}\in\Omega$ and σ such that the limit (7.98) holds when U is

---

[4] The following argument is valid for any jump discontinuity, but will be applied only to jumps in $B_{t,\varepsilon}$.

7.2. Uniqueness

defined by (7.97). Using a change of variables, we find that

$$\lim_{h \to 0^+} \frac{1}{h} \int_{x-\sigma^\star h}^{x+\sigma^\star h} |u(y, t+h) - U(y, t+h)| \, dy$$

$$= \lim_{h \to 0^+} \sigma^\star \left[\int_{-1-\lambda/\sigma^\star}^{0} |u(x + \lambda h + \sigma^\star h y, t+h) - u_l| \, dy \right.$$

$$\left. + \int_{0}^{1-\lambda/\sigma^\star} |u(x + \lambda h + \sigma^\star h y, t+h) - u_r| \, dy \right] = 0,$$

by Lemma 7.16. Hence for small positive h, we have that

$$\frac{1}{h} \int_{x-\sigma^\star h}^{x+\sigma^\star h} |u(y, t+h) - U(y, t+h)| \, dy \le \tilde{\varepsilon}, \tag{7.138}$$

for some small $\tilde{\varepsilon}$ to be determined later. By Lemma 7.14 we have $U(y, s) = \hat{u}(y, s-t)$, where $\hat{u}$ is defined by (7.115) with $\omega_0(y) = u(y,t)$, and U is defined by (7.97), in some forward neighborhood of (x,t). Then using (7.138) and (7.116) we obtain

$$\frac{1}{h} \int_{x-\sigma^\star h}^{x+\sigma^\star h} |(\mathcal{S}_h u(\,\cdot\,, t))(y) - u(y, t+h)| \, dy$$

$$\le \tilde{\varepsilon} + \frac{1}{h} \int_{x-\sigma^\star h}^{x+\sigma^\star h} |(\mathcal{S}_h u(\,\cdot\,, t))(y) - U(y, t+h)| \, dy$$

$$\le \tilde{\varepsilon} + \mathcal{O}(1) \, \text{T.V.}_{\langle x-2\sigma^\star h, x\rangle \cup \langle x, x+2\sigma^\star h\rangle} (u(\,\cdot\,, t))$$

$$\le 2\tilde{\varepsilon}, \tag{7.139}$$

for all h sufficiently small, since we compute the total variation on a shrinking interval excluding the jump in u at x.

Now we consider points (x,t) where u is either continuous or has a small jump discontinuity. Hence we can choose an interval $\langle c, d \rangle$ centered about x such that $B_{t,\varepsilon} \cap \langle c, d \rangle = \emptyset$. Recall that $B_{t,\varepsilon}$, defined in (7.102), is the set of points where $u(\,\cdot\,,t)$ has a jump larger than ε. Let the family of trapezoids Γ_h be defined by

$$\Gamma_h = \{ (y, s) \mid s \in [t, t+h], \; y \in \langle c + \sigma^\star(s-t), d - \sigma^\star(s-t) \rangle \}.$$

Now we claim that for h sufficiently small, we have that for all $(y,s) \in \Gamma_h$,

$$|u(y, s) - u(x, t)| \le 2\varepsilon + \text{T.V.}_{\langle c, d \rangle} (u(\,\cdot\,, t)). \tag{7.140}$$

To prove this, we argue as follows: By Lemma 7.14 discontinuities in u cannot propagate faster than $\sigma^\star$; hence $u(\,\cdot\,,t)$ is continuous in the lower corners of Γ_h, and the estimate surely holds for (y,s) located there. We must prove (7.140) for (y,s) in a region $[c+h', d-h'] \times [t, t+h]$ where h' is given and we can be free to choose h small. Now also $[c+h', d-h'] \cap B_{\varepsilon, t} = \emptyset$; hence for each $y \in [c+h', d-h']$ we can find ξ_y, h_y such that the estimate

(7.140) is valid for
$$(y, s) \in \langle y - \xi_y, y + \xi_y \rangle \times [t, t + h_y].$$

Now we can cover the compact interval $[c + h', d - h']$ with a finite number of intervals of the form $\langle y_i - \xi_{y_i}, y_i + \xi_{y_i} \rangle$, and choose
$$h = \min_i h_{y_i}.$$

Then we obtain (7.140) for (y, s) in $[c + h', d - h'] \times [t, t + h]$.

Now we must compare u and $\tilde{u}$ near (x, t). The eigenvectors of $\tilde{A} = df(u(x,t))$ are normalized according to (7.124). Observe that trivially
$$u = \sum_k \left(\tilde{l}_k \cdot u\right) \tilde{r}_k.$$

Then
$$\int_{c+\sigma^* h}^{d-\sigma^* h} |u(y, t+h) - \tilde{u}(y, t+h)| \, dy$$
$$\leq \sum_k \int_{c+\sigma^* h}^{d-\sigma^* h} \left| \tilde{l}_k \cdot \left(u(y - \tilde{\lambda}_k h, t) - u(y, t+h)\right) \right| dy. \tag{7.141}$$

To aid us here we use Lemma 7.17. Let $x_1 < x_2$ be in the interval $\langle c + \sigma^* h, d - \sigma^* h \rangle$. Then we shall estimate
$$E_k = \int_{x_1}^{x_2} \tilde{l}_k \cdot \left(u(y, t+h) - u(y - \tilde{\lambda}_k h, t)\right) dy.$$

If we integrate the conservation law over the region
$$\{(y, s) \mid y \in [x_1 - (s - (t+h))\tilde{\lambda}_k, x_2 + (s - (t+h))\tilde{\lambda}_k], \ s \in [t, t+h]\},$$
we find that
$$\int_{x_1}^{x_2} u(y, t+h) \, dy - \int_{x_1 - \tilde{\lambda}_k h}^{x_2 + \tilde{\lambda}_k h} u(y, t) \, dy$$
$$+ \int_t^{t+h} (f(u) - \tilde{\lambda}_k u)(x_2 + (s - (t+h))\tilde{\lambda}_k, s) \, ds$$
$$- \int_t^{t+h} (f(u) - \tilde{\lambda}_k u)(x_1 - (s - (t+h))\tilde{\lambda}_k, s) \, ds = 0.$$

Taking the inner product with $\tilde{l}_k$ we obtain
$$E_k = \int_t^{t+h} \tilde{l}_k \cdot (f(u) - \tilde{\lambda}_k u)(x_2 + (s - (t+h))\tilde{\lambda}_k, s) \, ds$$
$$- \int_t^{t+h} \tilde{l}_k \cdot (f(u) - \tilde{\lambda}_k u)(x_1 - (s - (t+h))\tilde{\lambda}_k, s) \, ds$$
$$= \int_t^{t+h} \tilde{l}_k \cdot \left(f(u_2) - f(u_1) - \tilde{\lambda}_k (u_2 - u_1)\right) ds, \tag{7.142}$$

7.2. Uniqueness

where we have defined
$$u_1 = u(x_1 - (s-(t+h))\tilde{\lambda}_k, s), \quad u_2 = u(x_2 + (s-(t+h))\tilde{\lambda}_k, s).$$

Let $A^\star$ denote the matrix
$$A^\star = \int_0^1 df(su_2 + (1-s)u_1)\, ds - \tilde{A}.$$

Then
$$\begin{aligned}\tilde{l}_k \cdot \big(f(u_2) - f(u_1) - \tilde{\lambda}_k(u_2-u_1)\big) &= \tilde{l}_k \cdot \big(A^\star(u_2-u_1) \\ &\quad + \tilde{A}(u_2-u_1) - \tilde{\lambda}_k(u_2-u_1)\big) \\ &= \tilde{l}_k \cdot A^\star(u_2-u_1).\end{aligned} \qquad (7.143)$$

Since
$$\|A^\star\| \leq \mathcal{O}(1)\,(|u_1 - u(x,t)| + |u_2 - u(x,t)|),$$

(7.142) and (7.143) yield
$$\begin{aligned}|E_k| &\leq \mathcal{O}(1) \int_t^{t+h} (|u_1 - u(x,t)| + |u_2 - u(x,t)|)\, |u_2-u_1|\, ds \\ &\leq \mathcal{O}(1) \sup_{(y,s)\in\Gamma_h} |u(y,s) - u(x,t)| \\ &\quad \times \int_t^{t+h} \mathrm{T.V.}_{\langle x_1-(s-(t+h))\tilde{\lambda}_k,\, x_2+(s-(t+h))\tilde{\lambda}_k\rangle}(u(\,\cdot\,,s))\, ds.\end{aligned}$$

Therefore,
$$\begin{aligned}\left|\int_{x_1}^{x_2} (u(y,t+h) - \tilde{u}(y,t+h))\, dy\right| &\leq \sum_k |E_k| \\ &\leq \mathcal{O}(1) \sup_{(y,s)\in\Gamma_h} |u(y,s) - u(x,t)| \\ &\quad \times \int_t^{t+h} \sum_k \mathrm{T.V.}_{\langle x_1-(s-(t+h))\tilde{\lambda}_k,\, x_2+(s-(t+h))\tilde{\lambda}_k\rangle}(u(\,\cdot\,,s))\, ds. \quad (7.144)\end{aligned}$$

Returning to (7.141) and using Lemma 7.17 we find that
$$\begin{aligned}\int_{c+\sigma^\star h}^{d-\sigma^\star h} |u(y,t+h) - \tilde{u}(y,t+h)|\, dy &\leq \mathcal{O}(1) \sup_{(y,s)\in\Gamma_h} |u(y,s) - u(x,t)| \\ &\quad \times \int_t^{t+h} \mathrm{T.V.}_{[c+\sigma^\star(s-t),\, d-\sigma^\star(s-t)]}(u(\,\cdot\,,s))\, ds.\end{aligned} \qquad (7.145)$$

Now we use (7.117), (7.145), and (7.140) to obtain

$$\int_{c+\sigma^*h}^{d-\sigma^*h} |(S_h u(\,\cdot\,,t))\,(y) - u(y,t+h)|\,dy$$

$$\leq \int_{c+\sigma^*h}^{d-\sigma^*h} \left(|(S_h u(\,\cdot\,,t))\,(y) - \tilde{u}(y,t+h)| + |\tilde{u}(y,t+h) - u(y,t+h)|\right)dy$$

$$\leq \mathcal{O}(1)\,h\,\left(\text{T.V.}_{\langle c,d \rangle}(u(\,\cdot\,,t))\right)^2$$
$$+ \mathcal{O}(1)\,(2\varepsilon + \text{T.V.}_{[c,d]}(u(\,\cdot\,,t)))$$
$$\times \int_t^{t+h} \text{T.V.}_{[c+\sigma^*(s-t),d-\sigma^*(s-t)]}(u(\,\cdot\,,s))\,ds. \qquad (7.146)$$

By Lemma 7.13 the set $B_{t,\varepsilon}$ contains only a finite number of points; $x_1 < x_2 < \cdots < x_N$ where $u(\,\cdot\,,t)$ has a discontinuity larger than ε. We can cover the set $[a,b] \setminus \cup_i \{x_i\}$ by a finite number of open intervals $\langle c_j, d_j \rangle$, $j = 1, \ldots, M$, such that:

(a) $x_i \notin \cup_j \langle c_j, d_j \rangle = \emptyset$ for $i = 1, \ldots, N$.

(b) $\text{T.V.}_{\langle c_j, d_j \rangle}(u(\,\cdot\,,t)) \leq 2\varepsilon$ for $j = 1, \ldots, M$.

(c) Every $x \in [a,b]$ is contained in at most two distinct intervals $\langle c_i, d_i \rangle$.

We have established that for sufficiently small h,

$$\frac{1}{h}\int_{x_i-\sigma^*h}^{x_i+\sigma^*h} |(S_h u(\,\cdot\,,t))\,(y) - u(y,t+h)| \leq \frac{\varepsilon}{N},$$

by (7.139) choosing $\tilde{\varepsilon} = \varepsilon/(2N)$. Also,

$$\int_{c_j+\sigma^*h}^{d_j-\sigma^*h} |(S_h u(\,\cdot\,,t))\,(y) - u(y,t+h)|\,dy$$

$$\leq \mathcal{O}(1)\,\varepsilon \int_t^{t+h} \text{T.V.}_{\langle c_j+\sigma^*(s-t), d_j-\sigma^*(s-t)\rangle}(u(\,\cdot\,,s))\,ds$$
$$+ \mathcal{O}(1)\,h\varepsilon \text{T.V.}_{\langle c_j, d_j \rangle}(u(\,\cdot\,,t))$$

for all i, j, and any $\varepsilon > 0$. Combining this we find that

$$\frac{1}{h}\int_a^b |(S_h u(\,\cdot\,,t))\,(y) - u(y,t+h)|\,dy$$

$$\leq \sum_i \frac{1}{h}\int_{x_i-\sigma^*h}^{x_i+\sigma^*h} |(S_h u(\,\cdot\,,t))\,(y) - u(y,t+h)|\,dy$$

$$+ \sum_j \int_{c_j+\sigma^\star h}^{d_j-\sigma^\star h} |(\mathcal{S}_h u(\,\cdot\,,t))\,(y) - u(y,t+h)|\; dy$$

$$\leq \varepsilon + \mathcal{O}\,(1)\,\frac{\varepsilon}{h}\int_t^{t+h} \text{T.V.}\,(u(\,\cdot\,,s))\; ds + \varepsilon \text{T.V.}\,(u(\,\cdot\,,t))$$

$$\leq \mathcal{O}\,(1)\,\varepsilon.$$

Since ε can be arbitrarily small, (7.137) holds, and we have proved the following theorem:

Theorem 7.20. *Let $f_j \in C^2(\mathbb{R}^n)$, $j = 1,\ldots,n$. Consider the strictly hyperbolic equation $u_t + f(u)_x = 0$. Assume that each wave family is either genuinely nonlinear or linearly degenerate. For every $u_0 \in \mathcal{D}$, defined by (7.94), the initial value problem*

$$u_t + f(u)_x = 0, \quad u(x,0) = u_0(x)$$

has a unique weak entropy solution satisfying conditions **A–C**, *pages 267–268. Furthermore, this solution can be found by the front-tracking construction.*

7.3 Notes

The material in Section 7.1 is taken almost entirely from the fundamental result of Bressan, Liu, and Yang [24]; there is really only an $\mathcal{O}\,(|e|)$ difference.

Stability of front-tracking approximations to systems of conservation laws was first proved by Bressan and Columbo in [19], in which they used a pseudopolygon technique to "differentiate" the front-tracking approximation with respect to the initial location of the fronts. This approach was since used to prove stability for many special systems; see [34], [4], [1], [2].

The same results as those in Section 7.1 of this chapter have also been obtained by Bressan, Crasta, and Piccoli, using a variant of the pseudopolygon approach [20]. This does lead to *many* technicalities, and [20] is heavy reading indeed!

The material in Section 7.2 is taken from the works of Bressan [17, 18] and coworkers, notably Lewicka [23], Goatin [21], and LeFloch [22].

There are few earlier results on uniqueness of solutions to systems of conservation laws; most notable are those by Bressan [14], where uniqueness and stability are obtained for Temple class system where every characteristic field is linearly degenerate, and in [16] for more general Temple class systems.

Continuity in L^1 with respect to the initial data was also proved by Hu and LeFloch [75] using a variant of Holmgren's technique. See also [57].

Stability for some non-strictly hyperbolic systems of conservation laws (these are really only "quasi-systems") has been proved by Winther and Tveito [142] and Klingenberg and Risebro [84].

We end this chapter with a suitable quotation:

> This is really easy:
>
> |what you have| ≤ |what you want|
>
> + |what you have − what you want|
>
> *Rinaldo Colombo, private communication*

Exercises

7.1 Show that the solution of the Cauchy problem obtained by the front-tracking construction of Chapter 6 is an entropy solution in the sense of conditions **A**–**C** on pages 267–268.

7.2 The proof of Theorem 7.8 is detailed only in the genuinely nonlinear case. Do the necessary estimates in the case of a linearly degenerate wave family.

Appendix A
Total Variation, Compactness, etc.

> I hate T.V. I hate it as much as peanuts.
> But I can't stop eating peanuts.
>
> *Orson Welles, The New York Herald Tribune (1956)*

A key concept in the theory of conservation laws is the notion of *total variation*, T.V. (u), of a function u of one variable. We define

$$\text{T.V.}(u) := \sup_i \sum |u(x_i) - u(x_{i-1})|. \tag{A.1}$$

The supremum in (A.1) is taken over all finite partitions $\{x_i\}$ such that $x_{i-1} < x_i$. The set of all functions with finite total variation on I we denote by $BV(I)$. Clearly, functions in $BV(I)$ are bounded. We shall omit explicit mention of the interval I if (we think that) this is not important, or if it is clear which interval we are referring to.

For any finite partition $\{x_i\}$ we can write

$$\sum_i |u(x_{i+1}) - u(x_i)| = \sum_i \max(u(x_{i+1}) - u(x_i), 0)$$
$$- \sum_i \min(u(x_{i+1}) - u(x_i), 0)$$
$$=: p + n.$$

Then the total variation of u can be written

$$\text{T.V.}(u) = P + N := \sup p + \sup n. \tag{A.2}$$

We call P the positive, and N the negative variation, of u. If for the moment we consider the finite interval $I = [a, x]$, and partitions with $a = x_1 < \cdots < x_n = x$, we have that
$$p_a^x - n_a^x = u(x) - u(a),$$
where we write p_a^x and n_a^x to indicate which interval we are considering. Hence
$$p_a^x \leq N_a^x + u(x) - u(a).$$
Taking the supremum on the left-hand side we obtain
$$P_a^x - N_a^x \leq u(x) - u(a).$$
Similarly, we have that $N_a^x - P_a^x \leq u(a) - u(x)$, and consequently
$$u(x) = P_a^x - N_a^x + u(a). \tag{A.3}$$
In other words, any function $u(x)$ in BV can be written as a difference between two increasing functions,[1]
$$u(x) = u_+(x) - u_-(x), \tag{A.4}$$
where $u_+(x) = u(a) + P_a^x$ and $u_-(x) = N_a^x$. Let ξ_j denote the points where u is discontinuous. Then we have that
$$\sum_j |u(\xi_j+) - u(\xi_j-)| \leq \text{T.V.}(u) < \infty,$$
and hence we see that there can be at most a countable set of points where $u(\xi+) \neq u(\xi-)$.

Equation (A.3) has the very useful consequence that if a function u in BV is also differentiable, then
$$\int |u'(x)|\, dx = \text{T.V.}(u). \tag{A.5}$$
This equation holds, since
$$\int |u'(x)|\, dx = \int \left(\frac{d}{dx}P_a^x + \frac{d}{dx}N_a^x\right) dx = P + N = \text{T.V.}(u).$$
We can also relate the total variation with the shifted L^1-norm. Define
$$\lambda(u, \varepsilon) = \int |u(x+\varepsilon) - u(x)|\, dx. \tag{A.6}$$
If $\lambda(u, \varepsilon)$ is a (nonnegative) continuous function in ε with $\lambda(u, 0) = 0$, we say that it is a *modulus of continuity* for u. More generally, we will use the name modulus of continuity for any continuous function $\lambda(u, \varepsilon)$ vanishing at $\varepsilon = 0$[2] such that $\lambda(u, \varepsilon) \geq \|u(\cdot + \varepsilon) - u\|_p$, where $\|\cdot\|_p$ is the L^p-norm. We

---

[1] This decomposition is often called the Jordan decomposition of u.
[2] This is *not* an exponent, but a footnote! Clearly, $\lambda(u, \varepsilon)$ is a modulus of continuity if and only if $\lambda(u, \varepsilon) = o(1)$ as $\varepsilon \to 0$.

Appendix A. Total Variation, Compactness, etc.

will need a convenient characterization of total variation (in one variable), which is described in the following lemma.

Lemma A.1. *Let u be a function in L^1. If $\lambda(u,\varepsilon)/|\varepsilon|$ is bounded as a function of ε, then u is in BV and*

$$\mathrm{T.V.}(u) = \lim_{\varepsilon \to 0} \frac{\lambda(u,\varepsilon)}{|\varepsilon|}. \tag{A.7}$$

Conversely, if u is in BV, then $\lambda(u,\varepsilon)/|\varepsilon|$ is bounded, and thus (A.7) holds. In particular, we shall frequently use

$$\lambda(u,\varepsilon) \le |\varepsilon|\,\mathrm{T.V.}(u) \tag{A.8}$$

if u is in BV.

Proof. Assume first that u is a smooth function. Let $\{x_i\}$ be a partition of the interval in question. Then

$$|u(x_i) - u(x_{i-1})| = \left| \int_{x_{i-1}}^{x_i} u'(x)\,dx \right| \le \lim_{\varepsilon \to 0} \int_{x_{i-1}}^{x_i} \left| \frac{u(x+\varepsilon) - u(x)}{\varepsilon} \right| dx.$$

Summing this over i we get

$$\mathrm{T.V.}(u) \le \liminf_{\varepsilon \to 0} \frac{\lambda(u,\varepsilon)}{|\varepsilon|}$$

for differentiable functions $u(x)$. Let u be an arbitrary bounded function in L^1, and u_k be a sequence of smooth functions such that $u_k(x) \to u(x)$ for almost all x, and $\|u_k - u\|_1 \to 0$. The triangle inequality shows that

$$|\lambda(u_k,\varepsilon) - \lambda(u,\varepsilon)| \le 2\|u_k - u\|_1 \to 0.$$

Let $\{x_i\}$ be a partition of the interval. We can now choose u_k such that $u_k(x_i) = u(x_i)$ for all i. Then

$$\sum |u(x_i) - u(x_{i-1})| \le \liminf_{\varepsilon \to 0} \frac{\lambda(u_k,\varepsilon)}{|\varepsilon|}.$$

Therefore,

$$\mathrm{T.V.}(u) \le \liminf_{\varepsilon \to 0} \frac{\lambda(u,\varepsilon)}{|\varepsilon|}.$$

Furthermore, we have

$$\begin{aligned}
\int |u(x+\varepsilon) - u(x)|\, dx &= \sum_j \int_{(j-1)\varepsilon}^{j\varepsilon} |u(x+\varepsilon) - u(x)|\, dx \\
&= \sum_j \int_0^\varepsilon |u(x+j\varepsilon) - u(x+(j-1)\varepsilon)|\, dx \\
&= \int_0^\varepsilon \sum_j |u(x+j\varepsilon) - u(x+(j-1)\varepsilon)|\, dx \\
&\leq \int_0^\varepsilon \text{T.V.}(u) \\
&= |\varepsilon|\, \text{T.V.}(u).
\end{aligned}$$

Thus we have proved the inequalities

$$\frac{\lambda(u,\varepsilon)}{|\varepsilon|} \leq \text{T.V.}(u) \leq \liminf_{\varepsilon \to 0} \frac{\lambda(u,\varepsilon)}{|\varepsilon|} \leq \limsup_{\varepsilon \to 0} \frac{\lambda(u,\varepsilon)}{|\varepsilon|} \leq \text{T.V.}(u), \quad (A.9)$$

which imply the lemma. $\square$

Observe that we trivially have

$$\tilde{\lambda}(u,\varepsilon) := \sup_{|\sigma| \leq |\varepsilon|} \lambda(u,\sigma) \leq |\varepsilon|\, \text{T.V.}(u). \qquad (A.10)$$

For functions in L^p care has to be taken as to which points are used in the supremum, since these functions in general are not defined pointwise. The right choice here is to consider only points x_i that are points of *approximate continuity*[3] of u. Lemma A.1 remains valid.

We include a useful characterization of total variation.

Theorem A.2. *Let u be a function in $L^1(I)$ where I is an interval. Assume $u \in BV(I)$. Then*

$$\text{T.V.}(u) = \sup_{\phi \in C_0^1(I),\, |\phi| \leq 1} \int_I u(x)\phi_x(x)\, dx. \qquad (A.11)$$

Conversely, if the right-hand side of (A.11) is finite for an integrable function u, then $u \in BV(I)$ and (A.11) holds.

---

[3] A function u is said to be approximately continuous at x if there exists a measurable set A such that $\lim_{r \to 0} |[x-r, x+r] \cap A| / |[x-r, x+r]| = 1$ (here $|B|$ denotes the measure of the set B), and u is continuous at x relative to A. (Every Lebesgue point is a point of approximate continuity.) The supremum (A.1) is then called the essential variation of the function. However, in the theory of conservation laws it is customary to use the name total variation in this case, too, and we will follow this custom here.

Appendix A. Total Variation, Compactness, etc.   293

Proof. Assume that u has finite total variation on I. Let ω be a nonnegative function bounded by unity with support in $[-1, 1]$ and unit integral. Define
$$\omega_\varepsilon(x) = \frac{1}{\varepsilon}\omega\left(\frac{x}{\varepsilon}\right),$$
and
$$u^\varepsilon = \omega_\varepsilon * u. \tag{A.12}$$
Consider points $x_1 < x_2 < \cdots < x_n$ in I. Then
$$\sum_i |u^\varepsilon(x_i) - u^\varepsilon(x_{i-1})|$$
$$\leq \int_{-\varepsilon}^{\varepsilon} \omega_\varepsilon(x) \sum_i |u(x_i - x) - u(x_{i-1} - x)|\, dx$$
$$\leq \text{T.V.}(u). \tag{A.13}$$
Using (A.5) and (A.13) we obtain
$$\int |(u^\varepsilon)'(x)|\, dx = \text{T.V.}(u^\varepsilon)$$
$$= \sup \sum_i |u^\varepsilon(x_i) - u^\varepsilon(x_{i-1})|$$
$$\leq \text{T.V.}(u).$$
Let $\phi \in C_0^1$ with $|\phi| \leq 1$. Then
$$\int u^\varepsilon(x)\phi'(x)\, dx = -\int (u^\varepsilon)'(x)\phi(x)\, dx$$
$$\leq \int |(u^\varepsilon)'(x)|\, dx$$
$$\leq \text{T.V.}(u),$$
which proves the first part of the theorem.
Now let u be such that
$$\|Du\| := \sup_{\substack{\phi \in C_0^1 \\ |\phi| \leq 1}} \int u(x)\phi_x(x)\, dx < \infty.$$
First we infer that
$$-\int (u^\varepsilon)'(x)\phi(x)\, dx = \int u^\varepsilon(x)\phi'(x)\, dx$$
$$= -\int (\omega_\varepsilon * u)(x)\phi'(x)\, dx$$
$$= -\int u(x)(\omega_\varepsilon * \phi)'(x)\, dx$$
$$\leq \|Du\|.$$

Using that (see Exercise A.1)
$$\|f\|_1 = \sup_{\substack{\phi \in C_0^1, \\ |\phi| \le 1}} \int f(x)\phi(x)\,dx,$$
we conclude that
$$\int |(u^\varepsilon)'(x)|\,dx \le \|Du\|. \tag{A.14}$$

Next we show that $u \in L^\infty$. Choose a sequence $u_j \in BV \cap C^\infty$ such that (see, e.g., [47, p. 172])
$$u_j \to u \text{ a.e.}, \quad \|u_j - u\|_1 \to 0, \quad j \to \infty, \tag{A.15}$$
and
$$\int |u_j'(x)|\,dx \to \|Du\|, \quad j \to \infty. \tag{A.16}$$

For any y, z we have
$$u_j(z) = u_j(y) + \int_y^z u_j'(x)\,dx.$$

Averaging over some bounded interval $J \subseteq I$ we obtain
$$|u_j| \le \frac{1}{|J|}\int_J |u_j(y)|\,dy + \int_I |u_j'(x)|\,dx, \tag{A.17}$$
which shows that the u_j are uniformly bounded, and hence $u \in L^\infty$. Thus
$$u^\varepsilon(x) \to u(x)$$
as $\varepsilon \to 0$ at each point of approximate continuity of u. Using points of approximate continuity $x_1 < x_2 < \cdots < x_n$ we conclude that
$$\sum_i |u(x_i) - u(x_{i-1})| = \lim_{\varepsilon \to 0} \sum_i |u^\varepsilon(x_i) - u^\varepsilon(x_{i-1})|$$
$$\le \limsup_{\varepsilon \to 0} \int |(u^\varepsilon)'(x)|\,dx$$
$$\le \|Du\|. \tag{A.18}$$
□

For a function u of two variables (x, y) the total variation is defined by
$$\text{T.V.}_{\cdot x, y}(u) = \int \text{T.V.}_{\cdot x}(u)(y)\,dy + \int \text{T.V.}_{\cdot y}(u)(x)\,dx. \tag{A.19}$$

The extension to functions of n variables is obvious.

Total variation is used to obtain compactness. The appropriate compactness statement is Kolmogorov's compactness theorem. We say that a subset M of a complete metric space X is *(strongly) compact* if any infinite subset

Appendix A. Total Variation, Compactness, etc.

of it contains a (strongly) convergent sequence. A set is *relatively compact* if its closure is compact. A subset of a metric space is called *totally bounded* if it is contained in a finite union of balls of radius ε for any $\varepsilon > 0$ (we call this finite union an ε-net). Our starting theorem is the following result.

Theorem A.3. *A subset M of a complete metric space X is relatively compact if and only if it is totally bounded.*

Proof. Consider first the case where M is relatively compact. Assume that there exists an ε_0 for which there is no finite ε_0-net. For any element $u_1 \in M$ there exists an element $u_2 \in M$ such that $\|u_1 - u_2\| \geq \varepsilon_0$. Since the set $\{u_1, u_2\}$ is not an ε_0-net, there has to be an $u_2 \in M$ such that $\|u_1 - u_3\| \geq \varepsilon_0$ and $\|u_2 - u_3\| \geq \varepsilon_0$. Continuing inductively construct a sequence $\{u_j\}$ such that

$$\|u_j - u_k\| \geq \varepsilon_0, \quad j \neq k,$$

which clearly cannot have a convergent subsequence, which yields a contradiction. Hence we conclude that there has to exist an ε-net for every ε.

Assume now that we can find a finite ε-net for M for every $\varepsilon > 0$, and let M_1 be an arbitrary infinite subset of M. Construct an ε-net for M_1 with $\varepsilon = \frac{1}{2}$, say $\{u_1^{(1)}, \ldots, u_{N_1}^{(1)}\}$. Now let $M_1^{(j)}$ be the set of those $u \in M_1$ such that $\|u - u_j^{(1)}\| \leq \frac{1}{4}$. At least one of $M_1^{(1)}, \ldots, M_1^{(N_1)}$ has to be infinite, since M_1 is infinite. Denote (one of) this by M_2 and the corresponding element u_2. On this set we construct an ε-net with $\varepsilon = \frac{1}{4}$. Continuing inductively we construct a nested sequence of subsets $M_{k+1} \subset M_k$ for $k \in \mathbb{N}$ such that M_k has an ε-net with $\varepsilon = 1/2^k$, say $\{u_1^{(k)}, \ldots, u_{N_k}^{(k)}\}$. For arbitrary elements u, v of M_k we have $\|u - v\| \leq \|u - u_k\| + \|u_k - v\| \leq 1/2^{k-1}$. The sequence $\{u_k\}$ with $u_k \in M_k$ is convergent, since

$$\|u_{k+m} - u_k\| \leq \frac{1}{2^{k-1}},$$

proving that M_1 contains a convergent sequence. □

A result that simplifies our argument is the following.

Lemma A.4. *Let M be a subset of a metric space X. Assume that for each $\varepsilon > 0$, there is a totally bounded set A such that $\mathrm{dist}(f, A) < \varepsilon$ for each $f \in M$. Then M is totally bounded.*

Proof. Let A be such that $\mathrm{dist}(f, A) < \varepsilon$ for each $f \in M$. Since A is totally bounded, there exist points $x_1, \ldots, x_n$ in X such that $A \subseteq \cup_{j=1}^n \mathcal{B}_\varepsilon(x_j)$, where

$$\mathcal{B}_\varepsilon(y) = \{z \in X \mid \|z - y\| \leq \varepsilon\}.$$

For any $f \in M$ there exists by assumption some $a \in A$ such that $\|a - f\| < \varepsilon$. Furthermore, $\|a - x_j\| < \varepsilon$ for some j. Thus $\|f - x_j\| < 2\varepsilon$, which proves
$$M \subseteq \bigcup_{j=1}^{n} \mathcal{B}_{2\varepsilon}(x_j).$$
Hence M is totally bounded. □

We can state and prove Kolomogorov's compactness theorem.

Theorem A.5 (Kolmogorov's compactness theorem). *Let M be a subset of $L^p(\Omega)$, $p \in [1, \infty)$, for some open set $\Omega \subseteq \mathbb{R}^n$. Then M is relatively compact if and only if the following three conditions are fulfilled:*

(i) *M is bounded in $L^p(\Omega)$, i.e.,*
$$\sup_{u \in M} \|u\|_p < \infty.$$

(ii) *We have*
$$\|u(\,\cdot\, + \varepsilon) - u\|_p \leq \lambda(|\varepsilon|)$$
for a modulus of continuity λ that is independent of $u \in M$ (we let u equal zero outside Ω).

(iii)
$$\lim_{\alpha \to \infty} \int_{\{x \in \Omega \mid |x| \geq \alpha\}} |u(x)|^p \, dx = 0 \text{ uniformly for } u \in M.$$

Remark A.6. In the case Ω is bounded, condition (iii) is clearly superfluous.

Proof. We start by proving that conditions (i)–(iii) are sufficient to show that M is relatively compact. Let φ be a nonnegative and continuous function such that $\varphi \leq 1$, $\varphi(x) = 1$ on $|x| \leq 1$, and $\varphi(x) = 0$ whenever $|x| \geq 2$. Write $\varphi_r(x) = \varphi(x/r)$. From condition (iii) we see that $\|\varphi_r u - u\| \to 0$ as $r \to \infty$. Using Lemma A.4 we see that it suffices to show that $M_r = \{\varphi_r u \mid u \in M\}$ is totally bounded. Furthermore, we see that M_r satisfies (i) and (ii). In other words, we need to prove only that (i) and (ii) together with the existence of some R so that $u = 0$ whenever $u \in M$ and $|x| \geq R$ imply that M is totally bounded. Let ω_ε be a mollifier, that is,
$$\omega \in C_0^\infty, \quad 0 \leq \omega \leq 1, \quad \int \omega \, dx = 1, \quad \omega_\varepsilon(x) = \frac{1}{\varepsilon^n} \omega\left(\frac{x}{\varepsilon}\right).$$
Then
$$\|u * \omega_\varepsilon - u\|_p^p = \int |u * \omega_\varepsilon(x) - u(x)|^p \, dx$$

$$= \int \left| \int_{\mathcal{B}_\varepsilon} (u(x-y) - u(x)) \omega_\varepsilon(y) \, dy \right|^p dx$$

$$\le \int \int_{\mathcal{B}_\varepsilon} |u(x-y) - u(x)|^p \, dy \, \|\omega_\varepsilon\|_q^p \, dx$$

$$= \varepsilon^{np/q-p} \|\omega\|_q^p \int_{\mathcal{B}_\varepsilon} \int |u(x-y) - u(x)|^p \, dx \, dy$$

$$\le \varepsilon^{np/q-p} \|\omega\|_q^p \int_{\mathcal{B}_\varepsilon} \max_{|z| \le \varepsilon} \lambda(|z|) \, dy$$

$$= \varepsilon^{n+np/q-p} \|\omega\|_q^p |\mathcal{B}_1| \max_{|z| \le \varepsilon} \lambda(|z|),$$

where $1/p + 1/q = 1$ and

$$\mathcal{B}_\varepsilon = \mathcal{B}_\varepsilon(0) = \{z \in \mathbb{R}^n \mid \|z\| \le \varepsilon\}.$$

Thus

$$\|u * \omega_\varepsilon - u\|_p \le \varepsilon^{n-1} \|\omega\|_q |\mathcal{B}_1|^{1/p} \max_{|z| \le \varepsilon} \lambda(|z|), \qquad (A.20)$$

which together with (ii) proves uniform convergence as $\varepsilon \to 0$ for $u \in M$. Using Lemma A.4 we see that it suffices to show that $N_\varepsilon = \{u * \omega_\varepsilon \mid u \in M\}$ is totally bounded for any $\varepsilon > 0$.

Hölder's inequality yields

$$|u * \omega_\varepsilon(x)| \le \|u\|_p \|\omega_\varepsilon\|_q,$$

so by (i), functions in N_ε are uniformly bounded. Another application of Hölder's inequality implies

$$|u * \omega_\varepsilon(x) - u * \omega_\varepsilon(y)| = \left| \int (u(x-z) - u(y-z)) \omega_\varepsilon(z) \, dz \right|$$

$$\le \|u(\cdot + x - y) - u\|_p \|\omega_\varepsilon\|_q,$$

which together with (ii) proves that N_ε is equicontinuous. The Arzela–Ascoli theorem implies that N_ε is relatively compact, and hence totally bounded in $C(\mathcal{B}_{R+r})$. Since the natural embedding of $C(\mathcal{B}_{R+r})$ into $L^p(\mathbb{R}^n)$ is bounded, it follows that N_ε totally bounded in $L^p(\mathbb{R}^n)$ as well. Thus we have proved that conditions (i)–(iii) imply that M is relatively compact.

To prove the converse, we assume that M is relatively compact. Condition (i) is clear. Now let $\varepsilon > 0$. Since M is relatively compact, we can find functions $u_1, \ldots, u_m$ in $L^p(\mathbb{R}^n)$ such that

$$M \subseteq \bigcup_{j=1}^m \mathcal{B}_\varepsilon(u_j).$$

Furthermore, since $C_0(\mathbb{R}^n)$ is dense in $L^p(\mathbb{R}^n)$, we may as well assume that $u_j \in C_0(\mathbb{R}^n)$. Clearly, $\|u_j(\cdot + y) - u_j\|_p \to 0$ as $y \to 0$, and so there is some $\delta > 0$ such that $\|u_j(\cdot + y) - u_j\|_p \leq \varepsilon$ whenever $|y| < \delta$. If $u \in M$ and $|y| < \delta$, then pick some j such that $\|u - u_j\|_p < \varepsilon$, and obtain

$$\begin{aligned}\|u(\cdot + z) - u\|_p &\leq \|u(\cdot + z) - u_j(\cdot + z)\|_p \\ &\quad + \|u_j(\cdot + z) - u_j\|_p + \|u_j - u\|_p \\ &= 2\|u_j - u\|_p + \|u_j(\cdot + z) - u_j\|_p \\ &\leq 3\varepsilon,\end{aligned}$$

proving (ii).

When r is large enough, $\chi_{\mathcal{B}_r} u_j = u_j$ for all j, and then, with the same choice of j as above, we obtain

$$\|\chi_{\mathcal{B}_r} u - u\|_p \leq \|\chi_{\mathcal{B}_r}(u - u_j)\|_p + \|u - u_j\|_p \leq 2\|u - u_j\|_p \leq 2\varepsilon,$$

which proves (iii). $\square$

Helly's theorem is a simple corollary of Kolmogorov's compactness theorem.

Corollary A.7 (Helly's theorem). *Let $\{h^\delta\}$ be a sequence of functions defined on an interval $[a,b]$, and assume that this sequence satisfies*

$$\mathrm{T.V.}\left(h^\delta\right) < M, \qquad \text{and} \qquad \left\|h^\delta\right\|_\infty < M,$$

where M is some constant independent of δ. Then there exists a subsequence h^{δ_n} that converges almost everywhere to some function h of bounded variation.

Proof. It suffices to apply (A.8) (for $p = 1$) together with the boundedness of the total variation to show that condition (ii) in Kolmogorov's compactness theorem is satisfied. $\square$

We remark that one can prove that the convergence in Helly's theorem is at every point, not only almost everywhere; see Exercise A.2.

The application of Kolmogorov's theorem in the context of conservation laws relies on the following result.

Theorem A.8. *Let $u_\eta \colon \mathbb{R}^n \times [0, \infty) \to \mathbb{R}$ be a family of functions such that for each positive T,*

$$|u_\eta(x,t)| \leq C_T, \quad (x,t) \in \mathbb{R}^n \times [0,T]$$

for a constant C_T independent of η. Assume in addition for all compact $B \subset \mathbb{R}^n$ and for $t \in [0,T]$ that

$$\sup_{|\xi| \leq |\rho|} \int_B |u_\eta(x+\xi,t) - u_\eta(x,t)|\, dx \leq \nu_{B,T}(|\rho|),$$

for a modulus of continuity ν. Furthermore, assume for s and t in $[0,T]$ that

$$\int_B |u_\eta(x,t) - u_\eta(x,s)|\, dx \leq \omega_{B,T}(|t-s|) \text{ as } \eta \to 0,$$

for some modulus of continuity ω_T. Then there exists a sequence $\eta_j \to 0$ such that for each $t \in [0,T]$ the function $\{u_{\eta_j}(t)\}$ converges to a function $u(t)$ in $L^1_{\text{loc}}(\mathbb{R}^n)$. The convergence is in $C([0,T]; L^1_{\text{loc}}(\mathbb{R}^n))$.

Proof. Kolmogorov's theorem implies that for each fixed $t \in [0,T]$ and for any sequence $\eta_j \to 0$ there exists a subsequence (still denoted by η_j) $\eta_j \to 0$ such that $\{u_{\eta_j}(t)\}$ converges to a function $u(t)$ in $L^1_{\text{loc}}(\mathbb{R}^n)$.

Consider now a dense countable subset E of the interval $[0,T]$. By possibly taking a further subsequence (which we still denote by $\{u_{\eta_j}\}$) we find that

$$\int_B |u_{\eta_j}(x,t) - u(x,t)|\, dx \to 0 \text{ as } \eta_j \to 0, \text{ for } t \in E.$$

Now let $\varepsilon > 0$ be given. Then there exists a positive δ such that $\omega_{B,T}(\tilde\delta) \leq \varepsilon$ for all $\tilde\delta \leq \delta$. Fix $t \in [0,T]$. We can find a $t_k \in E$ with $|t_k - t| \leq \delta$. Thus

$$\int_B |u_{\tilde\eta}(x,t) - u_{\tilde\eta}(x,t_k)|\, dx \leq \omega_{B,T}(|t-t_k|) \leq \varepsilon \text{ for } \tilde\eta \leq \eta$$

and

$$\int_B |u_{\eta_{j_1}}(x,t_k) - u_{\eta_{j_2}}(x,t_k)|\, dx \leq \varepsilon \text{ for } \eta_{j_1}, \eta_{j_2} \leq \eta \text{ and } t_k \in E.$$

The triangle inequality yields

$$\int_B |u_{\eta_{j_1}}(x,t) - u_{\eta_{j_2}}(x,t)|\, dx$$
$$\leq \int_B |u_{\eta_{j_1}}(x,t) - u_{\eta_{j_1}}(x,t_k)|\, dx + \int_B |u_{\eta_{j_1}}(x,t_k) - u_{\eta_{j_2}}(x,t_k)|\, dx$$
$$+ \int_B |u_{\eta_{j_2}}(x,t_k) - u_{\eta_{j_2}}(x,t)|\, dx$$
$$\leq 3\varepsilon,$$

proving that for each $t \in [0,T]$ we have that $u_\eta(t) \to u(t)$ in $L^1_{\text{loc}}(\mathbb{R}^n)$. The bounded convergence theorem then shows that

$$\sup_{t \in [0,T]} \int_B |u_\eta(x,t) - u(x,t)|\, dx\, dt \to 0 \text{ as } \eta \to 0,$$

thereby proving the theorem. $\square$

A.1 Notes

Extensive discussion about total variation can be found, e.g., in [47] and [148]. The proof of Theorem A.3 is taken from Sobolev [132, pp. 28 ff]. An alternative proof can be found in Yosida [147, p. 13]. The proof of Theorem A.2 is from [47, Theorem 1, p. 217]. Kolmogorov's compactness theorem, Theorem A.5, was first proved by Kolmogorov in 1931 [85] in the case where Ω is bounded, $p > 1$, and the translation $u(x + \varepsilon)$ of $u(x)$ is replaced by the spherical mean of u over a ball of radius ε in condition (ii). It was extended to the unbounded case by Tamarkin [135] in 1932 and finally extended to the case with $p = 1$ by Tulajkov [141] in 1933. M. Riesz [121] proved the theorem with translations. See also [49].

For other proofs of Kolmogorov's theorem, see, e.g., [132, pp. 28 ff], [27, pp. 69 f], [147, pp. 275 f], and [145, pp. 201f].

Exercises

A.1 Show that for any $f \in L^1$ we have
$$\|f\|_1 = \sup_{\substack{\phi \in C_0^1 \\ |\phi| \leq 1}} \int f(x)\phi(x)\, dx.$$

A.2 Show that in Helly's theorem, Corollary A.7, one can find a subsequence h^{δ_n} that converges for all x to some function h of bounded variation.

Appendix B
The Method of Vanishing Viscosity

> Details are the only things that interest.
>
> Oscar Wilde, *Lord Arthur Savile's Crime* (1891)

In this appendix we will give an alternative proof of existence of solutions of scalar multidimensional conservation laws based on the viscous regularization

$$u^\mu_t + \sum_{j=1}^m \frac{\partial}{\partial x_j} f_j(u^\mu) = \mu \Delta u^\mu, \quad u^\mu|_{t=0} = u_0, \tag{B.1}$$

where as usual Δu denotes the Laplacian $\sum_j u_{x_j x_j}$. Our starting point will be the following theorem:

Theorem B.1. *Let $u_0 \in L^1 \cap L^\infty \cap C^2$ with bounded derivatives and $f_j \in C^1$ with bounded derivative. Then the Cauchy problem (B.1) has a classical solution, denoted by u^μ, that satisfies*[1]

$$u^\mu \in C^2(\mathbb{R}^m \times \langle 0, \infty \rangle) \cap C(\mathbb{R}^m \times [0, \infty \rangle). \tag{B.2}$$

Furthermore, the solution satisfies the maximum principle

$$\|u^\mu(t)\|_\infty \leq \|u_0\|_\infty. \tag{B.3}$$

---

[1] The existence and regularity result (B.2) is valid for systems of equations in one spatial dimension as well.

Appendix B. The Method of Vanishing Viscosity

Let v^μ be another solution with initial data v_0 satisfying the same properties as u_0. Assume in addition that both u_0 and v_0 have finite total variation and are integrable. Then
$$\|u(\,\cdot\,,t) - v(\,\cdot\,,t)\|_{L^1(\mathbb{R}^m)} \leq \|u_0 - v_0\|_{L^1(\mathbb{R}^m)}, \tag{B.4}$$
for all $t \geq 0$.

Proof. We present the proof in the one-dimensional case only, that is, with $m = 1$. Let K denote the heat kernel, that is,
$$K(x,t) = \frac{1}{\sqrt{4\mu\pi t}} \exp\left(-\frac{x^2}{4\mu t}\right). \tag{B.5}$$

Define functions u^n recursively as follows: Let $u^{-1} = 0$, and define u^n to be the solution of
$$u_t^n + f(u^{n-1})_x = \mu u_{xx}^n, \quad u^n|_{t=0} = u_0, \quad n = 0, 1, 2, \ldots. \tag{B.6}$$

Then $u^n(t) \in C^\infty(\mathbb{R})$ for t positive. Applying Duhamel's principle we obtain
$$u^n(x,t) = \int K(x-y,t) u_0(y)\, dy$$
$$- \int\int_0^t K(x-y,t-s) f(u^{n-1}(y,s))_y\, ds\, dy$$
$$= u^0(x,t) - \int\int_0^t \frac{\partial}{\partial x} K(x-y,t-s) f(u^{n-1}(y,s))\, ds\, dy. \tag{B.7}$$

Define $v^n = u^n - u^{n-1}$. Then
$$v^{n+1}(x,t) = -\int\int_0^t \frac{\partial}{\partial x} K(x-y,t-s)\big(f(u^n(y,s)) - f(u^{n-1}(y,s))\big)\, ds\, dy.$$

Using Lipschitz continuity we obtain
$$\|v^{n+1}(t)\|_\infty \leq \|f\|_{\text{Lip}} \int_0^t \|v^n(s)\|_\infty \int \left|\frac{\partial}{\partial x} K(x,t-s)\right| dx\, ds$$
$$\leq \frac{\|f\|_{\text{Lip}}}{\sqrt{\pi\mu}} \int_0^t (t-s)^{-1/2} \|v^n(s)\|_\infty\, ds.$$

Assume that $|u_0| \leq M$ for some constant M. Then we claim that
$$\|v^n(t)\|_\infty \leq M \|f\|_{\text{Lip}}^n \frac{t^{n/2}}{\mu^{n/2} \Gamma(\frac{n+2}{2})}, \tag{B.8}$$
where we have introduced the gamma function defined by
$$\Gamma(p) = \int_0^\infty e^{-s} s^p\, ds.$$

We shall use the following properties of the gamma function. Let the beta function $B(p,q)$ be defined as

$$B(p,q) = \int_0^1 s^{p-1}(1-s)^{q-1}\,ds.$$

Then

$$B(p,q) = \frac{\Gamma(p)\Gamma(q)}{\Gamma(p+q)}.$$

After a change of variables the last equality implies that $\Gamma(\frac{1}{2}) = \sqrt{\pi}$. Equation (B.8) is clearly correct for $n = 0$. Assume it to be correct for n. Then

$$\left|v^{n+1}(x,t)\right| \leq M\|f\|_{\text{Lip}}^{n+1} \frac{1}{\sqrt{\pi}\mu^{(n+1)/2}\Gamma(\frac{n+2}{2})} \int_0^t (t-s)^{-1/2} s^{n/2}\,ds$$

$$= M\|f\|_{\text{Lip}}^{n+1} \frac{t^{(n+1)/2}}{\sqrt{\pi}\mu^{(n+1)/2}\Gamma(\frac{n+2}{2})} \int_0^1 (1-s)^{-1/2} s^{n/2}\,ds$$

$$= M\|f\|_{\text{Lip}}^{n+1} \frac{t^{(n+1)/2}}{\mu^{(n+1)/2}\Gamma(\frac{n+3}{2})}. \tag{B.9}$$

Hence we conclude that $\sum_n v^n$ converges uniformly on any bounded strip $t \in [0,T]$, and that

$$u = \lim_{n\to\infty} u^n = \lim_{n\to\infty} \sum_{j=0}^n v^j$$

exists. The convergence is uniform on the strip $t \in [0,T]$. It remains to show that u is a classical solution of the differential equation. We immediately infer that

$$u(x,t) = u^0(x,t) - \int\int_0^t \frac{\partial}{\partial x} K(x-y,t-s) f(u(y,s))\,ds\,dy. \tag{B.10}$$

It remains to show that (B.10) implies that u satisfies the differential equation

$$u_t + f(u)_x = \mu u_{xx}, \quad u|_{t=0} = u_0. \tag{B.11}$$

Next we want to show that u is differentiable. Define

$$M_n(t) = \sup_{x \in \mathbb{R}} \max_{0 \leq s \leq t} |u_x^n(x,s)|.$$

Clearly,

$$|u_x^n(x,t)| \leq \|f\|_{\text{Lip}} \frac{1}{\sqrt{\pi\mu}} \int_0^t (t-s)^{-1/2} M_{n-1}(s)\,ds + M_0(t).$$

Choose B such that $M^0 \leq B/2$. Then

$$M_n(t) \leq B\exp(Ct/\mu) \tag{B.12}$$

if C is chosen such that
$$\|f\|_{\text{Lip}}\frac{1}{\sqrt{\pi\mu}}\int_0^\infty s^{-1/2}e^{-Cs/\mu}\,ds \le \frac{1}{2}.$$

Inequality (B.12) follows by induction: It clearly holds for $n=0$. Assume that it holds for n. Then
$$\begin{aligned}|u_x^{n+1}(s,x)| &\le \|f\|_{\text{Lip}}\frac{1}{\sqrt{\pi\mu}}\int_0^t (t-s)^{-1/2}M_n(s)\,ds + B/2\\ &\le Be^{Ct/\mu}\left(\|f\|_{\text{Lip}}\frac{1}{\sqrt{\pi\mu}}\int_0^t s^{-1/2}e^{-Cs/\mu}\,ds + \frac{1}{2}\right)\\ &\le Be^{Ct/\mu}.\end{aligned}$$

Define
$$N_n(t) = \sup_{x\in\mathbb{R}}\max_{0\le s\le t}|u_{xx}^n(x,s)|.$$

Choose $\tilde B \ge \max\{2N^0, B^2+1\}$ and $\tilde C \ge C$ such that
$$2\tilde B(\|f'\|_\infty + \|f''\|_\infty)\frac{1}{\sqrt{\pi\mu}}\int_0^\infty s^{-1/2}e^{-2\tilde Cs/\mu} \le \frac{1}{2}.$$

Then we show inductively that
$$N_n(t) \le \tilde B e^{2\tilde Ct/\mu}.$$

The estimate is valid for $n=0$. Assume that it holds for n. Then
$$\begin{aligned}|u_{xx}^{n+1}(x,t)| &\le |u_{xx}^0(x,t)|\\ &\quad + \int_0^t (\|f''\|_\infty M_n(s)^2 + \|f'\|_\infty N_n(s))\int\left|\frac{\partial}{\partial x}K(y,t-s)\right|dy\,ds\\ &\le N_0\\ &\quad + (\|f'\|_\infty + \|f''\|_\infty)\frac{1}{\sqrt{\pi\mu}}\int_0^t (M_n(s)^2 + N_n(s))(t-s)^{-1/2}\,ds\\ &\le N_0\\ &\quad + (\|f'\|_\infty + \|f''\|_\infty)\frac{1}{\sqrt{\pi\mu}}\int_0^t (B^2e^{2Cs/\mu} + e^{\tilde Cs/\mu})(t-s)^{-1/2}\,ds\\ &\le \tilde B e^{2\tilde Ct/\mu}\left(1 + 2\tilde B(\|f'\|_\infty + \|f''\|_\infty)\frac{1}{\sqrt{\pi\mu}}\int_0^t e^{-2\tilde Cs/\mu}\,ds\right)\\ &\le \tilde B e^{2\tilde Ct/\mu}.\end{aligned}$$

We have now established that $u^n \to u$ uniformly and that u_x^n and u_{xx}^n both are uniformly bounded (in (x,t) and n). Lemma B.2 (proved after this theorem) implies that indeed u is differentiable and that u_x equals the uniform limit of u_x^n. Performing an integration by parts in (B.10) we find

that the limit u satisfies
$$u(x,t) = u^0(x,t) - \int\int_0^t K(x-y, t-s) f(u(y,s))_y \, ds \, dy.$$
Applying Lemma B.3 we conclude that u satisfies
$$u_t + f(u)_x = \mu u_{xx}, \quad u|_{t=0} = u_0,$$
with the required regularity.[2]

The proof of (B.3) is nothing but the maximum principle. Consider the auxiliary function
$$U(x,t) = u(x,t) - \eta(t + (\eta x)^2/2).$$
Since $U \to -\infty$ as $|x| \to \infty$, U obtains a maximum on $\mathbb{R} \times [0,T]$, say at the point (x_0, t_0). We know that
$$U(x_0, t_0) = u(x_0, t_0) - \eta(t_0 + (\eta x_0)^2/2) \geq u_0(0).$$
Hence
$$\eta^3 x_0^2 \leq 2u(x_0, t_0) - 2u_0(0) - 2\eta t_0 \leq \mathcal{O}(1) \quad \text{(B.13)}$$
independently of η, since u is bounded on $\mathbb{R} \times [0,T]$ by construction. Assume that $0 < t_0 \leq T$. At the maximum point we have
$$u_x(x_0, t_0) = \eta^3 x_0, \quad u_t(x_0, t_0) \geq \eta, \quad \text{and} \quad u_{xx}(x_0, t_0) \leq \eta^3,$$
which implies that
$$u_t(x_0, t_0) + f'(u(x_0, t_0)) u_x(x_0, t_0) - \mu u_{xx}(x_0, t_0) \geq \eta - \mathcal{O}(1) \eta^{3/2} - \mu \eta^3$$
$$> 0$$
if η is sufficiently small. We have used that $f'(u)$ is bounded and (B.13). This contradicts the assumption that the maximum was attained for t positive. Thus
$$u(x,t) - \eta(t + (\eta x)^2/2) \leq \sup_x U(x,0)$$
$$= \sup_x \left(u_0(x) - \eta^3 x^2/2 \right)$$
$$\leq \sup_x u_0(x),$$
which implies that $u \leq \sup u_0$. By considering η negative we find that $u \geq \inf u_0$, from which we conclude that $\|u\|_\infty \leq \|u_0\|_\infty$.

Lemma B.6 implies that any solution u satisfies the property needed for our uniqueness estimate, namely that if u_0 is in L^1, then $u(\,\cdot\,, t)$ is in L^1. This is so, since we have that
$$\|u(\,\cdot\,, t)\|_1 - \|u_0\|_1 \leq \|u(\,\cdot\,, t) - u_0\|_1 \leq Ct.$$

---

[2] The argument up this equality is valid for one-dimensional systems as well.

Furthermore, since u is of bounded variation (which is the case if u_0 is of bounded variation), u_x is in L^1, and thus $\lim_{|x|\to\infty} u_x(x,t) = 0$. Hence, if u_0 is in $L^1 \cap BV$, then we have that
$$\frac{d}{dt}\int u(x,t)\,dx = -\int (f(u)_x + \mu u_{xx})\,dx = 0.$$
Hence
$$\int u(x,t)\,dx = \int u_0(x)\,dx. \tag{B.14}$$

By the Crandall–Tartar lemma, Lemma 2.12, to prove (B.4) it suffices to show that if $u_0(x) \le v_0(x)$, then $u(x,t) \le v(x,t)$. To this end we first add a constant term to the viscous equation. More precisely, let u^δ denote the solution of (for simplicity of notation we we let $\mu = 1$ in this part of the argument)
$$u^\delta_t + f(u^\delta)_x = u^\delta_{xx} - \delta, \quad u^\delta|_{t=0} = u_0.$$
In integral form we may write (cf. (B.10))
$$u^\delta(x,t) = \int K(x-y,t)u_0(y)\,dy$$
$$- \int\int_0^t \frac{\partial}{\partial x}K(x-y,t-s)f(u^\delta(y,s))\,ds\,dy - \delta t.$$
Furthermore,
$$\left|u^\delta(x,t) - u(x,t)\right|$$
$$\le \int\int_0^t \left|\frac{\partial}{\partial x}K(x-y,t-s)\right|\,\left|f(u^\delta(y,s)) - f(u(y,s))\right|\,ds\,dy + |\delta|\,t$$
$$\le \|f\|_{\mathrm{Lip}} \int\int_0^t \left|\frac{\partial}{\partial x}K(x-y,t-s)\right|\,\left|u^\delta(y,s) - u(y,s)\right|\,ds\,dy + |\delta|\,t$$
$$\le \|f\|_{\mathrm{Lip}} \int_0^t \|u^\delta(s) - u(s)\|_\infty \frac{ds}{\sqrt{\pi\epsilon(t-s)}} + |\delta|\,t$$
$$\le \int_0^t \|u^\delta(s) - u(s)\|_\infty d\mu(s) + |\delta|\,t$$
with the new integrable measure $d\mu(s) = \|f\|_{\mathrm{Lip}}/\sqrt{\pi\epsilon(t-s)}$. Gronwall's inequality yields that
$$\|u^\delta(t) - u(t)\|_\infty \le t\,|\delta|\exp\left(\int_0^t d\mu(s)\right) = t\,|\delta|\exp\left(2\frac{\sqrt{t}\|f\|_{\mathrm{Lip}}}{\sqrt{\pi\epsilon}}\right),$$
which implies that $u^\delta \to u$ in L^∞ as $\delta \to 0$. Thus it suffices to prove the monotonicity property for u^δ and v^δ, where
$$v^\delta_t + f(v^\delta)_x = v^\delta_{xx} + \delta, \quad v^\delta|_{t=0} = v_0. \tag{B.15}$$

Let $u_0 \leq v_0$. We want to prove that $u^\delta \leq v^\delta$. Assume to the contrary that $u^\delta(x,t) > v^\delta(x,t)$ for some (x,t), and define
$$\hat{t} = \inf\{t \mid u^\delta(x,t) > v^\delta(x,t) \text{ for some } x\}.$$
Pick $\hat{x}$ such that $u^\delta(\hat{x}, \hat{t}) = v^\delta(\hat{x}, \hat{t})$. At this point we have
$$u^\delta_x(\hat{x},\hat{t}) = v^\delta_x(\hat{x},\hat{t}), \quad u^\delta_{xx}(\hat{x},\hat{t}) \leq v^\delta_{xx}(\hat{x},\hat{t}), \quad \text{and} \quad u^\delta_t(\hat{x},\hat{t}) \geq v^\delta_t(\hat{x},\hat{t}).$$
However, this implies the contradiction
$$-\delta = u^\delta_t + f'(u^\delta)u^\delta_x - u^\delta_{xx} \geq v^\delta_t + f'(v^\delta)v^\delta_x - v^\delta_{xx} \geq \delta \text{ at the point } (\hat{x},\hat{t})$$
whenever δ is positive.

Hence $u(x,t) \leq v(x,t)$ and the solution operator is monotone, and (B.4) holds. □

In the above proof we needed the following two results.

Lemma B.2. *Let $\phi_n \in C^2(I)$ on the interval I, and assume that $\phi_n \to \phi$ uniformly. If $\|\phi'_n\|_\infty$ and $\|\phi''_n\|_\infty$ are bounded, then ϕ is differentiable, and*
$$\phi'_n \to \phi'$$
uniformly as $n \to \infty$.

Proof. The family $\{\phi'_n\}$ is clearly equicontinuous and bounded. The Arzela–Ascoli theorem implies that a subsequence $\{\phi'_{n_k}\}$ converges uniformly to some function ψ. Then
$$\phi_{n_k} = \int^x \phi'_{n_k} \, dx \to \int^x \psi \, dx,$$
from which we conclude that $\phi' = \psi$. We will show that the sequence $\{\phi'_n\}$ itself converges to ψ. Assume otherwise. Then we have a subsequence $\{\phi'_{n_j}\}$ that does not converge to ψ. The Arzela–Ascoli theorem implies the existence of a further subsequence $\{\phi'_{n_{j'}}\}$ that converges to some element $\tilde{\psi}$, which is different from ψ. But then we have
$$\int^x \psi \, dx = \lim_{k\to\infty} \phi_{n_k} = \lim_{j'\to\infty} \phi_{n_{j'}} = \int^x \tilde{\psi} \, dx,$$
which shows that $\psi = \tilde{\psi}$, which is a contradiction. □

Lemma B.3. *Let $F(x,t)$ be a continuous function such that*
$$|F(x,t) - F(y,t)| \leq M |x-y|$$
uniformly in x,y,t. Define
$$u(x,t) = \int K(x-y,t)u_0(y)\,dy + \int\int_0^t K(x-y,t-s)F(y,s)\,ds\,dy.$$
Then u is in $C^2(\mathbb{R}^m \times \langle 0,\infty\rangle) \cap C(\mathbb{R}^m \times [0,\infty))$ and satisfies
$$u_t = u_{xx} + F(x,t), \quad u|_{t=0} = u_0.$$

Appendix B. The Method of Vanishing Viscosity

Proof. To simplify the presentation we assume that $u_0 = 0$. First we observe that

$$u(x,t) = \int_0^t F(x,s)\, ds + \int_0^t \int_0^t K(x-y, t-s)\big(F(y,s) - F(x,s)\big)\, ds\, dy.$$

The natural candidate for the time derivative of u is

$$u_t(x,t) = F(x,t) + \int_0^t \int \frac{\partial}{\partial t} K(x-y, t-s)\big(F(y,s) - F(x,s)\big)\, ds\, dy. \quad (B.16)$$

To show that this is well-defined we first observe that

$$\left|\frac{\partial}{\partial t} K(x-y, t-s)\right| \leq \frac{\mathcal{O}(1)}{t-s} K(x-y, 2(t-s)).$$

Thus

$$\int \int_0^t \left|\frac{\partial}{\partial t} K(x-y, t-s)\right| |F(y,s) - F(x,s)|\, ds\, dy$$

$$\leq M\mathcal{O}(1) \int_0^t \int \frac{1}{t-s} K(x-y, 2(t-s)) |y-x|\, dy\, ds$$

$$\leq M\mathcal{O}(1) \int_0^t \frac{1}{\sqrt{t-s}}\, ds \leq \mathcal{O}(1).$$

Consider now

$$\left|\frac{1}{\Delta t}\big(u(x, t+\Delta t) - u(x,t)\big) - u_t(x,t)\right|$$

$$\leq \left|\frac{1}{\Delta t}\int_t^{t+\Delta t} F(x,s)\, ds - F(x,t)\right|$$

$$+ \int \frac{1}{\Delta t} \int_t^{t+\Delta t} K(x-y, t+\Delta t - s) |F(y,s) - F(x,s)|\, ds\, dy$$

$$+ \int \int_0^t \left|\frac{1}{\Delta t}\big(K(x-y, t+\Delta t - s) - K(x-y, t-s)\big)\right.$$

$$\left. - \frac{\partial}{\partial t} K(x-y, t-s)\right| |F(y,s) - F(x,s)|\, dy\, ds$$

$$\leq \left|\frac{1}{\Delta t}\int_t^{t+\Delta t} F(x,s)\, ds - F(x,t)\right|$$

$$+ M\frac{1}{\Delta t}\int_t^{t+\Delta t} \int K(y, t+\Delta t - s) |y|\, dy\, ds$$

$$+ M \int \int_0^t \left|\frac{\partial}{\partial t} K(y, t + \theta\Delta t - s) - \frac{\partial}{\partial t} K(y, t-s)\right| |y|\, ds\, dy,$$

for some $\theta \in [0, 1]$. We easily see that the first two terms vanish in the limit when $\Delta t \to 0$. The last term can be estimated as follows (where $\delta > 0$):

$$\int\int_0^t \left|\frac{\partial}{\partial t}K(y, t+\theta\Delta t - s) - \frac{\partial}{\partial t}K(y, t - s)\right| |y|\, dy\, ds$$

$$\leq \int_0^{t-\delta}\int \left|\frac{\partial}{\partial t}K(y, t+\theta\Delta t - s) - \frac{\partial}{\partial t}K(y, t - s)\right| |y|\, dy\, ds$$

$$+ \int_{t-\delta}^{t}\int \left(\left|\frac{\partial}{\partial t}K(y, t+\theta\Delta t - s)\right| + \left|\frac{\partial}{\partial t}K(y, t - s)\right|\right) |y|\, dy\, ds$$

$$\leq \int_0^{t-\delta}\int \left|\frac{\partial}{\partial t}K(y, t+\theta\Delta t - s) - \frac{\partial}{\partial t}K(y, t - s)\right| |y|\, dy\, ds$$

$$+ \mathcal{O}(1) \int_{t-\delta}^{t}\int \left(\frac{1}{t+\theta\Delta t - s}K(y, 2(t+\theta\Delta t - s))\right.$$

$$\left. + \frac{1}{t-s}K(y, 2(t-s))\right) |y|\, dy\, ds.$$

Choosing δ sufficiently small in the second integral we can make that term less then a prescribed ϵ. For this fixed δ we choose Δt sufficiently small to make that integral less than ϵ. We conclude that indeed (B.16) holds. By using estimates

$$\left|\frac{\partial}{\partial x}K(x, t)\right| \leq \frac{\mathcal{O}(1)}{\sqrt{t}}K(x, 2t),$$

$$\left|\frac{\partial^2}{\partial x^2}K(x, t)\right| \leq \frac{\mathcal{O}(1)}{t}K(x, 2t),$$

we conclude that the spatial derivatives are given by

$$u_x(x, t) = \int\int_0^t \frac{\partial}{\partial x}K(x-y, t-s)F(y, s)\, ds\, dy,$$

$$u_{xx}(x, t) = \int\int_0^t \frac{\partial^2}{\partial x^2}K(x-y, t-s)F(y, s)\, ds\, dy, \qquad (B.17)$$

from which we conclude that

$$u_t(x, t) - u_{xx}(x, t)$$

$$= F(x, t) + \int\int_0^t \left(\frac{\partial}{\partial t}K(x-y, t-s) - \frac{\partial^2}{\partial x^2}K(x-y, t-s)\right)F(y, s)\, ds\, dy$$

$$= F(x, t). \qquad (B.18)$$

$\square$

Remark B.4. The lemma is obvious if F is sufficiently differentiable; see, e.g., [108, Theorem 3, p. 144].

Next, we continue by showing directly that as $\mu \to 0$, the sequence $\{u^\mu\}$ converges to the unique entropy solution of the conservation law (B.30). We

310 Appendix B. The Method of Vanishing Viscosity

remark that this convergence was already established in Chapter 3 when we considered error estimates.

In order to establish our estimates we shall need the following technical result.

Lemma B.5. *Let* $v \colon \mathbb{R}^m \to \mathbb{R}$ *such that* $v \in C^1(\mathbb{R}^m)$ *and* $|\nabla v| \in L^1(\mathbb{R}^m)$. *Then*

$$\int_{|v|\leq \eta} |\nabla v|\, dx \to 0 \text{ as } \eta \to 0.$$

Proof. By the inverse function theorem, the set

$$\left\{ x \mid v(x) = 0,\ \nabla v(x) \neq 0 \right\}$$

is a smooth $(m-1)$-dimensional manifold of $\mathbb{R}^m$. Thus

$$\int_{|v|\leq \eta} |\nabla v|\, dx = \int_{0<|v|\leq \eta} |\nabla v|\, dx.$$

The integrand (the norm of the gradient times the characteristic function of the region where $|v|$ is nonzero and less than η) tends pointwise to zero as $\eta \to 0$. The lemma follows using Lebesgue's dominated convergence theorem. □

The key estimates are contained in the next lemma.

Lemma B.6. *Assume that* $u_0 \in C^2(\mathbb{R}^m)$ *with bounded derivatives and finite total variation. Let* u^μ *denote the solution of equation* (B.1). *Then the following estimates hold:*

$$\text{T.V.}\,(u^\mu(t)) \leq \text{T.V.}\,(u^\mu(0)), \tag{B.19}$$

$$\|u^\mu(t) - u^\mu(s)\|_1 \leq C\,|t-s|. \tag{B.20}$$

Proof. We set $w^0 = \partial u^\varepsilon / \partial t$ and $w^i = \partial u^\varepsilon / \partial x_i$ for $i = 1, \ldots, m$. Then we find that

$$\frac{\partial w^i}{\partial t} + \sum_{j=1}^m \left(f'_j(u^\mu) w^i\right)_{x_j} = \mu \Delta w^i \tag{B.21}$$

for $i = 0, 1, \ldots, m$. Define the following continuous approximation to the sign function:

$$\text{sign}_\eta(x) = \begin{cases} 1 & \text{for } x \geq \eta, \\ x/\eta & \text{for } |x| < \eta, \\ -1 & \text{for } x \leq -\eta. \end{cases}$$

Multiply (B.21) by $\text{sign}_\eta(w^i)$ and integrate over $\mathbb{R}^m \times [0, T]$ for some T positive. This yields

$$\int_{\mathbb{R}^m} \int_0^T \frac{\partial w^i}{\partial t} \text{sign}_\eta(w^i)\, dt\, dx + \sum_{j=1}^m \int_{\mathbb{R}^m} \int_0^T \left(f'_j(u^\mu) w^i\right)_{x_j} \text{sign}_\eta(w^i)\, dt\, dx$$

$$= \int_{\mathbb{R}^m} \int_0^T \mu \Delta w^i \operatorname{sign}_\eta(w^i) \, dt \, dx. \tag{B.22}$$

The first term in (B.22) can be written

$$\int_{\mathbb{R}^m} \int_0^T \frac{\partial w^i}{\partial t} \operatorname{sign}_\eta(w^i) \, dt \, dx = \int_{\mathbb{R}^m} \int_0^T \left(w^i \operatorname{sign}_\eta(w^i) \right)_t dt \, dx$$
$$- \int_{\mathbb{R}^m} \int_0^T w^i \operatorname{sign}'_\eta(w^i) w^i_t \, dt \, dx.$$

Here we have that

$$\left| \int_{\mathbb{R}^m} \int_0^T w^i \operatorname{sign}'_\eta(w^i) w^i_t \, dt \, dx \right| = \frac{1}{\eta} \left| \int_{|w^i| \leq \eta} \int_{t \leq T} w^i \, w^i_t \, dt \, dx \right|$$
$$\leq \int_{|w^i| \leq \eta} \int_{t \leq T} \left| w^i_t \right| dt \, dx \to 0,$$

using Lemma B.5, which implies

$$\int_{\mathbb{R}^m} \int_0^T \frac{\partial w^i}{\partial t} \operatorname{sign}_\eta(w^i) \, dt \, dx \to \int_{\mathbb{R}^m} \int_0^T \frac{\partial}{\partial t} |w^i| \, dt \, dx$$
$$= \|w^i(T)\|_1 - \|w^i(0)\|_1, \tag{B.23}$$

as $\eta \to 0$. The second term in (B.22) reads

$$I := \sum_{j=1}^m \int_{\mathbb{R}^m} \int_0^T \left(f'_j(u^\mu) w^i \right)_{x_j} \operatorname{sign}_\eta(w^i) \, dt \, dx$$
$$= -\sum_{j=1}^m \int_{\mathbb{R}^m} \int_0^T f'_j(u^\mu) w^i \operatorname{sign}'_\eta(w^i) \frac{\partial w^i}{\partial x_j} \, dt \, dx$$
$$= -\frac{1}{\eta} \int_{\substack{|w^i| \leq \eta \\ t \leq T}} w^i \, f'(u^\mu) \cdot \nabla w^i \, dt \, dx,$$

where $f' = (f'_1, \ldots, f'_m)$. This can be estimated as follows:

$$|I| \leq \sup_u |f'(u)| \int_{|w^i| \leq \eta} \int_{t \leq T} |\nabla w^i| \, dt \, dx \to 0, \tag{B.24}$$

as $\eta \to 0$. Here the supremum is over $|u| \leq \|u^\mu(0)\|_\infty$. Finally,

$$\mu \int_{\mathbb{R}^m} \int_0^T \Delta w^i \operatorname{sign}_\eta(w^i) \, dt \, dx = -\mu \int_{\mathbb{R}^m} \int_0^T |\nabla w^i|^2 \operatorname{sign}'_\eta(w^i) \, dt \, dx \leq 0. \tag{B.25}$$

Using (B.23), (B.24), and (B.25) in (B.22) we obtain, when $\eta \to 0$,

$$\|w^i(T)\|_1 - \|w^i(0)\|_1 \leq 0. \tag{B.26}$$

For $i = 0$ this implies

$$\|u^\mu(t) - u^\mu(s)\|_1 = \int_{\mathbb{R}^m} \left| \int_s^t \frac{\partial u^\mu}{\partial t} \, dt \right| dx$$

$$\leq \int_{\mathbb{R}^m} \int_s^t \left| w^0(\tilde{t}) \right| d\tilde{t}\, dx$$

$$= \int_s^t \left\| w^0(\tilde{t}) \right\|_1 d\tilde{t}$$

$$\leq |t-s| \left\| w^0(0) \right\|_1.$$

For $i \geq 1$ we use (B.26) to prove (B.19). Recalling the results from Appendix A, we define

$$\lambda_i(u,\mu) = \int_{\mathbb{R}^m} |u(x+\mu e_i) - u(x)|\, dx \quad \text{and} \quad \lambda(u,\mu) = \sum_{i=1}^m \lambda_i(u,\mu).$$

Then the inequalities (A.10) hold. We have that

$$\lambda_i(u^\varepsilon(\,\cdot\,,t),\mu) = \int_{\mathbb{R}^m} |u^\varepsilon(x+\mu e_i, t) - u^\varepsilon(x,t)|\, dx$$

$$= \int_{\mathbb{R}^m} \left| \int_0^\mu w^i(x+\alpha e_i, t)\, d\alpha \right| dx$$

$$\leq \int_{\mathbb{R}^m} \int_0^\mu \left| w^i(x+\alpha e_i, t) \right| d\alpha\, dx$$

$$\leq \int_0^\mu \left\| w^i(\,\cdot\,,t) \right\|_1 d\alpha$$

$$= |\mu|\, \left\| w^i(\,\cdot\,,t) \right\|_1$$

$$\leq |\mu|\, \left\| w^i(\,\cdot\,,0) \right\|_1$$

$$= |\mu| \int_{\mathbb{R}^{m-1}} \text{T.V.}_{x_i}(u_0)\, dx_1 \cdots dx_{i-1}\, dx_{i+1} \cdots dx_m.$$

Thus we find that

$$\text{T.V.}(u^\varepsilon(\,\cdot\,,t)) = \liminf_{\mu \to 0} \frac{\lambda(u^\varepsilon(\,\cdot\,,t),\mu)}{|\mu|} \leq \text{T.V.}(u_0),$$

which proves (B.19). □

From the estimates in Lemma B.6 we may conclude, using Helly's theorem, Corollary A.7, and Theorem A.8, that there exists a (sub)sequence of $\{u^\mu\}$ that converges uniformly in $C([0,T]; L^1_{\text{loc}}(\mathbb{R}^m))$ to a function that we denote by u. It remains to show that u is an entropy solution of the conservation law.

Let k be in $\mathbb{R}$. Then

$$(u^\mu - k)_t + \nabla \cdot (f(u^\mu) - f(k)) = \mu \Delta (u^\mu - k). \tag{B.27}$$

Multiply (B.27) by $\text{sign}_\eta(u^\mu - k)$ times a nonnegative test function ϕ and integrate over $[0,T] \times \mathbb{R}^m$. We find, when we write $U = u^\mu - k$, that

$$0 = \iint \left(U_t \text{sign}_\eta(U) \phi \right.$$

$$+ \nabla \cdot (f(u^\mu) - f(k)) \operatorname{sign}_\eta(U)\phi - \mu \operatorname{sign}_\eta(U)\Delta U \phi \Big) \, dx \, dt$$

$$= \iint \Big(\big(U \operatorname{sign}_\eta(U)\big)_t \phi$$
$$- (f(u^\mu) - f(k)) \cdot \big(\operatorname{sign}_\eta(U)\nabla\phi + \phi \operatorname{sign}'_\eta(U)\nabla U\big)\Big) \, dx \, dt$$
$$+ \mu \iint \nabla U \cdot \nabla \big(\operatorname{sign}_\eta(U)\phi\big) \, dx \, dt - \iint U \operatorname{sign}'_\eta(U) U_t \, \phi \, dx \, dt$$

$$= -\iint \big(U \operatorname{sign}_\eta(U)\phi_t + \operatorname{sign}_\eta(U)\,(f(u^\mu) - f(k)) \cdot \nabla\phi\big) \, dx \, dt$$
$$- \int \big((U\phi)\big|_{t=0} - (U\phi)\big|_{t=T}\big) \, dx$$
$$- \iint \phi \operatorname{sign}'_\eta(U)\,(f(u^\mu) - f(k)) \cdot \nabla U \, dx \, dt$$
$$- \iint U\phi \operatorname{sign}'_\eta(U) U_t \, dx \, dt$$
$$- \mu \iint \operatorname{sign}_\eta(U)\nabla U \cdot \nabla\phi \, dx \, dt - \mu \iint |\nabla U|^2 \operatorname{sign}'_\eta(U)\phi \, dx \, dt.$$

All terms in the second line tend to zero as $\eta \to 0$, and the last term is nonpositive. Hence

$$\iint \big(|u^\mu - k|\phi_t + \operatorname{sign}(u^\mu - k)\,(f(u^\mu) - f(k)) \cdot \nabla\phi\big) \, dx \, dt$$
$$- \int (u^\mu(0) - k)\phi\big|_{t=0}^{t=T} \, dx \geq \mu \iint \operatorname{sign}(U)\,\nabla U \cdot \nabla\phi \, dx \, dt. \tag{B.28}$$

Taking $\mu \to 0$ we see that the right-hand side tends to zero, and we conclude that

$$\iint \big(|u - k|\phi_t + \operatorname{sign}(u - k)\,(f(u) - f(k)) \cdot \nabla\phi\big) \, dx \, dt$$
$$+ \int (u_0 - k)\phi|_{t=0} \, dx - \int (u(T) - k)\phi|_{t=T} \, dx \geq 0, \tag{B.29}$$

which is the Kružkov entropy condition. We have proved the following result.

Theorem B.7. *Let $u_0 \in C^2(\mathbb{R}^m) \cap L^\infty(\mathbb{R}^m)$ with bounded derivatives and finite total variation, and let $f_j \in C^1(\mathbb{R})$ with bounded derivative. Let u^μ be the unique solution of (B.1). Then there exists a convergent subsequence of $\{u^\mu\}$ that converges in $C([0,T]; L^1_{\text{loc}}(\mathbb{R}^m))$ to a function u that satisfies the Kružkov entropy condition (B.29), and hence is the unique solution of*

$$u_t + \sum_{j=1}^m \frac{\partial}{\partial x_j} f_j(u) = 0, \quad u|_{t=0} = u_0. \tag{B.30}$$

B.1 Notes

Our proof of Theorem B.1 is taken in part from [74], where a similar result is proved for an equation of the form

$$u_t + \sum_{j=1}^{m} \psi_j(x,t,u) u_{x_j} = \mu \Delta u.$$

We are grateful to H. Hanche-Olsen (private communication) for discussions on the proof of this theorem. We have also used [30]. Other proofs can be found; see, e.g., [107]. The conditions of Theorem B.1 can be weakened considerably. Alternative proofs of Theorem B.1 can be obtained using the dimensional splitting construction in Section 4.4. Lemma B.6 is familiar; see, e.g., [107]. Our presentation of Lemma B.6 and Theorem B.7 follows in part Bardos et al. [8].

Recently, Bianchini and Bressan [11, 12, 13] have published results concerning the vanishing viscosity method for general systems. More precisely, consider the solution u^ε of the system

$$u_t^\varepsilon + A(u^\varepsilon) u_x^\varepsilon = \varepsilon u_{xx}^\varepsilon, \quad u^\varepsilon|_{t=0} = u_0.$$

They prove that u^ε converges to u, the solution of

$$u_t + A(u) u_x = 0, \quad u|_{t=0} = u_0,$$

as $\varepsilon \to 0$. Their assumptions are the following: The matrices $A(u)$ are smooth and strictly hyperbolic in the neighborhood of a compact set K, the initial data u_0 has sufficiently small total variation, and $\lim_{x \to -\infty} u_0(x) \in K$. The proof uses an ingenious decomposition of the function u_x^ε in terms of gradients of viscous traveling waves selected by a center manifold technique.

Exercises

B.1 Consider the *system* of parabolic equations

$$u_t + f(u)_x = \mu\, u_{xx}, \quad u|_{t=0} = u_0, \qquad (B.31)$$

with $u_0(x) \in \mathbb{R}^n$.

  a. Show that there exists a solution u^μ of (B.31) that satisfies the regularity condition (B.2).
  b. Fix temporarily $\mu = 1$, and consider the equation

$$u_t + f(u)_x = u_{xx} - \delta, \quad u|_{t=0} = u_0, \qquad (B.32)$$

  with solution u^δ. Show that $\left\| u^\delta(t) - u(t) \right\|_1 \to 0$, where u solves (B.31) (with $\mu = 1$).

c. Assume that the flux function f satisfies
$$f(u_1, \ldots, u_{j-1}, u^*, u_{j+1}, \ldots, u_n) = \text{const}, \quad u_i \in \mathbb{R}, \; i \neq j,$$
for some j and some $u^* \in \mathbb{R}$. Assume that the jth component $u_{0,j}$ of u_0 satisfies $u_{0,j} \leq u^*$. Show that the jth component u_j of the solution u of (B.31) satisfies
$$u_j(x, t) \leq u^*.$$

d. Assume that there are constants $u_* < u^*$ and j such that
$$f(u_1, \ldots, u_{j-1}, u_*, u_{j+1}, \ldots, u_n) = \text{const}, \quad u_i \in \mathbb{R}, \; i \neq j,$$
$$f(u_1, \ldots, u_{j-1}, u^*, u_{j+1}, \ldots, u_n) = \text{const}, \quad u_i \in \mathbb{R}, \; i \neq j,$$
and that $u_* \leq u_{0,j} \leq u^*$. Show that
$$u_* \leq u_j(x, t) \leq u^*,$$
and hence that the region
$$\{u \in \mathbb{R}^n \mid u_* \leq u_j \leq u^*\}$$
is invariant for the solution of (B.31). Systems with this property appear, e.g., in multiphase flow in porous media and chemical chromatography. For more on invariant regions, see [71] and [65].

Appendix C
Answers and Hints

> ### HEALTH WARNING
> We do not claim any liability for these hints and answers. Use them at your own risk.
>
> THE AUTHORS

> The only way to get rid of a temptation is to yield to it. Resist it, and your soul grows sick with longing for the things it had forbidden itself.
>
> *Oscar Wilde, The Picture of Dorian Gray (1891)*

Appendix C. Answers and Hints

Chapter 1, pages 19–21.

1.1 The characteristics are given by
$$x = 2\arctan(x_0 e^\xi), \quad t = \xi + t_0, \quad z = \xi + z_0,$$
$$x = 2\arctan\left(\frac{x_0 e^\xi - 1}{x_0 e^\xi + 1}\right), \quad t = 2\arctan(t_0 e^\xi), \quad z = z_0,$$
$$x = 2\int_0^\xi \sin(z_0 e^\sigma)\,d\sigma + x_0, \quad t = \xi + t_0, \quad z = z_0 e^\xi,$$
$$x = \cos(z_0)\xi + x_0, \quad t = \sin(z_0)\xi + t_0, \quad z = z_0.$$

1.2 a.
$$u(x,y) = y \pm \sqrt{y^2 - x^2}.$$

b.
$$u(x,y) = \frac{1}{x}\bigl(1 + 2\sinh^2(y/2)\bigr).$$

c.
$$u(x,y) = h\bigl(\sqrt{x^2+y^2}\bigr)\exp\bigl(\arctan(y/x)\bigr).$$

d.
$$u(x,y) = (x+1)(y-1).$$

e.
$$u(x,y) = \exp\bigl(y + (1-x^2)/2\bigr) - 1.$$

f.
$$u(x,y) = (y - x^2)^2 \exp(x^2/2) - 1.$$

g.
$$x = (u - y^2)\exp(uy - 2y^3/3),$$
which determines u *implicitly* in terms of (x,y).

1.3 This is identical to the scalar case; work with each component f_i and u_i, etc.

1.4 Set $\mathbf{x} = (x,y)$ and $\mathbf{f} = (f,g)$. Then the conservation law reads $u_t + \nabla \cdot \mathbf{f} = 0$. Assume now that the position on the regular surface is $(\mathbf{x}, t)$ and let $\mathcal{B}_r = \bigl\{(\mathbf{y}, \tau) \mid |\mathbf{x} - \mathbf{y}|^2 + (t - \tau)^2 \leq r^2\bigr\}$. Let Γ_r be the intersection of the surface of discontinuity with $\mathcal{B}_r$, and choose a test function $\varphi \in C_0^\infty(\mathcal{B}_r)$. Then an application of Green's theorem yields
$$\int_{\Gamma_r} [(u_1, \mathbf{f}_1) - (u_2, \mathbf{f}_2)] \cdot (\sigma(s), \mathbf{n}(s))\,ds = 0,$$

where $\mathbf{n}(s)$ denotes the normal to the surface, u_1 and u_2 denote the limits of u, and σ is the speed of the discontinuity. Since φ is arbitrary, we obtain

$$\sigma(u_1 - u_2) = \mathbf{n} \cdot (\mathbf{f}_1 - \mathbf{f}_2).$$

1.5 Observe that we need only consider the extreme characteristics originating at $x = \mp 1$, since before these meet, the solution will be linear between these. These characteristics have speed ± 1, and hence the solution will be continuous until $t = 1$, and is given by $u(x,t) = u_0(x/(t-1))$. The solution of the linearized equation is given by $v(x,t) = u_0(\alpha e^{\alpha t} x)$. From this we find that

$$v_n\left(x, (m+1)/n\right) = v_n\left(\alpha_{m,n} e^{\alpha_{m,n}/n}, m/n\right),$$

and thus $\alpha_{m+1,n} = \alpha_{m,n} e^{\alpha_{m,n}/n}$. Set $1/n = \Delta t$. Assuming that the limit holds, we have $\alpha(m+1,n) = \bar{\alpha}(t + \Delta t)$, and thus

$$\frac{\ln(\bar{\alpha}(t + \Delta t)) - \ln(\bar{\alpha}(t))}{\Delta t} = \bar{\alpha}(t).$$

Letting Δt to zero, we find that $\bar{\alpha}'(t) = \bar{\alpha}^2(t)$, which gives the conclusion, since $\bar{\alpha}(0) = 1$. For $t \geq 1$, $\alpha_{m,n}$ diverges to $+\infty$, which incidentally gives us the correct solution.

1.6 The solutions in parts **a** and **b** are, respectively,

$$u(x,t) = \begin{cases} -1 & \text{for } x \leq t, \\ x/t & \text{for } |x| < t, \\ 1 & \text{for } x > t, \end{cases} \quad \text{and} \quad u(x,t) = u_0(x).$$

In the first case we directly verify that $u(x,t)$ also solves (1.26). In the second case the Rankine–Hugoniot condition is violated. Set $v = \frac{1}{2}u^2$. Then $v_t + \frac{2}{3}(v^{3/2})_x = 0$, and $v(x,0) = 1$. Hence $v(x,t) = 1$ is the correct solution.

1.7 The jumps satisfy the Rankine–Hugoniot condition. That's all.

Chapter 2, pages 57–61.

2.1 We find that

$$f'(u) = \frac{2u(1-u)}{(u^2 + (1-u)^2)^2},$$

and that the graph of f is "S-shaped" in the interval $[0,1]$ with a single inflection point for $u = \frac{1}{2}$. Hence the solution of the Riemann problem will be a rarefaction wave followed by a shock. The left limit of the shock, u_1, will solve the equation $f'(u_1) = (1 - f(u_1))/(1 - u_1)$, which gives $u_1 = 1 - \sqrt{2}/2$, and the speed of the shock will be $\sigma = (1 + \sqrt{2})/2$. For $u < u_1$ we must find the inverse of f', and after some manipulation this is found to be

$$(f')^{-1}(\xi) = \frac{1}{2}\left(1 - \sqrt{\frac{1}{\xi}(\sqrt{4\xi + 1} - 1)} - 1\right).$$

Hence the solution will be given by

$$u(x,t) = \begin{cases} 0 & \text{for } x \leq 0, \\ (f')^{-1}(x/t) & \text{for } 0 \leq x \leq t(1+\sqrt{2})/2, \\ 1 & \text{for } x > t(1+\sqrt{2})/2. \end{cases}$$

2.2 To show that the function in part **a** is a weak solution we check that the Rankine–Hugoniot condition holds. In part **b** we find that $u^\varepsilon(x,t) = u_0^\varepsilon(x\varepsilon/(t+\varepsilon))$. Then

$$\bar{u}(x,t) = \begin{cases} -1 & \text{for } x < t, \\ x/t & \text{for } |x| \leq t, \\ 1 & \text{for } x > t. \end{cases}$$

The solution found in part **a** does not satisfy the entropy condition, whereas $\bar{u}$ does.

2.3 Introduce coordinates (τ, y) by $\tau = t$, $y = x - at$. Then the resulting problem reads

$$u_\tau^\varepsilon = \varepsilon u_{yy}^\varepsilon.$$

The solution to this is found by convolution with the heat kernel and reads

$$u^\varepsilon(y, \tau) = u_l + \frac{u_r - u_l}{\sqrt{4\varepsilon\pi\tau}} \int_0^\infty \exp\bigl(-(y-z)^2/(4\varepsilon\tau)\bigr)\, dz.$$

The result follows from this formula.

2.4 Add (2.73) and (2.53); then choose ψ as (2.54).

2.5 a. The characteristics are given by

$$\frac{\partial t}{\partial \xi} = 1, \quad \frac{\partial x}{\partial \xi} = c(x)f'(z), \quad \frac{\partial z}{\partial \xi} = 0,$$

or

$$t = \xi + t_0, \quad \frac{\partial x}{\partial \xi} = c(x)f'(z_0), \quad z = z_0.$$

b. The Rankine–Hugoniot condition reads

$$s[\![u]\!] = c[\![f]\!].$$

c. The characteristics are given by

$$t = \xi + t_0, \quad x = \tan\left(z\xi + \arctan(x_0)\right), \quad z = z_0.$$

Using all values of z_0 between -1 and 1 for characteristics starting at the origin, and writing u in terms of (x, t), we obtain

$$u(x,t) = \begin{cases} -1 & \text{for } x \leq -\tan t, \\ \frac{\arctan x}{t} & \text{for } |x| < \tan t, \\ 1 & \text{for } x \geq \tan t. \end{cases}$$

d. One possibility is to approximate f by a continuous, piecewise linear flux function, and keep the function c. The characteristics will no longer be straight lines, and one will have to solve the ordinary differential equations that come from the jump condition. Another possibility is to approximate c by piecewise constant or piecewise linear functions.

e. The entropy condition reads

$$|u - k|_t + c(\operatorname{sign}(u - k)(f(u) - f(k)))_x \leq 0$$

weakly for all $k \in \mathbb{R}$.

2.6 a. The characteristics are given by

$$\frac{\partial t}{\partial \xi} = 1, \quad \frac{\partial x}{\partial \xi} = c(x)f'(z), \quad \frac{\partial z}{\partial \xi} = -c'(x)f(z). \qquad (C.1)$$

b. The entropy condition reads

$$|u - k|_t + (q(u,k)c(x))_x + \operatorname{sign}(u - k)f(k)c'(x) \leq 0, \qquad (C.2)$$

in the distributional sense.

c. Set $\eta(u,v) = |u - v|$ and $q(u,v) = \operatorname{sign}(u - v)(f(u) - f(v))$. Starting from the entropy condition (C.2) we get

$$\iint \Big[\eta(u,k)\varphi_t + q(u,k)c(x)\varphi_x - \operatorname{sign}(u - k)f(k)c'(x)\varphi\Big]\, dx\, dt \geq 0,$$

$$\iint \Big[\eta(v,k)\varphi_s + q(v,k)c(y)\varphi_y - \operatorname{sign}(v - k)f(k)c'(y)\varphi\Big]\, dx\, dt \geq 0.$$

We set $k = v$ in the first equation, $k = u$ in the second, and then add and integrate, obtaining

$$\iiiint \Big[\eta(u,v)(\varphi_t + \varphi_s) + q(u,v)(c(x)\varphi_x + c(y)\varphi_y)$$

$$- \operatorname{sign}(u - v)(f(u)c'(y) - f(v)c'(x))\varphi\Big]\, dx\, dt\, dy\, ds$$

$$+ \iiiint \operatorname{sign}(u - v)\Big[(f(u) - f(v))(c(y) - c(x))\varphi_y$$

$$- c'(x)f(v)\varphi + c'(y)f(u)\varphi\Big]\, dx\, dt\, dy\, ds \geq 0.$$

Now, $\varphi_y = \psi_y \omega + \psi \omega_y$. Therefore, the first term in the last integrand above can be split into

$$\text{sign}\,(u-v)\,[(f(u)-f(v))(c(y)-c(x))]\,\omega_y \psi$$
$$+ \text{sign}\,(u-v)\,[(f(u)-f(v))(c(y)-c(x))]\,\omega \psi_y.$$

The integral of the last term will vanish as $\varepsilon_1 \to 0$, since c is continuous. What remains is the integral of

$$\psi\,\text{sign}\,(u-v)\,\Big[(f(u)-f(v))(c(y)-c(x))\omega_y$$
$$- c'(x)f(v)\omega + c'(y)f(u)\omega\Big].$$

We have that

$$(c(y)-c(x))\omega_y + c'(y)\omega = \frac{\partial}{\partial y}\Big((c(y)-c(x))\omega\Big),$$

$$-(c(y)-c(x))\omega_y - c'(x)\omega = -\Big((c(x)-c(y))\omega_x + c'(x)\omega\Big)$$
$$= -\frac{\partial}{\partial x}\Big((c(x)-c(y))\omega\Big).$$

Thus the troublesome integrand can be written

$$\psi\,\text{sign}\,(u-v)\,\bigg(f(u)\frac{\partial}{\partial y}\Big((c(y)-c(x))\omega\Big)$$
$$- f(v)\frac{\partial}{\partial x}\Big((c(x)-c(y))\omega\Big)\bigg).$$

We add and subtract to find that this equals

$$\psi\,\text{sign}\,(u-v)(f(u)-f(v))\left(\frac{\partial}{\partial y}\Big((c(y)-c(x))\omega\Big)\right)$$
$$+ \psi\,\text{sign}\,(u-v)\,f(v)\,[c'(y)-c'(x)]\,\omega.$$

Upon integration, the last term will vanish in the limit, since c' is continuous. Thus, after a partial integration we are left with

$$\iiiint \frac{\partial}{\partial y}\Big(\psi q(u,v)\Big)(c(x)-c(y))\omega\,dx\,dt\,dy\,ds$$
$$\leq \|c'\|_\infty \varepsilon_1 \iiiint \left|\frac{\partial}{\partial y}\Big(\psi q(u,v)\Big)\right| \omega\,dx\,dt\,dy\,ds$$
$$\leq \text{const}\,\varepsilon_1\,(\text{T.V.}(v) + \text{T.V.}(\psi)).$$

By sending ε_0 and ε_1 to zero, we find that

$$\iint |u-v|\,\psi_t + \text{sign}\,(u-v)\,(f(u)-f(v))c(x)\psi_x\,dx\,dt \geq 0.$$

With this we can continue as in the proof of Proposition 2.10.

2.7 Mimic the proof the Rankine–Hugoniot condition by applying the computation (1.15).

2.8 The function q satisfies $q' = f'\eta'$. Thus $q = u^3/3$. The entropy condition reads
$$\int_\mathbb{R} \int_0^T \left(\frac{1}{2}u^2\phi_t + \frac{1}{3}u^3\phi_x\right) dt\, dx \geq -\frac{1}{2}\int_\mathbb{R}\left(u_0^2\phi|_{t=0} - (u^2\phi)|_{t=T}\right) dx.$$
Choose ϕ that approximate the identity function appropriately. Then
$$\int_\mathbb{R} u^2\, dx \leq \int_\mathbb{R} u_0^2.$$
Solutions of conservation laws are not contractive in the L^2-norm, in general, as the following counterexample shows. Let
$$u_0 = \begin{cases} 1 & \text{for } 0 < x < 1, \\ 0 & \text{otherwise,} \end{cases} \qquad v_0 = \begin{cases} \frac{1}{2} & \text{for } 0 < x < 1, \\ 0 & \text{otherwise.} \end{cases}$$
We find that
$$\|u(t) - v(t)\|_2^2 = \frac{1}{4} + \frac{5}{24}t$$
for $t < 2$.

2.9 a. The Rankine–Hugoniot relation is the same as before, viz.,
$$s[\![u]\!] = [\![f]\!].$$

b.
$$\int_\mathbb{R} \int_0^T \left(|u - k|\phi_t + q(u,k)\phi_x\right) dt\, dx$$
$$+ \int_\mathbb{R} \left((|u-k|\phi)|_{t=0} - (|u-k|\phi)|_{t=T}\right) dx$$
$$\geq \int_\mathbb{R} \int_0^T \operatorname{sign}(u - k)\, g(u)\, dt\, dx$$
for all $k \in \mathbb{R}$ and all nonnegative test functions $\phi \in C_0^\infty(\mathbb{R} \times [0,T])$. (Recall that $q(u,k) = \operatorname{sign}(u-k)(f(u) - f(k))$.)

2.10 First we note that the Rankine–Hugoniot condition implies that v is locally bounded and uniformly continuous. Assume now that $v - \varphi$ has a local maximum at (x_0, t_0), where $t_0 > 0$. Since p is piecewise differentiable, we can define the following limits:
$$\bar{p}_l = \lim_{x \to x_0-} p(x, t_0) \geq \varphi_x(x_0, t_0) \geq \bar{p}_r = \lim_{x \to x_0+} p(x, t_0).$$
The inequalities hold, since $v - \varphi$ has a maximum at (x_0, t_0) and where p is differentiable,
$$v_x = p + xp_x - tH(p)_x = p + \frac{x}{t}\dot{p} + tp_t = p + \frac{x}{t}\dot{p} - \frac{x}{t}\dot{p} = p.$$

Thus $\hat{\varphi}_x = \varphi_x(x_0, t_0)$ is between $\bar{p}_l$ and $\bar{p}_r$. We also take the upper convex envelope. Thus
$$H_l + \sigma(\hat{\varphi}_x - \bar{p}_l) \geq H(\hat{\varphi}_x),$$
$$H_r + \sigma(\hat{\varphi}_x - \bar{p}_r) \geq H(\hat{\varphi}_x),$$
where $H_{l,r} = H(\bar{p}_{l,r})$ and $\sigma = (H_l - H_r)/(\bar{p}_l - \bar{p}_r)$ if $p_l \neq p_r$ and $\sigma = H'(p_{l,r})$ otherwise. We add the two equations to find that
$$\sigma\hat{\varphi}_x \geq H(\hat{\varphi}_x) + \frac{\sigma}{2}(\bar{p}_l + \bar{p}_r) - \frac{1}{2}(H_l + H_r). \tag{C.3}$$
Now we find (x, t) close to (x_0, t_0) such that
$$\sigma = \frac{x_0 - x}{t_0 - t}.$$
Since $v - \varphi$ has a local maximum at (x_0, t_0), we have that
$$\frac{v(x_0, t_0) - v(x, t)}{t_0 - t} \geq \frac{\varphi(x_0, t_0) - \varphi(x, t)}{t_0 - t}.$$
If p is assumed to be left continuous, we can now use this to show that
$$\sigma\bar{p}_l - H_l \geq \hat{\varphi}_t + \sigma\hat{\varphi}_x.$$
Choosing (x, t) slightly to the right of the line $x = \sigma t$ we can also show that
$$\sigma\bar{p}_r - H_r \geq \hat{\varphi}_t + \sigma\hat{\varphi}_x,$$
and therefore
$$\frac{\sigma}{2}(\bar{p}_l + \bar{p}_r) - \frac{1}{2}(H_l + H_r) \geq \hat{\varphi}_t + \sigma\hat{\varphi}_x.$$
Using (C.3) we conclude that
$$\hat{\varphi}_t + H(\hat{\varphi}_x) \leq 0,$$
and v is a subsolution. To show that v is also a supsolution, proceed along similar lines.

2.11 We find that
$$\|f - f_\delta\|_{\text{Lip}} = \sup \frac{|(f(p) - f_\delta(p)) - (f(q) - f_\delta(q))|}{|p - q|}$$
$$\leq \sup |f'(p) - f'_\delta(p)|$$
$$= \sup_{\substack{j, p \\ j\delta \leq p \leq (j+1)\delta}} \left| f'(p) - \frac{f((j+1)\delta) - f(j\delta)}{\delta} \right|$$
$$= \sup_{\substack{p, q \\ |p-q| \leq \delta}} |f'(p) - f'(q)| \leq \delta \|f''\|_\infty.$$

2.12 See [113].

2.13 We first find the characteristics (parameterized using t)
$$x = x(\eta, t) = \begin{cases} \eta + (1 - e^{-t}) & \text{for } \eta \leq -\frac{1}{2}, \\ \eta(2e^{-t} - 1) & \text{for } -\frac{1}{2} < \eta < 0, \\ \eta & \text{for } \eta \geq 0, \end{cases}$$
$$u = u(\eta, t) = u_0(\eta)e^{-t}.$$

Characteristics with $\eta \in \left[-\frac{1}{2}, 0\right]$ collide at $t = \ln 2$. At that time a shock forms. The solution reads
$$u(x, t) = \begin{cases} e^{-t} & \text{for } x < \min\left(\frac{1}{2} - e^{-t}, \frac{1}{4} - \frac{1}{2}e^{-t}\right), \\ \frac{2x}{e^t - 2} & \text{for } \frac{1}{2} - e^{-t} \leq x \leq 0, \\ 0 & \text{for } x \geq \max\left(0, \frac{1}{4} - \frac{1}{2}e^{-t}\right). \end{cases}$$

2.14 The solution reads
$$u(x, t) = \begin{cases} 2 & \text{for } x < \frac{1}{2}(e^2 - 1)t, \\ 0 & \text{for } x \geq \frac{1}{2}(e^2 - 1)t. \end{cases}$$

2.15 a.
$$u(x, t) = \begin{cases} 1 & \text{for } x < t + 2, \\ 0 & \text{for } x \geq t + 2. \end{cases}$$

b.
$$u(x, 0) = \begin{cases} 0 & \text{for } x \leq 2, \\ 3\left(\frac{x}{t}\right)^2 & \text{for } 2 < x < \frac{t}{\sqrt{3}} + 2, \\ 1 & \text{for } x \geq \frac{t}{\sqrt{3}} + 2. \end{cases}$$

2.16 The solution reads
$$u(x, t) = \begin{cases} x/t & \text{for } 0 < x < t, \\ 1 & \text{for } t \leq x \leq 1 + t/2, \\ 0 & \text{otherwise,} \end{cases}$$
when $t \leq 2$, and
$$u(x, t) = \begin{cases} x/t & \text{for } 0 < x < \sqrt{2t}, \\ 0 & \text{otherwise,} \end{cases}$$
when $t > 2$.

2.17 In Figure C.1 you can see how the fronts are supposed to move, but you will have to work out the states yourself.

Appendix C. Answers and Hints

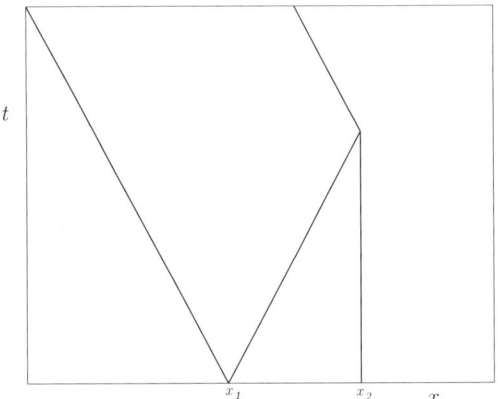

Figure C.1. The fronts for Exercise 2.17.

Chapter 3, pages 112–116.

3.1 We do the MacCormack method only; the Lax–Wendroff scheme is similar. It simplifies the computation to use repeatedly that

$$\phi(u + a\varepsilon + b\varepsilon^2 + \mathcal{O}\left(\varepsilon^3\right))$$
$$= \phi(u) + \phi'(u)a\varepsilon + \frac{\varepsilon^2}{2}(\phi''(u)a^2 + 2b\phi'(u)) + \mathcal{O}\left(\varepsilon^3\right)$$

as $\varepsilon \to 0$. Consider an exact classical (smooth) solution u of $u_t + f(u)_x = 0$, and compute (where S_M is the operator defined by the MacCormack scheme)

$$L_{\Delta t} = \frac{1}{\Delta t}\left(S(\Delta t)u - S_M(\Delta t)\right)$$
$$= \frac{1}{\Delta t}\Big\{u(x, t+\Delta t) - u(x,t)$$
$$+ \frac{\lambda}{2}\big[f\left(u(x,t) - \lambda(f(u(x+\Delta x, t)) - f(u(x,t)))\right)$$
$$- f\left(u(x-\Delta x, t) - \lambda(f(u(x,t)) - f(u(x-\Delta x, t)))\right)$$
$$+ f(u(x,t)) - f(u(x-\Delta x, t))\big]\Big\}$$
$$= \frac{1}{\Delta t}\Big\{(u_t + f(u)_x)\Delta t$$
$$+ \frac{\lambda^2}{2}\left[u_{tt} - 2f'(u)f''(u)u_x^2 - f'(u)^2 u_{xx}\right]\Delta x^2 + \mathcal{O}\left(\Delta x^3\right)\Big\}$$
$$= \mathcal{O}\left(\Delta x^2\right),$$

where we have used that a smooth solution of $u_t + f(u)_x = 0$ satisfies $u_{tt} - 2f'(u)f''(u)u_x^2 - f'(u)^2 u_{xx} = 0$ as well.

3.2 a. If $u = U_j^n$, $v = U_{j+1}^n$ and $w = U_{j-1}^n$, we have that

$$U_j^{n+1} = g(u, v, w)$$

$$= u - \lambda \left(\int_0^u f'(s) \vee 0 \, ds + \int_0^v f'(s) \wedge 0 \, ds \right.$$

$$\left. - \int_0^w f'(s) \vee 0 \, ds - \int_0^u f'(s) \wedge 0 \, ds \right)$$

$$= u - \lambda \left(\int_0^u |f'(s)| \, ds + \int_0^v f'(s) \wedge 0 \, ds - \int_0^w f'(s) \vee 0 \, ds \right).$$

Computing partial derivatives we find that

$$\frac{\partial g}{\partial u} = 1 - \lambda |f'(u)| \geq 0 \quad \text{if } \lambda |f'| \leq 1,$$

$$\frac{\partial g}{\partial v} = -\lambda f'(v) \wedge 0 \geq 0,$$

$$\frac{\partial g}{\partial w} = \lambda f'(w) \vee 0 \geq 0.$$

Consistency is easy to show.
b. If $f' \geq 0$, the scheme coincides with the upwind scheme; hence it is of first order.
c. For any number a we have

$$\left. \begin{array}{l} |a| = a \vee 0 - a \wedge 0, \\ a = a \vee 0 + a \wedge 0, \end{array} \right\} \Rightarrow \begin{cases} a \vee 0 = \dfrac{1}{2}(a + |a|), \\ a \wedge 0 = \dfrac{1}{2}(a - |a|). \end{cases}$$

Using this the form of the scheme easily follows.
d. We have that

$$\int |u| \, du = \text{sign}(u) \frac{u^2}{2}.$$

From this it follows that

$$f^{\text{EO}}(u, v) = \frac{1}{2} \left(\frac{u^2}{2} + \frac{v^2}{2} - \text{sign}(v) \frac{v^2}{2} + \text{sign}(u) \frac{u^2}{2} \right)$$

$$= \frac{1}{2} \left(\frac{u^2}{2}(1 + \text{sign}(u)) + \frac{v^2}{2}(1 - \text{sign}(v)) \right),$$

which is what we want to show. If f is convex with a unique minimum at $\bar{u}$, then

$$f^{\text{EO}}(u, v) = \frac{1}{2}(f(u \vee \bar{u}) + f(v \wedge \bar{u})).$$

3.3 The scheme is not monotone, since
$$\frac{\partial U_j^{n+1}}{\partial U_{j\pm 1}^n} = \mp \frac{\Delta t}{2\Delta x} f'(U_{j\pm 1}^n).$$

3.4 a. We find that
$$U_j^{n+1} = U_j^n/(1 + \lambda(U_j^n - U_{j-1}^n)),$$
provided that the denominator is nonzero. Thus
$$\frac{\partial U_j^{n+1}}{\partial U_j^n} = \lambda U_j^n/(1 + \lambda(U_j^n - U_{j-1}^n))^2,$$
$$\frac{\partial U_j^{n+1}}{\partial U_{j-1}^n} = (1 - \lambda U_j^n)/(1 + \lambda(U_j^n - U_{j-1}^n))^2.$$

Assume $\lambda < 1$. Considering U_j^{n+1} as a function of $U_j^n, U_{j-1}^n \in [0,1]$, we see that U_j^{n+1} takes on its largest value, namely one, when $U_j^n = U_{j-1}^n = 1$. Thus $U_j^{n+1} \in [0,1]$ for all n and j. The same computation shows that the scheme is monotone.

b. A constant is mapped into the same constant by this scheme, which therefore is consistent. A Taylor expansion around a smooth solution shows that the truncation error is of first order.

3.5 a. With the obvious notation we have that
$$V_j^{n+1} = V_j^n - \frac{\lambda}{\Delta x}\left(f_j^n - f_{j-1}^n + f_{j-2}^n - f_{j-1}^n\right),$$
and by a Taylor expansion about U_{j-1}^n,
$$f_j^n = f_{j-1}^n + (U_j^n - U_{j-1}^n)f'(U_{j-1}^n) + \frac{1}{2}(U_j^n - U_{j-1}^n)^2 f''(\eta_{j-1/2}),$$
$$f_{j-2}^n = f_{j-1}^n + (U_{j-2}^n - U_{j-1}^n)f'(U_{j-1}^n) + \frac{1}{2}(U_{j-2}^n - U_{j-1}^n)^2 f''(\eta_{j-3/2}).$$

Using this we get the desired result.

b. Assuming that $V_j^n \geq V_{j-1}^n \geq 0$ we find that
$$V_j^{n+1} \leq V_j^n - c\Delta t \left(V_j^n\right)^2 = g(V_j^n).$$

The function g has a maximum at $1/(2c\Delta t)$. Hence V_j^n is in an interval where g is increasing. Thus
$$V_j^{n+1} \leq g\left(\frac{1}{(2+n)c\Delta t}\right) = \frac{1}{(n+2)c\Delta t}\frac{n+1}{n+2} < \frac{1}{(n+3)c\Delta t}.$$

The case where $V_{j-1}^n > V_j^n$ is similar, and the case where $0 \geq V_j^n \vee V_{j-1}^n$ is trivial. Thus we have completed the induction. Hence for all

Appendix C. Answers and Hints   329

n, $\hat{V}^n$ will be in an interval where g is increasing, and
$$V_j^{n+1} \leq g\left(\hat{V}^n\right).$$
Taking the maximum over j and 0 on the left completes the claim.

c. Assuming that the claim holds for $n = 0$, we wish to show that it holds for any n by induction. Since $\hat{V}^n$ is in an interval where g is increasing, we find that
$$\hat{V}_j^{n+1} \leq \frac{\hat{V}_0}{1+cn\Delta t \hat{V}_0}\left(1 - \frac{\hat{V}_0}{1+cn\Delta t \hat{V}_0}\right)$$
$$= \frac{\hat{V}_0}{1+cn\Delta t \hat{V}_0} \frac{1+c\Delta t \hat{V}_0(n-1)}{1+c\Delta t \hat{V}_0 n},$$

so if
$$\frac{1+\hat{V}_0 c\Delta t(n-1)}{\left(1+\hat{V}_0 c\Delta tn\right)^2} \leq \frac{1}{1+\hat{V}_0 c\Delta t(n+1)},$$

we are ok. Set $k = \hat{V}_0 c\Delta t$. Since
$$(1+kn)^2 - k^2 < (1+kn)^2,$$
the claim follows.

d. Since $V_j^n \leq \hat{V}^n$, the claim follows by noting that
$$U_i^n - U_j^n = \sum_{k=j+1}^i V_j^n.$$

e. Since $\{U_j^n\}$ converges to the entropy solution u for almost every x (and y) and t, we find that the claim holds.

3.6 a. We find that $u(\cdot + p, t)$ is another entropy solution with the same initial condition; hence $u(\cdot + p, \cdot) = u$, and u is periodic.

b. Taking the infimum over y and the supremum over x we find that this holds.

c. Set $u_\varepsilon = u * w_\varepsilon$. Then u_ε is differentiable, and satisfies
$$\partial_t u_\varepsilon + \partial_x(f(u) * w_\varepsilon) = 0.$$

Thus
$$\frac{d}{dt}\int_0^p u_\varepsilon(x,t)\,dx = (f(u)*w_\varepsilon)(0,t) - (f(u)*w_\varepsilon)(p,t) = 0,$$

since also $f * w_\varepsilon$ is periodic with period p. Therefore,
$$\int_0^p u_\varepsilon(x,t)\,dx = \int_0^p u_{0,\varepsilon}(x)\,dx.$$

We know that u_ε converges to u in $L^1([0,p])$; hence

$$\int_0^p u(x,t)\,dx = \int_0^p u_0(x)\,dx.$$

Now, since $u(x,t) \to \bar{u}$ as t becomes large,

$$p\bar{u} = \int_0^p u_0(x)\,dx.$$

3.7 Using the Fourier representation, we find that

$$\int g(nx)\varphi(x)\,dx = \frac{a_0}{2}\int \varphi(x)\,dx$$

$$+ \sum_k a_k \int \cos(nkx)\varphi(x)\,dx + b_k \int \sin(nkx)\varphi(x)\,dx$$

if φ is in L^1. Since $\cos(nkx) \stackrel{*}{\rightharpoonup} 0$ and $\sin(nkx) \stackrel{*}{\rightharpoonup} 0$, part **a** follows. Now we have that $h(\sin(x))$ is a continuously differentiable function of period 2π, and by part **a**, $h(nx) \stackrel{*}{\rightharpoonup} a_0/2$, where

$$\frac{a_0}{2} = \frac{1}{2\pi}\int_0^{2\pi} h(\sin(x))\,dx = \frac{1}{\pi}\int_{-\pi/2}^{\pi/2} h(\sin(x))\,dx.$$

Introducing $\lambda = \sin(x)$ we find that

$$\frac{a_0}{2} = \frac{1}{\pi}\int_{-1}^{1} \frac{h(\lambda)}{\sqrt{1-\lambda^2}}\,d\lambda.$$

Hence for any set $E \subset \mathbb{R}$,

$$\nu(E) = \int_E \frac{\chi_{[-1,1]}}{\sqrt{1-\lambda^2}}\,d\lambda.$$

3.8 a. Observe first that v_k is constant and equal to one on the interval $[1,2]$. From the definition of F we see that

$$|F(\phi) - \phi| \le \frac{1}{3}|\phi(b) - \phi(a)|.$$

Thus

$$|v_{k+1} - v_k| = \sum_j |F(v_{j,k}) - v_{j,k}|\chi_{j,k} \le \left(\frac{2}{3}\right)^k,$$

and hence the limit

$$v(x) = \lim_{k\to\infty} v_k(x)$$

exists and is continuous.

b. Observe that T.V. $(v_{j,k}) = \left(\frac{2}{3}\right)^k$ (on $[0,1]$ and $[2,3]$), and that

$$\text{T.V.}(F(\phi)) = \frac{5}{3}|\phi(b) - \phi(a)| = \frac{5}{3}\text{T.V.}(\phi).$$

Thus

$$\text{T.V.}(v_{k+1}) = \sum_j \text{T.V.}(F(v_{j,k})) = \frac{5}{3}\sum_j \text{T.V.}(v_{j,k})$$

$$= \frac{5}{3}\left(\frac{2}{3}\right)^k \cdot 2 \cdot 3^k = \frac{10}{3} 2^k. \tag{C.4}$$

c. We see that

$$v(j/3^k) = v_k(j/3^k)$$

by construction.

d. Define the upwind scheme by

$$U_j^{n+1} = U_j^n + \lambda(f_j^n - f_{j-1}^n), \quad U_j^0 = v(j/3^k) = v_k(j/3^k). \tag{C.5}$$

From the assumptions on the flux function we know that the scheme is TVD with a CFL number at most one. Thus

$$\text{T.V.}(U^n) \leq \text{T.V.}(U^0) \leq \text{T.V.}(v_k) = \frac{10}{3} 2^k.$$

We apply Theorem 3.27, and consider

$$(\Delta x)^\beta \sum_n \sum_j |U_{j+1}^n - U_j^n| \Delta t \leq T 3^{-k\beta} \frac{10}{3} 2^k.$$

For this to be less than a constant $C(T)$, we need $2/3^\beta \leq 1$, or $\beta \geq \ln 2 / \ln 3 \approx 0.63$. For Theorem 3.27 to apply we note that (3.81) is satisfied with right-hand side zero.

Chapter 4, pages 159–163.

4.1 a. Set

$$\alpha_\varepsilon(t) = \int_0^t \left(\omega_\varepsilon(s - t_1) - \omega_\varepsilon(s - t_2)\right) ds,$$

and set $\psi(x,y,t) = \alpha_\varepsilon(t)\varphi(x,y)$ for some test function φ with $|\varphi(x,y)| \leq 1$. Then α_ε will tend to the characteristic function of the interval $[t_1, t_2]$ as $\varepsilon \to 0$. Hence, since u is a weak solution,

$$\iint \varphi(x,y)(u(x,y,t_1) - u(x,y,t_2))\, dx\, dy$$

$$+ \int_{t_1}^{t_2} \iint \left(f(u)\varphi_x + g(u)\varphi_y\right) dx\, dy\, dt = 0.$$

Then we have that

$$\|u(\cdot,\cdot,t_1) - u(\cdot,\cdot,t_2)\|_{L^1(\mathbb{R}^2)}$$
$$= \sup_{|\varphi|\leq 1} \iint \varphi(x,y)(u(x,y,t_1) - u(x,y,t_2))\,dx\,dy$$
$$\leq \int_{t_1}^{t_2} \sup_{|\varphi|\leq 1} \iint \left(f(u)\varphi_x + g(u)\varphi_y\right)\,dx\,dy\,dt$$
$$\leq \int_{t_1}^{t_2} \left(\int \text{T.V.}_{\cdot x}\left(f(u(\cdot,y,t))\right)\,dy + \int \text{T.V.}_{\cdot y}\left(g(u(x,\cdot,t))\right)\,dx\right)dt$$
$$\leq |t_1 - t_2|\left(\|f\|_{\text{Lip}} \vee \|g\|_{\text{Lip}}\right)\text{T.V.}(u_0).$$

See also Theorem 7.10.

b. Let $u_{\Delta t}$ and $v_{\Delta t}$ denote the dimensional splitting approximations to u and v, respectively. It is easy to show (using monotonicity for the one-dimensional solution operators) that if $u_0 \leq v_0$ a.e., then $u_{\Delta t} \leq v_{\Delta t}$ a.e. Hence

$$\iint \left[u_{\Delta t}(x,y,t) - v_{\Delta t}(x,y,t)\right] \vee 0\,dx\,dy = 0,$$

and since both $u_{\Delta t}$ and $v_{\Delta t}$ converge strongly in L^1 to u and v, respectively, it follows that

$$\iint \left[u(x,y,t) - v(x,y,t)\right] \vee 0\,dx\,dy = 0,$$

and thus $u \leq v$ a.e.

4.2 Let $u = S_t^j u_0$ denote the solution of $u_t + f_j(u)_x = 0$ with initial condition $u|_{t=0} = u_0$. Define $\{u^n\}$ by

$$u^0 = u_0, \quad u^{n+1/2} = S_{\Delta t}^1 u^n, \quad u^{n+1} = S_{\Delta t}^2 u^{n+1/2}.$$

Interpolate by defining $u_{\Delta t} = S_{2(t-t_n)}^1 u^n$ if $t_n \leq t \leq t_{n+1/2}$ and $u_{\Delta t} = S_{2(t-t_{n+1/2})}^1 u^{n+1/2}$ whenever $t_{n+1/2} \leq t \leq t_{n+1}$. (Here $t_n = n\Delta t$.) By mimicking the multidimensional case, one concludes that (i) $\|u_{\Delta t}\|_\infty \leq C$; (ii) $\text{T.V.}(u_{\Delta t}(t)) \leq \text{T.V.}(u_0)$; and (iii) $\|u_{\Delta t}(t) - u_{\Delta t}(s)\|_1 \leq C|t-s|$. Theorem A.8 shows that $u_{\Delta t}$ has a limit u as $\Delta t \to 0$. Write the Kružkov entropy condition for $u_{\Delta t}$ for each time interval $[t_n, t_{n+1/2}]$ (for f_1) and $[t_n, t_{n+1/2}]$ (for f_2), add the results, and let $Dt \to 0$. As in the multidimensional case, the limit is the Kružkov entropy condition for u and the original initial value problem (4.93). The analysis in Section 4.3 applies concerning convergence rates.

4.3 For simplicity, we assume that $m = 2$. Using the heat kernel we can write

$$u^{n+1/2}(x,y) = \frac{1}{\sqrt{4\pi\Delta t}} \int \exp\left(-\frac{(x-z)^2}{4\Delta t}\right) u^n(z,y)\, dz$$

$$u^{n+1}(x,y) = \frac{1}{\sqrt{4\pi\Delta t}} \int \exp\left(-\frac{(y-w)^2}{4\Delta t}\right) u^{n+1/2}(x,w)\, dw$$

$$= \frac{1}{4\pi\Delta t} \iint \exp\left(-\frac{(x-z)^2 + (y-w)^2}{4\Delta t}\right) u^n(z,w)\, dz\, dw.$$

From this we see that $u^{n+1}(x,y)$ is the exact solution of the heat equation with initial data $u^n(x,y)$ after a time Δt. If we let $u(x,y,t)$ denote the exact solution of the original heat equation, we therefore see that

$$u(x,y,t_n) = u_{\Delta t}(x,y,t_n), \quad n = 0, 1, 2, \ldots.$$

We have that u and $u_{\Delta t}$ are L^1 continuous in t; hence it follows that $u_{\Delta t} \to u$ in L^1 as $\Delta t \to 0$.

If we want a rate of this convergence, we first assume that $u(\cdot, \cdot, t_n)$ is uniformly continuous. For $t \in \langle t_n, t_{n+1/2}\rangle$ we have that

$$u_{\Delta t}(x,y,t) - u(x,y,t)$$

$$= \frac{1}{4\pi(t-t_n)} \iint \exp\left(-\frac{(x-z)^2 + (y-w)^2}{4(t-t_n)}\right)$$
$$\times \left[u(z,y,t_n) - u(z,w,t_n)\right] dw\, dz$$

$$= \frac{1}{\sqrt{4\pi(t-t_n)}} \int \exp\left(-\frac{(x-z)^2}{4(t-t_n)}\right)$$
$$\times \int \frac{1}{\sqrt{4\pi(t-t_n)}} \exp\left(-\frac{(y-w)^2}{4(t-t_n)}\right)$$
$$\times \left[u(z,y,t_n) - u(z,w,t_n)\right] dw\, dz$$

$$= \frac{1}{\sqrt{4\pi(t-t_n)}} \int \exp\left(-\frac{(x-z)^2}{4(t-t_n)}\right) \left[\eta(z,y,t) - \eta(z,y,t_n)\right] dz,$$

where $\eta(z,w,t)$ denotes the solution of

$$\eta_t = \eta_{ww}, \quad \eta(z,w,t_n) = u^n(z,w).$$

If $u^n(z,w)$ is uniformly continuous, then

$$|\eta(z,y,t) - \eta(z,y,t_n)| \leq C\sqrt{\Delta t},$$

and hence

$$|u_{\Delta t}(x,y,t) - u(x,y,t)| \leq C\sqrt{\Delta t},$$

and an identical estimate is available if $t \in \langle t_{n+1/2}, t_{n+1}\rangle$. Hence if $u_0(x,y)$ is uniformly continuous, then
$$\|u_{\Delta t}(\cdot,\cdot,t) - u(\cdot,\cdot,t)\|_{L^\infty(\mathbb{R}^2)} \le C\sqrt{\Delta t}.$$

If $u_0(x,y)$ is not assumed to be continuous, but merely of bounded variation, we must use Kružkov's interpolation lemma to conclude that
$$\|\eta(z,\cdot,t) - \eta(z,\cdot,t_n)\|_{L^1_{\text{loc}}(\mathbb{R})} \le C\sqrt{\Delta t}.$$

Using this we find that
$$\|u_{\Delta t}(\cdot,\cdot,t) - u(\cdot,\cdot,t)\|_{L^1_{\text{loc}}(\mathbb{R}^2)} \le C\sqrt{\Delta t}.$$

4.4 a. The scheme will be monotone if $U_j^{n+1/2}(U^n)$ and $U_j^n(U^{n+1/2})$ are monotone in all arguments. This is the case if
$$\lambda\|f'\|_1 \le 1, \quad \text{and} \quad \mu \le \frac{1}{2}.$$

b. We see that if $\lambda|f'| \le 1$, then
$$\min_j U_j^n \le U_j^{n+1/2} \le \max_j U_j^n,$$
and if $\mu \le \frac{1}{2}$,
$$\min_j U_j^{n+1/2} \le U_j^{n+1} \le \max_j U_j^{n+1/2}.$$

Therefore, the sequence $\{u_{\Delta t}\}$ is uniformly bounded. Now let V_j^n be another solution with initial data V_j^0. Set $W_j^n = U_j^n - V_j^n$. Then
$$W_j^{n+1/2} = \frac{1}{2}(1 - f'(\eta_{j+1}))W_{j+1}^n + \frac{1}{2}(1 + f'(\eta_{j-1}))W_{j-1}^n,$$
where η_j is between U_j^n and V_j^n. By the CFL condition, the coefficients of $W_{j\pm 1}^n$ are positive; hence
$$\left|W_j^{n+1/2}\right| \le \frac{1}{2}(1 - f'(\eta_{j+1}))\left|W_{j+1}^n\right| + \frac{1}{2}(1 + f'(\eta_{j-1}))\left|W_{j-1}^n\right|.$$

Summing over j we find that
$$\sum_j \left|W_j^{n+1/2}\right| \le \sum_j |W_j^n|.$$

Similarly, we find that
$$\left|W_j^{n+1}\right| \le (1 - 2\mu)\left|W_j^{n+1/2}\right| + \mu\left|W_{j+1}^{n+1/2}\right| + \mu\left|W_{j-1}^{n+1/2}\right|,$$
so that
$$\sum_j |W_j^{n+1}| \le \sum_j \left|W_j^{n+1/2}\right| \le \sum_j |W_j^n|.$$

Setting $V_j^n = U_{j-1}^n$ we see that T.V. $(u_{\Delta t})$ is uniformly bounded, and setting $V_j^0 = 0$ we see that $\|u_{\Delta t}\|_1$ is also uniformly bounded. To apply Theorem A.8 we need to use Kružkov's interpolation lemma, Lemma 4.10, to find a temporal modulus of continuity. Now we find that

$$\left|\int \phi(x)\left(u_{\Delta t}(x, n\Delta t) - u_{\Delta t}(x, m\Delta t)\right) dx\right|$$
$$= \left|\sum_{k=n+1}^{m} \sum_j \int_{x_{j-1/2}}^{x_{j+1/2}} \phi(x)\, dx \left(U_j^{k+1} - U_j^k\right)\right|.$$

If we set $\bar\phi_j = \int_{x_{j-1/2}}^{x_{j+1/2}} \phi(x)\, dx$ and $D_j^k = U_j^k - U_{j-1}^k$, then the the sum over j can be written

$$\sum_j \bar\phi_j \mu \left(D_{j+1}^{k+1/2} - D_j^{k+1/2}\right) + \sum_j \bar\phi_j \frac{1}{2}\left(D_{j+1}^k - D_j^k\right)$$
$$+ \sum_j \bar\phi_j \lambda \left(f_{j+1}^n - f_{j-1}^n\right). \tag{C.6}$$

We do a partial summation in the first sum to find that it equals

$$-\mu \sum_j D_j^{k+1/2} \left(\bar\phi_j - \bar\phi_{j-1}\right).$$

Now

$$\left|\bar\phi_j - \bar\phi_{j-1}\right| = \left|\int_{x_{j-1/2}}^{x_{j+1/2}} \int_{x-\Delta x}^{x} \phi'(y)\, dy\, dx\right| \le \|\phi'\|_1 \Delta x^2.$$

Since $\mu = \Delta t/\Delta x^2$, the first sum in (C.6) is bounded by

$$\Delta t \|\phi'\|_1 \text{T.V.}(u_{\Delta t}).$$

Similarly, the second term is bounded by

$$\Delta t \frac{\|\phi'\|_1}{2} \frac{\Delta x}{\lambda} \text{T.V.}(u_{\Delta t}).$$

Finally, since $\lambda = \Delta t/\Delta x$, the last term in (C.6) is bounded by

$$\Delta t \|\phi\|_1 \text{T.V.}(u_{\Delta t}).$$

Therefore,

$$\left|\int \phi(x)\left(u_{\Delta t}(x, n\Delta t) - u_{\Delta t}(x, m\Delta t)\right) dx\right|$$
$$\le (n-m)\Delta t\, \text{const}\, \left(\|\phi\|_1 + \|\phi'\|_1\right).$$

Then by Lemma 4.10 and the bound on the total variation,

$$\|u_{\Delta t}(\,\cdot\,, t_1) - u_{\Delta t}(\,\cdot\,, t_2)\|_1 \le \text{const}\, \sqrt{|t_1 - t_2|}.$$

Hence we can conclude, using Theorem A.8, that a subsequence of $\{u_{\Delta t}\}$ converges strongly in L^1.

To show that the limit is a weak solution we have to do a long and winding calculation. We give only an outline here. Firstly,

$$\iint \left(u_{\Delta t}\varphi_t + f(u_{\Delta t})\varphi_x + u_{\Delta t}\varphi_{xx}\right) dx\, dt$$

$$= \sum_{n,j} \int_{x_{j-1/2}}^{x_{j+1/2}} \int_{t_n}^{t_{n+1}} \left(U_j^n \varphi_t + f_j^n \varphi_x + U_j^n \varphi_{xx}\right) dx\, dt.$$

After a couple of partial summations this equals

$$-\sum_{n,j} \int_{x_{j-1/2}}^{x_{j+1/2}} \varphi^{n+1}(x)\, dx \left[U_j^{n+1} - U_j^n\right] \tag{C.7}$$

$$-\sum_{n,j} \int_{t_n}^{t_{n+1}} \varphi_{j+1/2}(t)\, dt \left[f_j^n - f_{j-1}^n\right] \tag{C.8}$$

$$-\sum_{n,j} \int_{t_n}^{t_{n+1}} \varphi'_{j+1/2}(t)\, dt D_j^{n+1/2} \tag{C.9}$$

$$-\sum_{n,j} \int_{t_n}^{t_{n+1}} \left(\varphi'_{j+1/2}(t) - \varphi'_{j-1/2}(t)\right) dt \left[U_j^{n+1/2} - U_j^n\right]. \tag{C.10}$$

Now split (C.7) by writing the term in the square brackets as

$$\left[\mu\left(D_j^{n+1/2} - D_{j-1}^{n+1/2}\right) + \frac{1}{2}\left(D_j^n - D_{j-1}^n\right) - \frac{\lambda}{2}\left(f_{j+1}^n - f_{j-1}^n\right)\right]. \tag{C.11}$$

The trick is now to pair the term in (C.11) with μ with (C.9), and the term with λ with (C.8). The limits of the sums of these are zero, as is the limit of (C.10), and the remaining part of (C.11) also tends to zero as Δt approaches zero. If you carry out all of this, you will find that the limit is a weak solution.

4.5 a. We multiply the inequality by $e^{-\gamma t}$ and find that

$$\frac{d}{dt}\left(e^{-\gamma t} u(t)\right) \leq 0.$$

Thus

$$u(t)e^{-\gamma t} \leq u(0).$$

b. Multiply the inequality by e^{-Ct} to find that

$$\frac{d}{dt}\left(u(t)e^{-Ct}\right) \leq Ce^{-Ct}.$$

We integrate from 0 to t:

$$u(t)e^{-Ct} - u(0) \leq \int_0^t Ce^{-Cs}\,ds = 1 - e^{-Ct}.$$

After some rearranging, this is what we want.

c. Multiplying by $\exp(-\int_0^t c(s)ds)$ we find that

$$u(t) \leq u(0)e^{\int_0^t c(s)ds} + \int_0^t d(s)e^{\int_s^t c(\tau)d\tau}\,ds,$$

which implies the claim.

d. Set $U(t) = \int_0^t u(t)dt$. Then

$$U'(t) \leq C_1 U(t) + C_2,$$

and an application of part **c** gives that

$$\int_0^t u(s)\,ds \leq -\frac{C_2}{C_1}(1 - e^{C_1 t}).$$

Inserting this in the original inequality yields the claim.

e. Introduce $w = u/f$. Then

$$w(t) \leq 1 + \int_0^t \frac{f(s)}{f(t)}g(s)w(s)\,ds \leq 1 + \int_0^t g(s)w(s)\,ds.$$

Let U be the right-hand side of the above inequality, that is,

$$U = 1 + \int_0^t g(s)w(s)\,ds.$$

Clearly, $U(0) = 1$ and $U'(t) = g(t)w(t) \leq g(t)U(t)$. Applying part **c** we find that

$$u(t) = w(t)f(t) \leq f(t)U(t) \leq f(t)\exp\left(\int_0^t g(s)\,ds\right).$$

(If f is differentiable, we see that $U(t) = f(t) + \int_0^t g(s)u(s)\,ds$ satisfies $U'(t) \leq f'(t) + g(t)U(t)$, and hence we could have used part **c** directly.)

4.6 a. To verify the left-hand side of the inequality, we proceed as before. When doing this we have a right-hand side given by

$$\iiiint \text{sign}(u-v)(g(u)-g(v))\psi \omega_{\varepsilon_0}\omega_{\varepsilon_1}\,dx\,dt\,dy\,ds.$$

Now we can send ε_0 and ε_1 to zero and obtain the answer.

b. By choosing the test function as before, we find that the double integral on the left-hand side is less than zero. Therefore, we find that

$$h(0) - h(T) \geq \iint \text{sign}(u-v)(g(u)-g(v))\psi(x,t)\,dxdt,$$

and therefore

$$h(T) \leq h(0) + \int_0^T \int |g(u) - g(v)| \psi \, dx \, dt$$

$$\leq h(0) + \gamma \int_0^T \int |u - v| \psi \, dx \, dt.$$

c. By making M arbitrarily large we obtain the desired inequality.

4.7 a. We set

$$u_{\Delta t}(x, 0) = u_0(x),$$

$$u_{\Delta t}(x, t)$$
$$= \begin{cases} S\left(2(t - t_n)\right) u_n(x), & t \in \langle t_n, t_{n+1/2}], \\ R(x, 2(t - t_{n+1/2}) + t_n, t_{n+1/2}) u_{n+1/2}(x), & t \in \langle t_{n+1/2}, t_n]. \end{cases}$$

b. A bound on $\|u_{\Delta t}\|_\infty$ is obtained as before. To obtain a bound on the total variation we first note that

$$\text{T.V.}\left(u^{n+1/2}\right) \leq \text{T.V.}\left(u^n\right).$$

Let $u(x, t)$ and $v(y, t)$ be two solutions of the ordinary differential equation in the interval $[t_n, t_{n+1}]$. Then as before we find that

$$|u(x, t) - v(x, t)|_t \leq |g(x, t, u(x, t)) - g(x, t, v(y, t))|$$
$$+ |g(x, t, v(y, t)) - g(y, t, v(y, t))|$$
$$\leq \gamma |u(x, t) - v(y, t)|$$
$$+ |g(x, t, v(y, t)) - g(y, t, v(y, t))|.$$

Setting $w = u - v$ we obtain

$$|w(t)| \leq e^{\gamma(t - t_{n+1})} \left(|w(0)| + \int_{t_n}^{t_{n+1}} |g(x, t, v(y, t)) - g(y, t, v(y, t))| \, dt \right).$$

If now $u(0) = u^{n+1/2}(x)$ and $v(0) = u^{n+1/2}(y)$, this reads

$$\left| u^{n+1}(x) - u^{n+1}(y) \right|$$
$$\leq e^{\gamma \Delta t} \Bigg(\left| u^{n+1/2}(x) - u^{n+1/2}(y) \right|$$
$$+ \int_{t_n}^{t_{n+1}} \left| g\left(x, t, u_{\Delta t}\left(y, \frac{t - t_n}{2} + t_{n+1/2}\right)\right) \right.$$
$$\left. - g\left(y, t, u_{\Delta t}\left(y, \frac{t - t_n}{2} + t_{n+1/2}\right)\right) \right| dt \Bigg).$$

By the assumption on g this implies that

$$\text{T.V.}\left(u^{n+1}\right) \leq e^{\gamma \Delta t} \left(\text{T.V.}\left(u^{n+1/2}\right) + \int_{t_n}^{t_{n+1}} b(t) \, dt \right),$$

and thus
$$\text{T.V.}(u_{\Delta t}) \leq e^{\gamma T} \left(\text{T.V.}(u_0) + \|b\|_1 \right).$$
We can now proceed as before to see that
$$\int_I \left| u^{n+1}(x) - u^n(x) \right| dx \leq \text{const } \Delta t.$$
Hence we conclude, using Theorem A.8, that a subsequence of $\{u_{\Delta t}\}$ converges strongly in L^1 to a function u of bounded variation.

c. To show that u is an entropy solution, we can use the same argument as before.

4.8 Write $v = u_t$ and differentiate the heat equation $u_t = \Delta u$ with respect to t. Thus
$$v_t = \varepsilon \Delta v.$$
Let sign_η be a smooth sign function, namely,
$$\text{sign}_\eta(x) = \begin{cases} 1 & \text{for } x \geq \eta, \\ x/\eta & \text{for } |x| < \eta, \\ -1 & \text{for } x \leq -\eta. \end{cases}$$
If we multiply
$$v_t = \varepsilon \Delta v$$
by $\text{sign}_\eta(v)$ and integrate over $\mathbb{R}^m \times [0, t]$, we obtain
$$\int_{\mathbb{R}^m} \int_0^t v_t \, \text{sign}_\eta(v) \, dt \, dx = \varepsilon \int_{\mathbb{R}^m} \int_0^t \Delta v \, \text{sign}_\eta(v) \, dt \, dx$$
$$= -\varepsilon \int_{\mathbb{R}^m} \int_0^t |\nabla v|^2 \, \text{sign}'_\eta(v) \, dt \, dx$$
$$\leq 0.$$
For the left-hand side we obtain
$$\int_{\mathbb{R}^m} \int_0^t v_t \, \text{sign}_\eta(v) \, dt \, dx = \int_{\mathbb{R}^m} \int_0^t \left((v \, \text{sign}_\eta(v)) \right)_t dt \, dx$$
$$- \int_{\mathbb{R}^m} \int_0^t v \, v_t \, \text{sign}'_\eta(v) \, dt \, dx.$$
The last term vanishes in the limit $\eta \to 0$; see Lemma B.5. Thus we conclude that as $\eta \to 0$,
$$\|v(t)\|_1 - \|v(0)\|_1 \leq 0.$$
From this we obtain
$$\|u(t) - u_0\|_1 = \int_{\mathbb{R}^m} \left| \int_0^t v(\tilde{t}) \, d\tilde{t} \right| dx \leq \int_0^t \|v(\tilde{t})\|_1 \, d\tilde{t} \leq \|v(0)\|_1 \, t.$$

Chapter 5, pages 203–206.

5.1 The eigenvalues are $\lambda = \pm\sqrt{-p'(v)}$. Hence we need that p' is negative.

5.2 The shock curves are given by
$$S_1: u = u_l + (v - v_l)/\sqrt{vv_l}, \quad v \leq v_l,$$
$$S_2: u = u_l - (v - v_l)/\sqrt{vv_l}, \quad v \geq v_l,$$
and the rarefaction curves read
$$R_1: u = u_l + \ln(v/v_l), \quad v \leq v_l,$$
$$R_2: u = u_l - \ln(v/v_l), \quad v \geq v_l.$$
We see that for $v_l > 0$ the wave curves extend to infinity. Given a right state (v_r, u_r) and a left state (v_l, u_l) we see that the forward slow curves (i.e., the S_1 and the R_1 curves above) intersect the backward fast curves from (v_r, u_r) (i.e., the curves of left states that can be connected to the given right state (v_r, u_r)) in a unique point (v_m, u_m). Hence the Riemann problem has a unique solution for all initial data in the half-plane $v > 0$.

5.3 The shock curves are given by
$$S_1: u = u_l - \sqrt{(v - v_l)(p(v_l) - p(v))}, \quad v \leq v_l,$$
$$S_2: u = u_l + \sqrt{(v - v_l)(p(v_l) - p(v))}, \quad v \geq v_l,$$
and the rarefaction curves read
$$R_1: u = u_l + \int_{v_l}^{v} \sqrt{-p'(y)}\, dy, \quad v \leq v_l,$$
$$R_2: u = u_l - \int_{v_l}^{v} \sqrt{-p'(y)}\, dy, \quad v \geq v_l.$$
If $\int_{v_l}^{\infty} \sqrt{-p'(y)}\, dy < \infty$, then there are points (v_r, u_r) for which the Riemann problem does not have a solution. For further details, see [130, pp. 306 ff].

5.4 The solution consists of a slow rarefaction wave connecting the left state $(h_l, 0)$ and an intermediate state (h_m, v_m) with
$$v_m = -2(\sqrt{h_m} - \sqrt{h_l}),$$
and there is a fast shock connecting (h_m, v_m) and $(h_r, 0)$, where
$$v_m = \frac{1}{\sqrt{2}}(h_r - h_m)\sqrt{\frac{1}{h_m} + \frac{1}{h_r}}.$$
The intermediate state is determined by the relation
$$2\sqrt{2}(\sqrt{h_m} - \sqrt{h_l}) = (h_r - h_m)\sqrt{\frac{1}{h_m} + \frac{1}{h_r}},$$

which has a unique solution (the left-hand side is increasing in h_m, ending up at zero for h_l, while the right-hand side is decreasing in h_m, starting at zero for h_r).

5.5 **a.** Set $f(w) = \varphi w$ and compute

$$df = \begin{pmatrix} \varphi_u + \varphi & \varphi_v u \\ \varphi_u v & \varphi_v v + \varphi \end{pmatrix} = \begin{pmatrix} \varphi_u u & \varphi_v u \\ \varphi_u v & \varphi_v v \end{pmatrix} + \begin{pmatrix} \varphi & 0 \\ 0 & \varphi \end{pmatrix} = A + \varphi I.$$

Hence an eigenvector of A with eigenvalue μ is also an eigenvector of df with corresponding eigenvalue $\mu + \varphi$. Using this we find that

$$\lambda_1 = \varphi, \quad r_1 = \begin{pmatrix} \varphi_v \\ -\varphi_u \end{pmatrix}, \quad \lambda_2 = \varphi + (\varphi_u u + \varphi_v v), \quad r_2 = \begin{pmatrix} u \\ v \end{pmatrix}.$$

b. In this case $\varphi_u = u$ and $\varphi_v = v$, and hence

$$\lambda_1 = \frac{1}{2}(u^2 + v^2), \quad r_1 = \begin{pmatrix} v \\ -u \end{pmatrix}, \quad \lambda_2 = \frac{3}{2}(u^2 + v^2), \quad r_2 = \begin{pmatrix} u \\ v \end{pmatrix}.$$

We find that the contours of φ, i.e., circles about the origin, are contact discontinuities with associated speed equal to half the radius squared. These correspond to the eigenvalue λ_1. The rarefaction curves of the eigenvalue λ_2 are half-lines starting at (u_l, v_l) given by

$$\frac{u}{u_l} = \frac{v}{v_l}$$

such that $u^2 + v^2 \geq u_l^2 + v_l^2$. For the shock part of the solution the Rankine–Hugoniot relation gives

$$(\varphi - \varphi_l)(vu_l - v_l u) = 0.$$

If $\varphi = \varphi_l$, we are on the contact discontinuity, so we must have

$$\frac{u}{v} = \frac{u_l}{v_l}.$$

This is again a straight line through the origin and through u_l, v_l. What parts of this line can we use? The shock speed is given by

$$\sigma = \frac{u_l^2 + v_l^2}{2}\left[\left(\frac{v}{v_l}\right)^2 + \left(\frac{v}{v_l}\right) + 1\right]$$

$$= \frac{u_l^2 + v_l^2}{2}\left[\left(\frac{u}{u_l}\right)^2 + \left(\frac{u}{u_l}\right) + 1\right]$$

$$= \frac{u_l^2 + v_l^2}{2u_l v_l}\left[uv + \sqrt{uu_l vv_l} + u_l v_l\right].$$

Using polar coordinates $r = 2\varphi$, $u = r\cos(\theta)$ and $v = r\sin(\theta)$, we see that

$$\sigma = \varphi_l\left[\left(\frac{r}{r_l}\right)^2 + \left(\frac{r}{r_l}\right) + 1\right].$$

We want (this is an extra condition!) to have σ decreasing along the shock path, so we define the admissible part of the Hugoniot locus to be the line segment bounded by (u_l, v_l) and the origin.

Now the solution of the Riemann problem is found by first using a contact discontinuity from (u_l, v_l) to some (u_m, v_m), where

$$\varphi(u_l, v_l) = \varphi(u_m, v_m) \quad \text{and} \quad \frac{u_m}{u_r} = \frac{v_m}{v_r},$$

and (u_m, v_m) is on the same side of the origin as (u_r, v_r). Finally, (u_m, v_m) is connected to (u_r, v_r) with a 2-wave.

c. In this case

$$\varphi_u = \varphi_v = \frac{-1}{(1+u+v)^2},$$

and therefore

$$\lambda_2 = \frac{1}{(1+u+v)^2}.$$

The calculation regarding the Hugoniot loci remains valid; hence the curves $\varphi = \text{const}$ are contact discontinuities. These are the lines given by $v = c - u$ and $u, v > 0$. The rarefaction curves of the second family remain straight lines through the origin, as are the shocks of the second family. On a Hugoniot curve of the second family, the speed is found to be

$$\sigma = \frac{1}{(1+u_l+v_l)(1+u+v)},$$

and if we want the Lax entropy condition to be satisfied, we must take the part of the Hugoniot locus pointing away from the origin. In this case both families of wave curves are straight lines. Such systems are often called *line fields*, and this system arises in chromatography.

Observe that the family denoted by subscript 2 now has the smallest speed. Hence when we find the solution of the Riemann problem, we first use the second family, then the contact discontinuity.

5.6 Integrate the exact solution u_j^n over the rectangle $[(j-1)\Delta x, (j+1)\Delta x] \times [n\Delta t, (n+1)\Delta t]$. The CFL condition yields that no waves cross the lines $(j \pm 1)\Delta x$. Thus

$$0 = \int_{(j-1)\Delta x}^{(j+1)\Delta x} \int_{n\Delta t}^{(n+1)\Delta t} \left((u_j^n)_t + f(u_j^n)_x \right) dx\, dt$$

$$= \int_{(j-1)\Delta x}^{(j+1)\Delta x} u_j^n \, dx \Big|_{n\Delta t}^{(n+1)\Delta t} + \int_{n\Delta t}^{(n+1)\Delta t} f(u_j^n) \, dt \Big|_{(j-1)\Delta x}^{(j+1)\Delta x}$$

$$= \int_{(j-1)\Delta x}^{(j+1)\Delta x} u_j^n(x, \Delta t) \, dt - \Delta x \left(U_{j+1}^n + U_{j-1}^n \right)$$

Appendix C. Answers and Hints 343

$$+ \Delta t \left(f(U^n_{j+1}) - f(U^n_{j-1}) \right)$$
$$= \int_{(j-1)\Delta x}^{(j+1)\Delta x} u^n_j(x, \Delta t)\, dt - 2\Delta x U^{n+1}_j.$$

5.7 a. Choose a hypersurface M transverse to $r_k(u_0)$. Then the linear first-order equation $\nabla w(u)\cdot r_k(u) $ with w equal to some regular function w_0 on M has a unique solution locally, using, e.g., the method of characteristics. On M we may choose $n-1$ linearly independent functions, say $w^1_0,\ldots,w^{n-1}_0$. The corresponding solutions w^j, $j=1,\ldots,n-1$, will have linearly independent gradients. See [128, p. 117] for more details, and [130, p. 321] for a different argument.
b. The kth rarefaction curve is the integral curve of the kth right eigenvector field, and hence any k-Riemann invariant is constant along that rarefaction curve.
c. The set of equations (5.136) yields $n(n-1)$ equations to determine n scalar functions w_j. See [42, pp. 127 ff] for more details.
d. For the shallow-water equations we have
$$w_1 = \frac{q}{h} - 2\sqrt{h}, \quad w_2 = \frac{q}{h} + 2\sqrt{h}.$$

5.8 a. Recall from Exercise 5.2 that the rarefaction curves are given by
$$\ln(v/v_l) = \mp(u - u_l).$$
Therefore, the Riemann invariants are given by
$$w_-(v,u) = \ln(v) + u, \quad w_+(v,u) = \ln(v) - u.$$
b.–e. Introducing $\mu = w_-$ and $\tau = w_+$, we find that
$$v = \exp\left(\frac{\mu + \tau}{2}\right), \quad u = \frac{\mu - \tau}{2}.$$
The Hugoniot loci are given by
$$u - u_l = \mp\left(\sqrt{\frac{v}{v_l}} - \sqrt{\frac{v_l}{v}}\right),$$
which in (μ,τ) coordinates reads
$$\frac{1}{2}[(\mu - \mu_l) - (\tau - \tau_l)]$$
$$= \mp\left(\exp\left(\frac{\mu - \mu_l + \tau - \tau_l}{4}\right) - \exp\left(\frac{\mu_l - \mu + \tau_l - \tau}{4}\right)\right).$$
Using $\Delta\mu$ and $\Delta\tau$, this relation says that
$$\Delta\mu - \Delta\tau = \mp 4 \sinh\left(\frac{\Delta\mu + \Delta\tau}{4}\right). \qquad\text{(C.12)}$$

Hence the Hugoniot loci are translation-invariant. The rarefaction curves are coordinate lines, and trivially translation-invariant. To show that H_- (the part of the Hugoniot locus with the minus sign) is the reflection of H_+, we note that such a reflection maps

$$\Delta\mu \mapsto \Delta\tau \quad \text{and} \quad \Delta\tau \mapsto \Delta\mu.$$

Hence a reflection leaves the right-hand side of (C.12) invariant, and changes the sign of the left-hand side. Thus the claim follows.

Recalling that the "−" wave corresponds to the eigenvalue $+1/u$, we must use the "+" waves first when solving the Riemann problem. Therefore, we now term the "+" waves 1-waves, and the "−" waves 2-waves. Hence the lines parallel to the μ-axis are the rarefaction curves of the first family. Then the 1-wave curve is given by

$$\begin{cases} u - u_l = \ln\left(\frac{v}{v_l}\right) & \text{for } v > v_l, \\ u - u_l = \sqrt{\frac{v}{v_l}} - \sqrt{\frac{v_l}{v}} & \text{for } u < u_l, \end{cases}$$

which in (μ, τ) coordinates reads

$$\begin{cases} \tau = \tau_l & \text{for } \mu > \mu_l, \\ \tau = \tau(\mu) & \text{for } \mu < \mu_l. \end{cases}$$

Here, the curve $\tau(\mu)$ is given implicitly by (C.12), and we find that

$$\tau'(\mu) = \frac{1 - \cosh(\cdots)}{1 + \cosh(\cdots)}.$$

From this we conclude that

$$0 \geq \frac{d\tau}{d\mu} \geq -1, \quad \frac{d^2\tau}{d\mu^2} \geq 0,$$

and $\tau(\mu_l) = \tau_l$ and $\lim_{\mu \to -\infty} \tau'(\mu) = -1$. By the reflection property we can also find the second wave curve in a similar manner.

5.9 a. We obtain (cf. (2.10))

$$\begin{aligned} 0 &= \iint (u_t + f(u)_x) \nabla_u \eta(u) \phi \, dx \, dt \\ &= \iint \eta(u)_t \phi \, dx \, dt + \iint df(u) u_x \nabla_u \eta(u) \phi \, dx \, dt \\ &= -\iint \eta(u) \phi_t \, dx \, dt + \iint \nabla_u q(u) u_x \phi \, dx \, dt \\ &= -\iint \left(\eta(u) \phi_t + q(u) \phi_x \right) dx \, dt. \end{aligned}$$

b. The right-hand side of (5.137) is the gradient of a vector-valued function q if

$$d^2\eta(u) df(u) = df(u)^t d^2\eta(u), \tag{C.13}$$

where $df(u)^t$ denotes the transpose of $df(u)$, and $d^2\eta(u)$ denotes the Hessian of η. The relation (C.13) follows from the fact that mixed derivatives need to be equal, i.e.,

$$\frac{\partial^2 q}{\partial u_j \partial u_k} = \frac{\partial^2 q}{\partial u_k \partial u_j}.$$

Equation (C.13) imposes $n(n-1)/2$ conditions on the scalar function η.

c. We obtain

$$\eta(u) = \frac{1}{2}u^2 - \int^v p(y)\,dy, \quad q(u) = up(v).$$

d. Equation (C.13) reduces in the case of the shallow water equations to one hyperbolic equation,

$$\left(\frac{q^2}{2h} + h\right)\eta_{qq} = \eta_{hh} + 2\frac{q}{h}\eta_{hq},$$

with solution

$$\eta(u) = \frac{q^2}{2h} + \frac{1}{2}h^2, \quad Q(u) = \frac{q^3}{2h^2} + hq$$

(where for obvious reasons we have written the entropy flux with capital Q rather than q).

5.10 a. Let r_k and l_k denote the right and left eigenvectors, respectively. Thus

$$Ar_k = \lambda_k r_k, \quad l_k A = \lambda_k l_k, \quad l_j \cdot r_k = \delta_{j,k}.$$

Decompose the left and right states in terms of right eigenvectors as

$$u_l = \sum_{k=1}^n \alpha_k r_k, \quad u_r = \sum_{k=1}^n \beta_k r_k.$$

Then we can write the solution of the Riemann problem as (see [98, pp. 64 ff])

$$u(x,t) = u_l + \sum_{\substack{k \\ \lambda_k < x/t}} (\beta_k - \alpha_k) r_k = u_r - \sum_{\substack{k \\ \lambda_k > x/t}} (\beta_k - \alpha_k) r_k.$$

b. A more compact form, valid for the general Cauchy problem as well, is

$$u(x,t) = \sum_{k=1}^n l_k \cdot u_0(x - \lambda_k t) r_k.$$

Here the initial data is u_0, and the left eigenvectors are normalized. Using that

$$u_t = \sum_{k=1}^{n}(-\lambda_k)l_k \cdot u_0'(x - \lambda_k t)r_k,$$

$$u_x = \sum_{k=1}^{n} l_k \cdot u_0'(x - \lambda_k t)r_k,$$

we see that

$$u_t + Au_x = \sum_{k=1}^{n}\left(-\lambda_k l_k \cdot u_0'(x - \lambda_k t)r_k + l_k \cdot u_0'(x - \lambda_k t)Ar_k\right) = 0.$$

Decomposing the initial data

$$u_0(x) = \sum_{k=1}^{n} \alpha_k(x)r_k,$$

we see that

$$\alpha_k(x) = l_k \cdot u_0(x),$$

and hence $u(x, 0) = u_0(x)$.

c. In the case of the wave equation we find that

$$A = \begin{pmatrix} 0 & -1 \\ -c^2 & 0 \end{pmatrix}, \quad c > 0,$$

$$\lambda_1 = -c, \quad r_1 = \frac{1}{\sqrt{1+c^2}}\begin{pmatrix} 1 \\ c \end{pmatrix}, \quad l_1 = \frac{1}{\sqrt{1+c^2}}(c, 1),$$

$$\lambda_2 = c, \quad r_2 = \frac{1}{\sqrt{1+c^2}}\begin{pmatrix} 1 \\ -c \end{pmatrix}, \quad l_2 = \frac{1}{\sqrt{1+c^2}}(-c, 1),$$

when

$$u = \begin{pmatrix} \phi_x \\ \phi_t \end{pmatrix}.$$

The initial data

$$\phi(x, 0) = f(x), \quad \phi_t(x, 0) = g(x)$$

translates into

$$u_0(x, 0) = \begin{pmatrix} f'(x) \\ g(x) \end{pmatrix}.$$

We conclude that

$$\phi(x, t) = \frac{1}{2}(f(x + ct) + f(x - ct)) + \frac{1}{2c}\int_{x-ct}^{x+ct} g(\xi)\,d\xi,$$

the familiar d'Alembert's formula for the solution of the linear wave equation.

Chapter 6, pages 232–233.

6.1 d. Up to second order in ϵ, the wave curves are given by
$$S_i(\epsilon) = \exp(\epsilon r_i).$$
For an interaction between an i-wave and a j-wave $(j < i)$ we have that (up to second order)
$$u_r = \exp(\epsilon_i r_i) \exp(\epsilon_j) u_l.$$
After the interaction,
$$u_r = \exp(\epsilon'_n r_n) \cdots \exp(\epsilon'_j r_j) \cdots \exp(\epsilon'_i r_i) \cdots \exp(\epsilon'_1 r_1) u_l$$
up to second order. Comparing these two expressions and using part **c** and the fact that $\{r_i\}$ are linearly independent, we find that
$$\epsilon'_k = \delta_{kj}\epsilon_j + \delta_{ki}\epsilon_i + \text{second-order terms}.$$

6.2 a. The definition of the front-tracking algorithm follows from the general case and the grid in the (μ, τ) plane. In this case we do not need to remove any front of high generation. To show that T_n is nonincreasing in n, we must study interactions. In this case we have a left wave ϵ_l separating states (μ_l, τ_l) and (μ_m, τ_m), colliding with a wave ϵ_r separating (μ_m, τ_m) and (μ_r, τ_r). Now, the family of ϵ_l must be greater than or equal to the family of ϵ_r. The claim follows by studying each case, and recalling that (consult Exercise 5.8) $0 \geq \tau'(\mu) > -1$.
b. If the total variation of μ_0 and τ_0 is finite, then we have enough to show that front tracking produces a compact sequence. Hence if
$$\text{T.V.}(\ln(u_0)) \quad \text{and} \quad \text{T.V.}(v_0)$$
are finite, then the sequence produced by front tracking is compact. A reasonable condition could then be that $u_0(x) \geq c > 0$ almost everywhere, $\text{T.V.}(u_0) < M$ and $\text{T.V.}(v_0) < M$.
c. To show that the limit is a weak solution, we proceed as in the general case.

Chapter 7, page 288.

7.1 We have already established **A** and **B** in Theorem 6.6. To prove **C**, let T be defined by (6.19) and Q by (6.18), and assume that $t = \gamma(x)$ is a curve with a Lipschitz constant L that is smaller than the inverse of the largest characteristic speed, i.e.,
$$L < \frac{1}{\max_{u \in \mathcal{D}, k} |\lambda_k(u)|}.$$

For such a curve we have that all fronts in u_δ will cross γ from below. Let $\gamma_t(x)$ be defined by
$$\gamma_t(x) = \min\{t, \gamma(x)\}.$$
Since all fronts will cross γ, and hence also γ_t from below, we can define $T|_{\gamma_t}$. Then we have that
$$T|_{\gamma_t} \leq (T + kQ)|_{\gamma_t} \leq (T + kQ)(t) \leq (T + kQ)(0).$$
Since $\gamma = \lim_{t\to\infty} \gamma_t$, it follows that the total variation of u_δ on γ is finite, independent of δ. Hence the total variation on γ of the limit u is also finite.

7.2 In the linearly degenerate case the Hugoniot loci coincide with the rarefaction curves. If you (meticulously) recapitulate all cases you will find that some of the estimates are easier because of this, while others are identical. If you do this exercise, you have probably mastered the material in Section 7.1!

Appendix A, page 300.

A.1 Observe first that
$$\left|\int \phi(x)f(x)\,dx\right| \leq \int |f(x)|\,dx,$$
for any function $|\phi| \leq 1$. Now let ω_ε be a standard mollifier, and define
$$\phi_\varepsilon = \omega_\varepsilon * \text{sign}(f).$$
Clearly, $f\phi_\varepsilon \to |f|$ pointwise, and a simple application of dominated convergence implies that indeed,
$$\int \phi_\varepsilon(x) f(x)\,dx \to \int |f(x)|\,dx \text{ as } \varepsilon \to 0.$$

A.2 Write as usual $h^\delta = v^\delta - w^\delta$, where v^δ and w^δ both are nonincreasing functions. Both sequences $\{v^\delta\}$ and $\{w^\delta\}$ satisfy the conditions of the theorem, and hence we may pass to a subsequence such that both v^δ and w^δ converge in L^1. After possibly taking yet another subsequence we conclude that we may obtain pointwise convergence almost everywhere. Let v be the limit of v^δ, which we may assume to be nondecreasing (by possibly redefining it on a set of measure zero). Write $v(x\pm)$ for the right, respectively left, limits of v at x. Fix $x \in \langle a, b \rangle$. Let $\epsilon > 0$ and $\eta > 0$ be such that $v(y) < v(x+) + \epsilon$ whenever $x < y < x + \eta$. If $v^\delta \geq v(x+) + 2\epsilon$, then
$$\|v^\delta - v\|_1 \geq \int_x^{x+\delta} (v^\delta(y) - v(y))\,dy > \eta\epsilon,$$

and so, since $\left\|v^\delta - v\right\|_1 \to 0$ as $\delta \to 0$, we must conclude that $v^\delta(x) < v(x+) + 2\epsilon$ for δ sufficiently small. Similarly, $v^\delta(x) > v(x-) - 2\epsilon$. In particular, $v^\delta(x) \to v(x)$ whenever v is continuous at x, thus at all but a countable set of points x. In the same way we show that $w^\delta(x) \to w(x)$ for all but at most a countable set of points x. A diagonal argument shows that we can pass to a subsequence such that both v^δ and w^δ converge pointwise for all x in $[a, b]$.

Appendix B, page 314.

B.1 For more details on this exercise including several applications, as well as how to extend the result to hyperbolic systems in the $\mu \to 0$ limit, please consult [71].

a. Just follow the argument in Theorem B.1.

b. Mimic the argument starting with (B.15). (Let δ in (B.15) be negative.)

c. Redo the previous point, assuming that $u_{0,j} \geq u_*$.

References

[1] D. Amadori and R. M. Colombo. Continuous dependence for 2×2 conservation laws with boundary. *J. Differential Equations*, 138:229–266, 1997.

[2] D. Amadori and R. M. Colombo. Viscosity solutions and standard Riemann semigroup for conservation laws with boundary. *Rend. Semin. Mat. Univ. Padova*, 99:219–245, 1998.

[3] K. A. Bagrinovskiĭ and S. K. Godunov. Difference methods for multidimensional problems. *Dokl. Akad. Nauk SSSR*, 115:431–433, 1957. In Russian.

[4] P. Baiti and A. Bressan. The semigroup generated by a Temple class system with large data. *Differ. Integral Equ.*, 10:401–418, 1997.

[5] P. Baiti and H. K. Jenssen. On the front tracking algorithm. *J. Math. Anal. Appl.*, 217:395–404, 1997.

[6] J. M. Ball. A version of the fundamental theorem for Young measures. In M. Rascle, D. Serre, and M. Slemrod, editors, *PDE's and Continuum Models of Phase Transitions*, pages 241–259, Berlin, 1989. Springer.

[7] D. B. Ballou. Solutions to nonlinear hyperbolic Cauchy problems without convexity conditions. *Trans. Amer. Math. Soc.*, 152:441–460, 1970.

[8] C. Bardos, A. Y. LeRoux, and J. C. Nedelec. First order quasilinear equations with boundary conditions. *Comm. Partial Differential Equations*, 4:1017–1034, 1979.

[9] L. M. Barker. SWAP — a computer program for shock wave analysis. Technical report, Sandia National Laboratory, Albuquerque, New Mexico, 1963. SC4796(RR).

[10] H. Bateman. Some recent researches on the motion of fluids. *Monthly Weather Review*, 43:163–170, 1915.

[11] S. Bianchini and A. Bressan. BV estimates for a class of viscous hyperbolic systems. *Indiana Univ. Math. J.*, 49:1673–1713, 2000.

[12] S. Bianchini and A. Bressan. A case study in vanishing viscosity. *Discrete Contin. Dynam. Systems*, 7:449–476, 2001.

[13] S. Bianchini and A. Bressan. Vanishing viscosity solutions of nonlinear hyperbolic systems. Preprint 86/2001/M, S.I.S.S.A., Trieste, Italy, 2001.

[14] A. Bressan. Contractive metrics for nonlinear hyperbolic systems. *Indiana Univ. Math. J.*, 37:409–421, 1988.

[15] A. Bressan. Global solutions of systems of conservation laws by wave-front tracking. *J. Math. Anal. Appl.*, 170:414–432, 1992.

[16] A. Bressan. A contractive metric for systems of conservation laws with coinciding shock and rarefaction waves. *J. Differential Equations*, 106:332–366, 1993.

[17] A. Bressan. The unique limit of the Glimm scheme. *Arch. Rat. Mech.*, 130:205–230, 1995.

[18] A. Bressan. *Hyperbolic Systems of Conservation Laws*. Oxford Univ. Press, Oxford, 2000.

[19] A. Bressan and R. M. Colombo. The semigroup generated by 2×2 conservation laws. *Arch. Rat. Mech.*, 133:1–75, 1995.

[20] A. Bressan, G. Crasta, and B. Piccoli. Well-posedness of the Cauchy problem for $n \times n$ systems of conservation laws. *Mem. Amer. Math. Soc.*, 694:1–134, 2000.

[21] A. Bressan and P. Goatin. Oleinik type estimates for $n \times n$ conservation laws. *J. Differential Equations*, 156:26–49, 1998.

[22] A. Bressan and P. LeFloch. Uniqueness of weak solutions to systems of conservation laws. *Arch. Rat. Mech.*, 140:301–317, 1997.

[23] A. Bressan and M. Lewicka. A uniqueness condition for hyperbolic systems of conservation laws. *Discrete Contin. Dynam. Systems*, 6:673–682, 2000.

[24] A. Bressan, T.-P. Liu, and T. Yang. L^1 stability estimates for $n \times n$ conservation laws. *Arch. Rat. Mech.*, 149:1–22, 1999.

[25] J. M. Burgers. Applications of a model system to illustrate some points of the statistical theory of free turbulence. *Proc. Roy. Neth. Acad. Sci. (Amsterdam)*, 43:2–12, 1940.

[26] J. M. Burgers. *The Nonlinear Diffusion Equation*. Reidel, Dordrecht, 1974.

[27] G. Buttazzo, M. Gianquinta, and S. Hildebrandt. *One-dimensional Variational Problems*. Oxford Univ. Press, Oxford, 1998.

[28] W. Cho-Chun. On the existence and uniqueness of the generalized solutions of the Cauchy problem for quasilinear equations of first order without the convexity condition. *Acta Math. Sinica*, 4:561–577, 1964.

[29] A. J. Chorin and J. E. Marsden. *A Mathematical Introduction to Fluid Mechanics*. Springer, New York, third edition, 1993.

[30] K. N. Chueh, C. C. Conley, and J. A. Smoller. Positively invariant regions for systems of nonlinear diffusion equations. *Indiana Univ. Math. J.*, 26:373–392, 1977.

[31] B. Cockburn and P.-A. Gremaud. Error estimates for finite element methods for scalar conservation laws. *SIAM J. Numer. Anal.*, 33:522–554, 1996.

[32] B. Cockburn and P.-A. Gremaud. A priori error estimates for numerical methods for scalar conservation laws. Part I: The general approach. *Math. Comp.*, 65:533–573, 1996.

[33] J. D. Cole. On a quasi-linear parabolic equation occurring in aerodynamics. *Quart. Appl. Math.*, 9:225–236, 1951.

[34] R. M. Colombo and N. H. Risebro. Continuous dependence in the large for some equations of gas dynamics. *Comm. Partial Differential Equations*, 23:1693–1718, 1998.

[35] F. Coquel and P. Le Floch. Convergence of finite difference schemes for conservation laws in several space dimensions: The corrected antidiffusive flux approach. *Math. Comp.*, 57:169–210, 1991.

[36] F. Coquel and P. Le Floch. Convergence of finite difference schemes for conservation laws in several space dimensions: A general theory. *SIAM J. Num. Anal.*, 30:675–700, 1993.

[37] R. Courant and K. O. Friedrichs. *Supersonic Flow and Shock Waves.* Springer, New York, 1976.

[38] M. G. Crandall and A. Majda. The method of fractional steps for conservation laws. *Numer. Math.*, 34:285–314, 1980.

[39] M. G. Crandall and A. Majda. Monotone difference approximations for scalar conservation laws. *Math. Comp.*, 34:1–21, 1980.

[40] M. G. Crandall and L. Tartar. Some relations between nonexpansive and order preserving mappings. *Proc. Amer. Math. Soc.*, 78:385–390, 1980.

[41] C. M. Dafermos. Polygonal approximation of solutions of the initial value problem for a conservation law. *J. Math. Anal. Appl.*, 38:33–41, 1972.

[42] C. M. Dafermos. *Hyperbolic Conservation Laws in Continuum Physics.* Springer, New York, 2000.

[43] R. J. DiPerna. Global existence of solutions to nonlinear systems of conservation laws. *J. Differential Equations*, 20:187–212, 1976.

[44] R. J. DiPerna. Compensated compactness and general systems of conservation laws. *Trans. Amer. Math. Soc.*, 292:383–420, 1985.

[45] R. J. DiPerna. Measure-valued solutions to conservation laws. *Arch. Rat. Mech.*, 88:223–270, 1985.

[46] B. Engquist and S. Osher. One-sided difference approximations for nonlinear conservation laws. *Math. Comp.*, 36:321–351, 1981.

[47] L. C. Evans and R. F. Gariepy. *Measure Theory and Fine Properties of Functions.* CRC Press, Boca Raton, Florida, USA, 1992.

[48] A. R. Forsyth. *Theory of Differential Equations, Part IV — Partial Differential Equations.* Cambridge Univ. Press, 1906.

[49] M. Fréchet. Sur les ensembles compacts de fonctions de carrés sommables. *Acta Szeged Sect. Math.*, 8:116–126, 1937.

[50] I. M. Gel'fand. Some problems in the theory of quasilinear equations. *Amer. Math. Soc. Transl.*, 29:295–381, 1963.

[51] J.-F. Gerbeau and B. Perthame. Derivation of viscous Saint–Venant system for laminar shallow water; numerical validation. *Discrete Contin. Dynam. Systems Ser. B*, 1:89–102, 2001.

[52] J. Glimm. Solutions in the large for nonlinear hyperbolic systems of equations. *Comm. Pure Appl. Math.*, 18:697–715, 1965.

[53] J. Glimm, J. Grove, X. L. Li, K.-M. Shyue, Y. Zeng, and Q. Zhang. Three-dimensional front tracking. *SIAM J. Sci. Comput.*, 19:703–727, 1998.

[54] J. Glimm, J. Grove, X. L. Li, and N. Zhao. Simple front tracking. In G.-Q. Chen and E. DiBenedetto, editors, *Nonlinear Partial Differential Equations*, volume 238 of *Contemp. Math.*, pages 33–149, Providence, 1999. Amer. Math. Soc.

[55] J. Glimm, E. Isaacson, D. Marchesin, and O. McBryan. Front tracking for hyperbolic systems. *Adv. in Appl. Math.*, 2:91–119, 1981.

[56] J. Glimm, C. Klingenberg, O. McBryan, B. Plohr, D. Sharp, and S. Yaniv. Front tracking and two-dimensional Riemann problems. *Adv. in Appl. Math.*, 6:259–290, 1985.

[57] P. Goatin and P. G. LeFloch. Sharp L^1 continuous dependence of solutions of bounded variation for hyperbolic systems of conservation laws. *Arch. Rat. Mech.*, 157:35–73, 2001.

[58] E. Godlewski and P.-A. Raviart. *Hyperbolic Systems of Conservation Laws*. Mathématiques et Applications, Ellipses, Paris, 1991.

[59] E. Godlewski and P.-A. Raviart. *Numerical Approximation of Hyperbolic Systems of Conservation Laws*. Springer, Berlin, 1996.

[60] S. K. Godunov. Finite difference method for numerical computation of discontinuous solutions of the equations of fluid dynamics. *Math. Sbornik*, 47:271–306, 1959. In Russian.

[61] R. Haberman. *Mathematical Models*. Prentice-Hall, Englewood Cliffs, 1977.

[62] A. Harten, J. M. Hyman, and P. D. Lax. On finite-difference approximations and entropy conditions for shocks. *Comm. Pure Appl. Math.*, 29:297–322, 1976.

[63] G. W. Hedstrom. Some numerical experiments with Dafermos's method for nonlinear hyperbolic equations. In R. Ansorge and W. Törnig, editors, *Numerische Lösung nichtlinearer partieller Differential- und Integrodifferentialgleichungen, Lecture Notes in Mathematics*, pages 117–138. Springer, Berlin, 1972.

[64] G. W. Hedstrom. The accuracy of Dafermos' method for nonlinear hyperbolic equations. In *Proceedings of Equadiff III (Third Czechoslovak Conf. Differential Equations and Their Applications, Brno, 1972)*, volume 1, pages 175–178, Brno, 1973. Folia Fac. Sci. Natur. Univ. Purkynianae Brunensis, Ser. Monograph.

[65] D. Hoff. Invariant regions for systems of conservation laws. *Trans. Amer. Math. Soc.*, 289:591–610, 1985.

[66] H. Holden and L. Holden. On scalar conservation laws in one-dimension. In S. Albeverio, J. E. Fenstad, H. Holden, and T. Lindstrøm, editors, *Ideas and Methods in Mathematics and Physics*, pages 480–509. Cambridge Univ. Press, Cambridge, 1992.

[67] H. Holden, L. Holden, and R. Høegh-Krohn. A numerical method for first order nonlinear scalar conservation laws in one-dimension. *Comput. Math. Applic.*, 15:595–602, 1988.

[68] H. Holden, K.-A. Lie, and N. H. Risebro. An unconditionally stable method for the Euler equations. *J. Comput. Phys.*, 150:76–96, 1999.

[69] H. Holden and N. H. Risebro. A method of fractional steps for scalar conservation laws without the CFL condition. *Math. Comp.*, 60:221–232, 1993.

[70] H. Holden and N. H. Risebro. Conservation laws with a random source. *Appl. Math. Optim.*, 36:229–241, 1997.

[71] H. Holden, N. H. Risebro, and A. Tveito. Maximum principles for a class of conservation laws. *SIAM J. Appl. Math.*, 55:651–661, 1995.

[72] E. Hölder. Historischer Überblick zur mathematischen Theorie von Unstetigkeitswellen seit Riemann und Christoffel. In P. L. Butzer and P. Feher, editors, *E. B. Christoffel*, pages 412–434. Birkhäuser, Basel, 1981.

[73] E. Hopf. The partial differential equation $u_t + uu_x = \mu u_{xx}$. *Comm. Pure Appl. Math.*, 3:201–230, 1950.

[74] L. Hörmander. *Lectures on Nonlinear Hyperbolic Differential Equations*. Springer, Berlin, 1997.

[75] Jiaxin Hu and P. G. LeFloch. L^1 continuous dependence property for hyperbolic systems of conservation laws. *Arch. Rat. Mech.*, 151:45–93, 2000.

[76] H. Hugoniot. Sur un théorème général relatif à la propagation du mouvement dans les corps. *C. R. Acad. Sci. Paris Sér. I Math.*, 102:858–860, 1886.

[77] H. Hugoniot. Mémoire sur la propagation du mouvement dans les corps et spécialement dans les gaz parfaits. *J. l'Ecoles Polytechn.*, 57:3–97, 1887. *ibid.* 58:1–125, 1887. Translated into English in [79], pp. 161–243, 245–358.

[78] H. Hugoniot. Mémoire sur la propagation du mouvement dans un fluid indéfini. *J. Math. Pures Appl.*, 3:477–492, 1887. *ibid.* 4:153–167, 1887.

[79] J. N. Johnson and R. Chéret, editors. *Classical Papers in Shock Compression Science*. Springer, New York, 1998.

[80] K. Hvistendahl Karlsen. On the accuracy of a dimensional splitting method for scalar conservation laws. Master's thesis, University of Oslo, 1994.

[81] K. Hvistendahl Karlsen. On the accuracy of a numerical method for two-dimensional scalar conservation laws based on dimensional splitting and front tracking. Preprint Series 30, Department of Mathematics, University of Oslo, 1994.

[82] K. Hvistendahl Karlsen and N. H. Risebro. An operator splitting method for convection–diffusion equations. *Numer. Math.*, 77:365–382, 1997.

[83] J. Kevorkian. *Partial Differential Equations. Analytical Solution Techniques*. Wadsworth & Brooks/Cole, Pacific Grove, 1990.

[84] C. Klingenberg and N. H. Risebro. Stability of a resonant system of conservation laws modeling polymer flow with gravitation. *J. Differential Equations*, 170:344–380, 2001.

[85] A. N. Kolmogorov. On the compactness of sets of functions in the case of convergence in mean. *Nachr. Ges. Wiss. Göttingen, Math.-Phys. Kl. I*, (9):60–63, 1931. Translated and reprinted in: *Selected Works of A. N. Kolomogorov, Vol. I* (V. M. Tikhomirov. ed.), Kluwer, Dordrecht, 1991, pp. 147–150.

[86] D. Kröner. *Numerical Schemes for Conservation Laws*. Wiley-Teubner, Chichester, 1997.

[87] S. N. Kružkov. Results concerning the nature of the continuity of solutions of parabolic equations, and some of their applications. *Math. Notes*, 6:517–523, 1969.

[88] S. N. Kružkov. First order quasi-linear equations in several independent variables. *Math. USSR Sbornik*, 10:217–243, 1970.

[89] N. N. Kuznetsov. On stable methods of solving a quasi-linear equation of first order in a class of discontinuous functions. *Soviet Math. Dokl.*, 16:1569–1573, 1975.

[90] N. N. Kuznetsov. Accuracy of some approximative methods for computing the weak solutions of a first-order quasi-linear equation. *USSR Comput. Math. and Math. Phys. Dokl.*, 16:105–119, 1976.

[91] J. O. Langseth. On an implementation of a front tracking method for hyperbolic conservation laws. *Advances in Engineering Software*, 26:45–63, 1996.

[92] J. O. Langseth, N. H. Risebro, and A. Tveito. A conservative front tracking scheme for 1D hyperbolic conservation laws. In A. Donato and F. Oliveri, editors, *Nonlinear Hyperbolic Problems: Theoretical, Applied and Computational Aspects*, pages 385–392, Braunschweig, 1992. Vieweg.

[93] J. O. Langseth, A. Tveito, and R. Winther. On the convergence of operator splitting applied to conservation laws with source terms. *SIAM J. Appl. Math.*, 33:843–863, 1996.

[94] P. D. Lax. Weak solutions of nonlinear hyperbolic equations and their numerical computation. *Comm. Pure Appl. Math.*, 7:159–193, 1954.

[95] P. D. Lax. Hyperbolic systems of conservation laws. II. *Comm. Pure Appl. Math.*, 10:537–566, 1957.

[96] P. D. Lax. *Hyperbolic Systems of Conservation Laws and the Mathematical Theory of Shock Waves*. CBMS–NSF Regional Conference Series in Applied Mathematics. SIAM, Philadelphia, 1973.

[97] P. D. Lax and B. Wendroff. Systems of conservation laws II. *Comm. Pure Appl. Math.*, 13:217–237, 1960.

[98] R. J. LeVeque. *Numerical Methods for Conservation Laws*. Birkhäuser, Basel, second edition, 1992.

[99] K.-A. Lie. *Front Tracking and Operator Splitting for Convection Dominated Problems*. PhD thesis, NTNU, Trondheim, 1998.

[100] K.-A. Lie, V. Haugse, and K. Hvistendahl Karlsen. Dimensional splitting with front tracking and adaptive grid refinement. *Numerical Methods for Part. Diff. Eq.*, 14:627–648, 1997.

[101] M. J. Lighthill and G. B. Whitham. On kinematic waves. II. Theory of traffic flow on long crowded roads. *Proc. Roy. Soc. London. Ser. A*, 229:317–345, 1955.

[102] T.-P. Liu. The deterministic version of the Glimm scheme. *Comm. Math. Phys.*, 57:135–148, 1977.

[103] T.-P. Liu. Large time behavior of solutions of initial and initial-boundary value problems of a general system of hyperbolic conservation laws. *Comm. Math. Phys.*, 55:163–177, 1977.

[104] T.-P. Liu. *Hyperbolic and Viscous Conservation Laws*. CBMS-NSF Regional Conference Series in Applied Mathematics. SIAM, Philadelphia, 1999.

[105] T.-P. Liu and J. Smoller. On the vacuum state for isentropic gas dynamics. *Adv. Pure Appl. Math.*, 1:345–359, 1980.

[106] B. J. Lucier. A moving mesh numerical method for hyperbolic conservation laws. *Math. Comp.*, 46:59–69, 1986.

[107] J. Málek, J. Nečas, M. Rokyta, and M. Ruzička. *Weak and Measure-valued Solutions to Evolutionary PDEs*. Chapman & Hall, London, 1996.

[108] R. McOwen. *Partial Differential Equations*. Prentice-Hall, Upper Saddle River, New Jersey, USA, 1996.

[109] T. Nishida and J. Smoller. Solutions in the large for some nonlinear hyperbolic conservation laws. *Comm. Pure Appl. Math*, 26:183–200, 1973.

[110] O. A. Oleĭnik. Discontinuous solutions of non-linear differential equations. *Amer. Math. Soc. Transl. Ser.*, 26:95–172, 1963.

[111] O. A. Oleĭnik. Uniqueness and stability of the generalized solution of the Cauchy problem for a quasi-linear equation. *Amer. Math. Soc. Transl. Ser.*, 33:285–290, 1963.

[112] O. A. Oleĭnik and S. N. Kružkov. Quasi-linear second-order parabolic equations with many independent variables. *Russian Math. Surveys*, 16(5):105–146, 1961.

[113] S. Osher. Riemann solvers, the entropy condition, and difference approximations. *SIAM J. Num. Anal.*, 21:217–235, 1984.

[114] S. Osher and J. A. Sethian. Fronts propagating with curvature dependent speed: algorithms based on Hamilton–Jacobi formulations. *J. Comp. Phys.*, 79:12–49, 1988.

[115] B. Keyfitz Quinn. Solutions with shocks: An example of an L_1-contractive semigroup. *Comm. Pure Appl. Math.*, 24:125–132, 1971.

[116] W. J. M. Rankine. On the thermodynamic theory of waves of finite longitudinal disturbances. *Phil. Trans. Roy. Soc.*, 160:277–288, 1870. Reprinted in [79], pp. 133–147.

[117] M. Renardy and R. C. Rogers. *An Introduction to Partial Differential Equations*. Springer, New York, 1993.

[118] P. I. Richards. Shock waves on the highway. *Oper. Res.*, 4:42–51, 1956.

[119] G. F. B. Riemann. Selbstanzeige: Ueber die Fortpflanzung ebener Luftwellen von endlicher Schwingungsweite. *Göttinger Nachrichten*, (19):192–197, 1859.

[120] G. F. B. Riemann. Ueber die Fortpflanzung ebener Luftwellen von endlicher Schwingungsweite. *Abh. König. Gesell. Wiss. Göttingen*, 8:43–65, 1860. Translated into English in [79], pp. 109–128.

[121] M. Riesz. Sur les ensembles compacts de fonctions sommables. *Acta Szeged Sect. Math.*, 6:136–142, 1933.

[122] N. H. Risebro. The partial differential equation $u_t + \nabla f(u) = 0$. A numerical method. Master's thesis, University of Oslo, 1987.

[123] N. H. Risebro. A front-tracking alternative to the random choice method. *Proc. Amer. Math. Soc.*, 117:1125–1129, 1993.

[124] N. H. Risebro and A. Tveito. Front tracking applied to a nonstrictly hyperbolic system of conservation laws. *SIAM J. Sci. Stat. Comput.*, 12:1401–1419, 1991.

[125] N. H. Risebro and A. Tveito. A front tracking method for conservation laws in one dimension. *J. Comp. Phys.*, 101:130–139, 1992.

[126] B. L. Roždestvenskiĭ and N. N. Janenko. *Systems of Quasilinear Equations and Their Applications to Gas Dynamics*. Amer. Math. Soc., Providence, 1983.

[127] M. Schatzman. Glimm functionals and uniqueness of solutions of the Riemann problem. *Indiana Univ. Math. J.*, 34:533–589, 1985.

[128] D. Serre. *Systems of Conservation Laws. Volume 1*. Cambridge Univ. Press, Cambridge, 1999.

[129] D. Serre. *Systems of Conservation Laws. Volume 2*. Cambridge Univ. Press, Cambridge, 2000.

[130] J. Smoller. *Shock Waves and Reaction–Diffusion Equations*. Springer, New York, second edition, 1994.

[131] J. Smoller and B. Temple. Global solutions of the relativistic Euler equations. *Comm. Math. Phys.*, 156:67–99, 1993.

[132] S. L. Sobolev. *Applications of Functional Analysis in Mathematical Physics*. Amer. Math. Soc., Providence, 1963.

[133] B. K. Swartz and B. Wendroff. AZTEK: A front tracking code based on Godunov's method. *Appl. Num. Math.*, 2:385–397, 1986.

[134] A. Szepessy. An existence result for scalar conservation laws using measure valued solutions. *Comm. Partial Differential Equations*, 14:1329–1350, 1989.

[135] J. D. Tamarkin. On the compactness of the space L_p. *Bull. Amer. Math. Soc.*, 32:79–84, 1932.

[136] Pan Tao and Lin Longwei. The global solution of the scalar nonconvex conservation law with boundary condition. *J. Partial Differential Equations*, 8:371–383, 1995.

[137] B. Temple. Systems of conservation laws with invariant submanifolds. *Trans. Amer. Math. Soc.*, 280:781–795, 1983.

[138] Z.-H. Teng. On the accuracy of fractional step methods for conservation laws in two dimensions. *SIAM J. Num. Anal.*, 31:43–63, 1994.

[139] J. W. Thomas. *Numerical Partial Differential Equations. Conservation Laws and Elliptic Equations*. Springer, New York, 1999.

[140] E. Toro. *Riemann Solvers and Numerical Methods for Fluid Mechanics*. Springer, Berlin, 1997.

[141] A. Tulajkov. Zur Kompaktheit im Raum L_p für $p = 1$. *Nachr. Ges. Wiss. Göttingen, Math.-Phys. Kl. I*, (39):167–170, 1933.

[142] A. Tveito and R. Winther. Existence, uniqueness, and continuous dependence for a system of hyperbolic conservation laws modeling polymer flooding. *SIAM J. Math. Anal.*, 22:905–933, 1991.

[143] A. I. Vol'pert. The spaces BV and quasi-linear equations. *Math. USSR Sbornik*, 2:225–267, 1967.

[144] E. T. Whittaker and G. N. Watson. *A Course of Modern Analysis*. Cambridge Univ. Press, Cambridge, fourth edition, 1999.

[145] J. Wloka. *Functionalanalysis und Anwendungen*. de Gruyter, Berlin, 1971.

[146] W.-A. Yong. A simple approach to Glimm's interaction estimates. *Appl. Math. Lett.*, 12:29–34, 1999.

[147] K. Yosida. *Functional Analysis*. Springer, Berlin, 1995. Reprint of the 1980 edition.

[148] W. P. Ziemer. *Weakly Differentiable Functions*. Springer, New York, 1989.

Index

approximate δ distribution, 45
approximately continuous, 292
Arzela–Ascoli theorem, 297, 307

Borel measure, 275
bores, 177
breakpoint, 34
Buckley–Leverett equation, 57, 67
Burgers' equation (inviscid), 3, 4, 8, 11, 18, 21, 36, 56, 57, 59, 113
Burgers' equation (viscous), 18

characteristic equation, 3
characteristics, 3, 16, 17, 19
chromatography, 342
Cole–Hopf transformation, 18
conservation laws, 1
conservative method, 65
consistent method, 65
contact discontinuity, 173
Courant–Friedrichs–Lewy(CFL) condition, 66
Crandall–Tartar's lemma, 51, 55, 57, 74, 306

Dafermos' method, 56
dam breaking, 199

dimensional splitting, 117, 158
directional derivative, 194
distributions, 6
  derivatives of, 6
downwind scheme, 66
Duhamel's principle, 302

Engquist–Osher's scheme, 112
entropy condition, 14
  Kružkov, 26, 119
  Lax, 188
  Oleĭnik, 114
  traveling wave, 24
  vanishing viscosity, 301–314
entropy/entropy flux pair, 59, 76, 205
ε-net, 295
essential variation, 292
Euler's equations, 169, 203

finite speed of propagation, 37
first-order convergence, 54
flux density, 2
flux function, 2
fractional steps method, 117, 158
front
  generation of, 218
  strength of, 211

front tracking, 17, 36, 48, 56, 127
  two-dimensional, 158
front tracking in a box
  scalar case, 40
  systems, 220
fronts
  scalar case, 40, 41
  systems, 210

generalized functions, 6
generation
  of front, 218
genuinely nonlinear, 171, 172
Glimm's functional, 217
Godunov's scheme, 66, 112
Green's theorem, 10, 32
Gronwall's inequality, 156, 161, 306

Hamilton–Jacobi equation, 57, 60
heat kernel, 147, 302
Helly's theorem, 57, 298, 300
Hugoniot locus, 177
hybrid methods, 66
hyperbolic, 170
hyperfast, 41, 218

implicit function theorem, 179, 203, 205
interaction estimate, 213
interaction potential, 215
invariant regions, 315

Jordan decomposition, 290

Kolmogorov's compactness theorem, 296, 300
Kružkov's interpolation lemma, 149, 159
Kružkov entropy condition, 26, 34, 36, 43, 53, 76, 154
  multidimensional, 119
Kružkov entropy solution, 28, 57
Kružkov function, 27
Kuznetsov's lemma, 82, 136
Kuznetsov's theory, 57

Lagrangian coordinates, 36
Lax entropy condition, 188, 202, 226
Lax inequalities, 187

Lax shock
  definition of, 189
Lax's theorem, 196
Lax–Friedrichs' scheme, 65, 71, 74, 112, 160, 203
Lax–Wendroff's scheme, 66, 112, 158
Lax–Wendroff's theorem, 71, 112
Lebesgue point, 292
level set methods, 57
Lie bracket, 214, 232
line fields, 342
linearly degenerate, 36, 171, 173
linearly degenerate wave, 173
Lipschitz constant, 45
Lipschitz continuity in time, 55
Lipschitz continuous, 33, 45
Lipschitz seminorm, 45
local truncation error, 69
L^1-contractive, 55, 73
lower convex envelope, 30

MacCormack's scheme, 66, 112
maximum principle, 40, 55, 305
measure-valued solutions, 99
model equation, 71
modulus of continuity, 290
  in time, 81
mollifier, 45
monotone method, 73
monotone scheme, 77, 88, 111, 112
monotonicity, 55, 58
monotonicity preserving, 55, 73
Moses' problems
  first, 200
  seond, 201
multi-index, 149

negative variation, 290
numerical entropy flux, 77
numerical flux, 65

Oleĭnik entropy condition, 114
operator splitting methods, 117

p-system, 169, 203
parabolic regularization, 159
positive variation, 290

quasilinear equation, 3, 19

Rademacher's theorem, 273
Radon measure, 275
random choice method, 158, 231
Rankine–Hugoniot condition, 10, 19, 20, 25, 58, 59, 177
rarefaction front, 219
rarefaction wave, 16, 34, 171
real state, 229
relatively compact, 295
Richtmyer two-step Lax–Wendroff scheme, 66
Riemann invariants, 204
  coordinate system of, 204
Riemann problem, 13, 14, 16, 17, 19, 30, 33, 60

shallow-water equations, 166, 170, 174, 177, 182, 191, 198, 202, 203
  bores, 177
  conservation of
    energy, 183
    mass, 166
    momentum, 166
  hydrostatic balance, 167
  pressure, 167
  traveling wave, 185
  vacuum, 199
shock, 34
  admissible, 188
shock line, 227
smoothing method, 84
strictly hyperbolic, 170
strongly compact, 294
structure theorem
  for BV functions, 269

Temple class systems, 237
test functions, 6
total variation, 49, 289, 300
total variation diminishing(TVD), 55, 73
total variation stable, 73
totally bounded, 295
traffic flow, 11, 18, 34
traffic hydrodynamics, 18
traveling wave, 25, 185
traveling wave entropy condition, 24, 25

TVD, *see* total variation diminishing

upper concave envelope, 33
upwind scheme, 65, 66

vanishing viscosity method, 87
viscosity solution, 60
viscous profile, 25
viscous regularization, 24, 56

wave
  family of, 171
wave curve, 190
wave equation, 168
  d'Alembert's formula, 346
waves
  systems, 210
weak solution, 8

Young's theorem, 100

Applied Mathematical Sciences

(continued from page ii)

60. *Ghil/Childress:* Topics in Geophysical Dynamics: Atmospheric Dynamics, Dynamo Theory and Climate Dynamics.
61. *Sattinger/Weaver:* Lie Groups and Algebras with Applications to Physics, Geometry, and Mechanics.
62. *LaSalle:* The Stability and Control of Discrete Processes.
63. *Grasman:* Asymptotic Methods of Relaxation Oscillations and Applications.
64. *Hsu:* Cell-to-Cell Mapping: A Method of Global Analysis for Nonlinear Systems.
65. *Rand/Armbruster:* Perturbation Methods, Bifurcation Theory and Computer Algebra.
66. *Hlaváček/Haslinger/Necasl/Lovísek:* Solution of Variational Inequalities in Mechanics.
67. *Cercignani:* The Boltzmann Equation and Its Applications.
68. *Temam:* Infinite-Dimensional Dynamical Systems in Mechanics and Physics, 2nd ed.
69. *Golubitsky/Stewart/Schaeffer:* Singularities and Groups in Bifurcation Theory, Vol. II.
70. *Constantin/Foias/Nicolaenko/Temam:* Integral Manifolds and Inertial Manifolds for Dissipative Partial Differential Equations.
71. *Catlin:* Estimation, Control, and the Discrete Kalman Filter.
72. *Lochak/Meunier:* Multiphase Averaging for Classical Systems.
73. *Wiggins:* Global Bifurcations and Chaos.
74. *Mawhin/Willem:* Critical Point Theory and Hamiltonian Systems.
75. *Abraham/Marsden/Ratiu:* Manifolds, Tensor Analysis, and Applications, 2nd ed.
76. *Lagerstrom:* Matched Asymptotic Expansions: Ideas and Techniques.
77. *Aldous:* Probability Approximations via the Poisson Clumping Heuristic.
78. *Dacorogna:* Direct Methods in the Calculus of Variations.
79. *Hernández-Lerma:* Adaptive Markov Processes.
80. *Lawden:* Elliptic Functions and Applications.
81. *Bluman/Kumei:* Symmetries and Differential Equations.
82. *Kress:* Linear Integral Equations, 2nd ed.
83. *Bebernes/Eberly:* Mathematical Problems from Combustion Theory.
84. *Joseph:* Fluid Dynamics of Viscoelastic Fluids.
85. *Yang:* Wave Packets and Their Bifurcations in Geophysical Fluid Dynamics.
86. *Dendrinos/Sonis:* Chaos and Socio-Spatial Dynamics.
87. *Weder:* Spectral and Scattering Theory for Wave Propagation in Perturbed Stratified Media.
88. *Bogaevski/Povzner:* Algebraic Methods in Nonlinear Perturbation Theory.
89. *O'Malley:* Singular Perturbation Methods for Ordinary Differential Equations.
90. *Meyer/Hall:* Introduction to Hamiltonian Dynamical Systems and the N-body Problem.
91. *Straughan:* The Energy Method, Stability, and Nonlinear Convection.
92. *Naber:* The Geometry of Minkowski Spacetime.
93. *Colton/Kress:* Inverse Acoustic and Electromagnetic Scattering Theory, 2nd ed.
94. *Hoppensteadt:* Analysis and Simulation of Chaotic Systems, 2nd ed.
95. *Hackbusch:* Iterative Solution of Large Sparse Systems of Equations.
96. *Marchioro/Pulvirenti:* Mathematical Theory of Incompressible Nonviscous Fluids.
97. *Lasota/Mackey:* Chaos, Fractals, and Noise: Stochastic Aspects of Dynamics, 2nd ed.
98. *de Boor/Höllig/Riemenschneider:* Box Splines.
99. *Hale/Lunel:* Introduction to Functional Differential Equations.
100. *Sirovich (ed):* Trends and Perspectives in Applied Mathematics.
101. *Nusse/Yorke:* Dynamics: Numerical Explorations, 2nd ed.
102. *Chossat/Iooss:* The Couette-Taylor Problem.
103. *Chorin:* Vorticity and Turbulence.
104. *Farkas:* Periodic Motions.
105. *Wiggins:* Normally Hyperbolic Invariant Manifolds in Dynamical Systems.
106. *Cercignani/Illner/Pulvirenti:* The Mathematical Theory of Dilute Gases.
107. *Antman:* Nonlinear Problems of Elasticity.
108. *Zeidler:* Applied Functional Analysis: Applications to Mathematical Physics.
109. *Zeidler:* Applied Functional Analysis: Main Principles and Their Applications.
110. *Diekmann/van Gils/Verduyn Lunel/Walther:* Delay Equations: Functional-, Complex-, and Nonlinear Analysis.
111. *Visintin:* Differential Models of Hysteresis.
112. *Kuznetsov:* Elements of Applied Bifurcation Theory, 2nd ed.
113. *Hislop/Sigal:* Introduction to Spectral Theory: With Applications to Schrödinger Operators.
114. *Kevorkian/Cole:* Multiple Scale and Singular Perturbation Methods.
115. *Taylor:* Partial Differential Equations I, Basic Theory.
116. *Taylor:* Partial Differential Equations II, Qualitative Studies of Linear Equations.

(continued on next page)

Applied Mathematical Sciences

(continued from previous page)

117. *Taylor:* Partial Differential Equations III, Nonlinear Equations.
118. *Godlewski/Raviart:* Numerical Approximation of Hyperbolic Systems of Conservation Laws.
119. *Wu:* Theory and Applications of Partial Functional Differential Equations.
120. *Kirsch:* An Introduction to the Mathematical Theory of Inverse Problems.
121. *Brokate/Sprekels:* Hysteresis and Phase Transitions.
122. *Gliklikh:* Global Analysis in Mathematical Physics: Geometric and Stochastic Methods.
123. *Le/Schmitt:* Global Bifurcation in Variational Inequalities: Applications to Obstacle and Unilateral Problems.
124. *Polak:* Optimization: Algorithms and Consistent Approximations.
125. *Arnold/Khesin:* Topological Methods in Hydrodynamics.
126. *Hoppensteadt/Izhikevich:* Weakly Connected Neural Networks.
127. *Isakov:* Inverse Problems for Partial Differential Equations.
128. *Li/Wiggins:* Invariant Manifolds and Fibrations for Perturbed Nonlinear Schrödinger Equations.
129. *Müller:* Analysis of Spherical Symmetries in Euclidean Spaces.
130. *Feintuch:* Robust Control Theory in Hilbert Space.
131. *Ericksen:* Introduction to the Thermodynamics of Solids, Revised ed.
132. *Ihlenburg:* Finite Element Analysis of Acoustic Scattering.
133. *Vorovich:* Nonlinear Theory of Shallow Shells.
134. *Vein/Dale:* Determinants and Their Applications in Mathematical Physics.
135. *Drew/Passman:* Theory of Multicomponent Fluids.
136. *Cioranescu/Saint Jean Paulin:* Homogenization of Reticulated Structures.
137. *Gurtin:* Configurational Forces as Basic Concepts of Continuum Physics.
138. *Haller:* Chaos Near Resonance.
139. *Sulem/Sulem:* The Nonlinear Schrödinger Equation: Self-Focusing and Wave Collapse.
140. *Cherkaev:* Variational Methods for Structural Optimization.
141. *Naber:* Topology, Geometry, and Gauge Fields: Interactions.
142. *Schmid/Henningson:* Stability and Transition in Shear Flows.
143. *Sell/You:* Dynamics of Evolutionary Equations.
144. *Nédélec:* Acoustic and Electromagnetic Equations: Integral Representations for Harmonic Problems.
145. *Newton:* The N-Vortex Problem: Analytical Techniques.
146. *Allaire*: Shape Optimization by the Homogenization Method.
147. *Aubert/Kornprobst:* Mathematical Problems in Image Processing: Partial Differential Equations and the Calculus of Variations.
148. *Peyret:* Spectral Methods for Incompressible Viscous Flow.
149. *Ikeda/Murota:* Imperfect Bifurcation in Structures and Materials.
150. *Skorokhod/Hoppensteadt/Salehi:* Random Perturbation Methods with Applications in Science and Engineering.
151. *Bensoussan/Frehse:* Topics on Nonlinear Partial Differential Equations and Applications.
152. *Holden/Risebro:* Front Tracking for Hyperbolic Conservation Laws.
153. *Osher/Fedkiw:* Level Sets and Dynamic Implicit Surfaces.
154. *Bluman/Anco:* Symmetry and Integration Methods for Differential Equations.